FIBRE CHANNEL FOR SANS

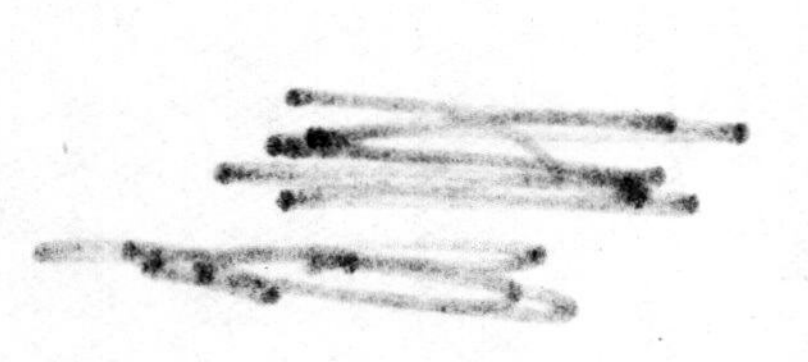

Fibre Channel for SANs

Alan F. Benner

McGraw-Hill
New York • Chicago • San Francisco
Lisbon • London • Madrid • Mexico City
Milan • New Delhi • San Juan • Seoul
Singapore • Sydney • Toronto

Cataloging-in-Publication Data is on file with the Library of Congress

McGraw-Hill

A Division of The ***McGraw-Hill*** *Companies*

1 2 3 4 5 6 7 8 9 0 DOC/DOC 0 9 8 7 6 5 4 3 2 1

ISBN 0-07-137413-2

The sponsoring editor of this book was Steve Chapman. The editing supervisor was Steven Melvin, and the production supervisor was Sherri Souffrance.

Printed and bound by R. R. Donnelley & Sons, Inc.

This book is printed on recycled, acid-free paper containing a minimum of 50% recycled, de-inked fiber.

Table of Contents

Contents

Contents

List of Figures

Figures

Figures

Preface

In the preface to the first edition of this book I wrote about the difficulty of keeping up with the flow of information in a book format, since the state of the technology changes so much during the time between when a book is started and when it actually appears. Since then, the pace of change in Fibre Channel, Storage Area Networks (SANs), and server networking has, if anything, accelerated. At least a dozen new standards have been approved since the first edition, and several more are either under development or have been superseded by better work.

Fortunately, however, most of the current progress is up in the network management and data management layers, so that the basic operation of the network, which is the primary subject of this book, has pretty well stabilized. It's possible, therefore, to expect that the subject matter here will stay stable for a reasonable period of time.

The most fundamental change in Fibre Channel since the first edition of this book, however, has been in where it gets used. When Fibre Channel was first created, it was not clear exactly which problems it would be most effective at solving, since it was such a widely applicable technology. Since then, Fibre Channel has become the de facto standard interconnect technology for storage networking. In fact, the term Storage Area Network, newly defined since that time, has currently become essentially synonymous with Fibre Channel.

Since Fibre Channel has become so tightly linked with storage networking, much of the progress in Fibre Channel has been towards solving specific storage networking problems. In this book, however, I've tried to keep a general focus on the whole of Fibre Channel technology, since restricting the discussion to only those aspects important in current-day SANs runs the risk of ignoring parts of the technology that will be extremely important in the future.

The intention of this book is two-fold. First, it's intended to be an overview guide to the concepts, the structures, and the goals of the Fibre Channel architecture at a fairly detailed level. A dedicated reader should be able to use it to understand most of the details of Fibre Channel before referring to the ANSI materials for authoritative information. Second, and perhaps more importantly, it's an attempt to show the reasons why the network is designed the way it is. Networking technology is consistently becoming more and more important, and new technologies are being developed regularly. As new networking technologies are invented, I hope that this book will help people exploit the goods features in the already-existing networks.

A few words on notation. Determining a consistent notation in this type of work is not trivial, since the subject matter bridges both computer and communications arenas, which have traditionally used slightly different notations. For example, communications data rates are generally measured in megabits per second, where "mega" means 10^6, while computer data is measured in megabytes, where "mega" means 2^{20}. The text mixes both somewhat, using "b" to represent bits, as in Gbps, and "B" to represent bytes, as in MBps. In recognition of the communications-oriented nature of the subject, the prefixes "mega (M)" and "giga (G)" will mean 10^6 and 10^9 here, rather than 2^{20} and 2^{30}. All numbers in the book are written in binary (b'0110 0101'), hexadecimal (x'FF FFFD'), or decimal (65,532) formats. Single bits are written as 1 or 0.

A number of common words, such Sequence, Exchange, and Connection, have specific meanings in Fibre Channel that are quite distinct from their common usage. In this book, words with specific Fibre Channel meanings are generally capitalized to distinguish them from the common usage. This capitalization generally matches the format used in the ANSI standards documents. Information provided here is in the public domain, through generally available books, articles, ANSI documents, or other reference material.

Several terms used in this book, such as ATM and HIPPI, are taken from other architectures. Any trademarks used are properties of their rightful owners. Ethernet is a trademark of the Xerox Corporation. ESCON, FICON, and SBCON are trademarks or registered trademark of the IBM Corporation in the United States or other countries or both. InfiniBand is a registered trademark of the InfiniBand Trade Association.

The book is organized as follows. The first few chapters give an overview of the features and goals for the Fibre Channel architecture, along with an example of how data is transmitted under a Fibre Channel network.

The middle chapters cover the concepts and structures of Fibre Channel in a fair amount of detail. These include chapters on all of the Fibre Channel physical components and logical constructs, supported functions, flow control, and error recovery.

The final chapters cover configuration and operation of the Arbitrated Loop topology, mapping of Fibre Channel constructs to upper level protocols such as SCSI and the IP level of TCP/IP. These chapters show how Fibre Channel fits in with currently existing software and operating system levels. In the last chapter I have taken the opportunity to make some predictions for the future of Fibre Channel, SANs, and server networking, as far as I dare.

Thanks are due to far too many people to list here, but I'll try — I apologize in advance to those I may have missed. Many thanks to Carl Zeitler, Ki Won Lee, Mike Yang, Dan Eisenhower, Ron Cash, Roger Weekly, Giles Frazier, Jerry Chapman, Jerry Rouse, Jonathan Thatcher, Bill George, Al Widmer, Tom McConathy, Casey Cannon, Gary Nutt, R. Bryan Cook, Paul Green, Dal Allan, Martin Sachs, Horst Truestedt, Richard Taborek, Schelto Van Doorn, and Roger Cummings, who helped especially during the writing of the first edition. Thanks to Herman Presby, Ivan Kaminow, and Jon Sauer, for helping me get involved in optical networking. Thanks to Frank Kampf, Bob Stucke, Harish Sethu, Doug Joseph and Bob Cypher for helping me understand some of the many issues in computer communications. Thanks also to Steve Chapman and Caroline Levine for help in putting the book together and getting the project finished, and to Steve again for help the second time around.

I particularly want to thank Joe Mathis, for teaching me a lot about how to actually get work of this kind done successfully, and Renato Recio, for always demonstrating how to do it well. I also want to thank the InfiniBand crowd, you know who you are, for the chance to do some great work with great people. Hopefully we'll get to do more in the future.

I hope it's clear from this book that I love figuring out and explaining how to build better communications systems — I hope it's helpful to people.

This book is dedicated to Micah, for putting up with the time I spend on this stuff when I should be doing things more important, and to Ashley and Denali, who are busy building their own futures — it's wonderful to watch it happening.

Chapter 1

Fibre Channel and Storage Area Networks

Introduction

Fibre Channel technology is over a decade old. How successful has it been?

Here is an illustration. The first edition of this book included a section called "The Unification of LAN and Channel technologies," which described how Fibre Channel would be part of a trend towards convergence between LANs and channels. LANs (Local Area Networks) are used for computer-to-computer communications, and channels are high-efficiency, high-performance links between computers and their long-term storage devices (disk and tape drives), and other I/O devices.

Since then, the prediction has come true, in three quite different ways.

- Most important has been the introduction and widespread use of the term "Storage Area Network," or SAN, describing a network which is highly optimized for transporting traffic between servers and storage devices.
- At the physical layer, the LAN and Fibre Channel technologies have become nearly identical — Gigabit Ethernet and Fibre Channel share common signaling and data encoding mechanisms, and the future 10 Gb/s Ethernet and Fibre Channel are expected to share nearly the same data rate.
- The management methods for Fibre Channel SANs have steadily approached the traditional methods used for LAN management, although the current level of management effort required for Fibre Channel SANs is still higher than for LANs.

Interestingly, however, although the LAN and SAN types of computer data communications have converged at a technology level, they have so far stayed quite different in how they are used and how they are managed. That is, systems are usually built with the SAN storage traffic separated on separate networks from the LAN traffic, so that the management, topologies, and provisioning of each network can be optimized for the types of traffic traversing them.

The trends that originally motivated the creation of Fibre Channel have continued or accelerated. The speed of processors, the capacities of memory, disks, and tapes, and the use of switched communications networks have all been doubling every 18 to 24 months, and the doubling period has in many cases even been steadily shortening slightly. However, the rate of I/O improvement has been much slower, so that devices are even more I/O limited. The continuing observation is that computers usually appear nearly instantaneous, except when doing I/O (e.g., downloading web pages), or managing stored data (e.g., backing up file systems).

Fibre Channel, and Storage Area Networks, are focused at (a) optimizing the movement of data between server and storage systems, and (b) managing the data and the access to the data, so that communications are optimized as much as possible, while continuously and reliably providing access to data, for whoever needs it.

Fibre Channel Features

Following is a list of the major features that Fibre Channel provides:

- Unification of networking and I/O channel data communications: This was described in detail above, and allows storage to be decoupled from servers and managed separately. Similarly, many servers can directly access the data as if it were their own, as long as they are coordinated to manage it coherently.
- Bandwidth: The base definition of Fibre Channel provides better than 100 MBps for I/O and communications on current architectures, with speeds defined up to 4 times this rate, for implementation as market and applications dictate.
- Inexpensive implementation: Fibre Channel uses an 8B/10B encoding for all data transmission, which, by limiting low-frequency components, allows design of AC-coupled gigabit receivers using inexpensive CMOS VLSI technology
- Low overhead: The very low 10^{-12} bit error rate achievable using a combination of reliable hardware and 8B/10B encoding allows very low extra overhead in the protocol, providing efficient usage of the transmission bandwidth and saving effort in implementation of low-level error recovery mechanisms.
- Low-level control: Local operations depend very little on global information. This means, for example, that the actions that one Port takes are only minimally affected by actions taking place on other Ports, and that individual computers need to maintain very little information about the rest of the network. This feature minimizes the amount of work to do at the higher levels.
  - For example, hardware-controlled flow control alleviates the host processors from the burden of managing much of the flow control overhead.
  - Similarly, the low-level hardware does sophisticated error detection and deletion, so that it can assure delivery of data intact or not at all. Upper layer protocols don't have to do as much error detection, and can be more efficient.
- Flexible topology: Physical connection topologies are defined for (1) point-to-point links, (2) shared-media loop topologies, and (3) packet-switching network topologies. Any of these can be built using the same

hardware, allowing users to match physical topology to the required connectivity characteristics.

- Distance: 50 m in a room simplifies wiring, more important is 10 km, which allows remote copy without WAN infrastructure. Consider a high performance disk drive attached to a computer over an optical fiber. The access time for the disk drive (to rotate the disk and move the head over the data) would be roughly 5 ms. The speed of light in optical fiber is about 124 mi/ms. This means that the time to reach an optically connected disk drive located a mile away would be only 0.008 ms more than the time to reach a disk drive in the same enclosure.
- Availability: More capability to attach to multiple servers allows the data to be accessed through many paths, which enhances availability in case one of those paths fails.
- Flexible transmission service: Mechanisms are defined for multiple Classes of services, including (1) dedicated bandwidth between Port pairs at the full hardware capacity, (2) multiplexed transmission with multiple other source or destination Ports, with acknowledgment of reception, and (3) best-effort multiplexed datagram transmission without acknowledgment, for more efficient transmission in environments where error recovery is handled at a higher level, (4) dedicated connections with configurable quality of service guarantees on transmission bandwidth and latency, and (5) reliable multicast, with a dedicated connection at the full hardware capacity.
- Standard protocol mappings: Fibre Channel can operate as a data transport mechanism for multiple Upper Level Protocols, with mappings defined for IP, SCSI-3, IPI-3 Disk, IPI-3 Tape, HIPPI, the Single Byte Channel Command set for ESCON, the AAL5 mapping of ATM for computer data, and VIA or Virtual Interface Architecture. The most commonly used of these currently are the mapping to SCSI-3, which is termed "FCP," and the mapping to ESCON, which is termed either "FICON," or "SBCON," depending on context.
- Wide industry support: Most major computer, disk drive, and adapter manufacturers are currently developing hardware and/or software components based on the Fibre Channel ANSI standard.

These improvements to traditional channels don't actually provide much real benefit when a single server is used to process the data on a single storage device. However, when multiple servers act together (for better reliability, or higher throughput, or better pipelining, etc.) to work with the data on multiple storage devices of different types, then the advantages of Fibre Channel can become very important.

Storage Area Networks

What is a Storage Area Network, and how is it different from the various other types of networks that are built?

Here is a definition of a Storage Area Network, from one of the leaders in the industry:

> A Storage Area Network (SAN) is a dedicated, centrally managed, secure information infrastructure, which enables any-to-any interconnection of servers and storage systems.

This definition is unfortunately not particularly instructive as to, for example, the difference between SANs and LANs, or MANs, or even WANs, all of which, in some applications, could fit this description.

The difference between SANs and other types of networks can perhaps best be understood by considering the difference between the storage and networking ports on a desktop computer. Every computer has access to some kind of long-term storage, and almost every computer has access to some way of communicating with other computers. The storage interface is highly optimized, tightly controlled (in laptops and most desktop machines, it may not even be visible outside the box), and not shared with any other computers — which helps make it highly predictable, efficient, and fast. Network interfaces, on the other hand, are much slower, less efficient (you have to wait for them), and have higher overhead, but they allow access to any other machine that it knows how to communicate with.

Storage Area Networks are built to incorporate the best of both storage and networking interfaces: fast, efficient communications, optimized for efficient movement of large amounts of data, but with access to a wide range of other servers and storage devices on the network.

The primary difference then between a Storage Area Network and the other types of networks mentioned is that, in a SAN, communication within the network is well-managed, very well-controlled, and predictable. Therefore, each entity on the network can almost operate is if it has sole access to whichever partner on the network that it is currently communicating with.

A primary reason for this has been the idea of decoupling the servers from their storage, and allowing multiple servers to access the same data at the same time. The key here is that client systems often access their through servers, which assure consistency, security, and authorization for the data access. Clients, however, don't particularly care which server is used to access the data, and the data is the same no matter which server is accessing it. This three-tiered system of clients displaying the data, servers processing and managing the data, and storage subsystems holding the data, is tied together with networks — LANs and SANs — between each layer.

Fibre Channel overlaps very little with Ethernet, except in very specific applications. For general-purpose communications, Ethernet is very difficult to compete with (particularly since the Ethernet community tends to adopt the best networking innovations every time there is a new generation, which is regularly).

Fibre Channel does, however, overlap very closely with the storage technologies such as IDE and SCSI. In fact, to a file system or higher-level device, Fibre Channel may appear almost exactly like SCSI — the SCSI command set is transported across a Fibre Channel link, just as it would be across a SCSI bus.

The preceding picture is generally valid for on mid-range machines. On high-end machines, the networking interface is usually still Ethernet (although Token Ring, FDDI, HiPPI, and others have all been important), but the storage interface has, for the last 10 years or so, been a channel protocol. The primary one in the early '90s was called ESCON, for Enterprise System Connections. ESCON was the first real SAN, since it allowed multiple servers to access multiple storage units through a high-performance, switched fabric. In fact, currently the ESCON protocols are still transmitted over a high-performance, switched fabric, but now the fabric is Fibre Channel, and the name has changed to FICON or SBCON.

SAN topologies

A typical topology for a large-scale system using both a Fibre Channel-based Storage Area Network and a Local Area Network is shown in Figure 1.1.

This configuration allows a number of advantages, vs. a system with storage devices tightly integrated with each separate server.

- Networked Access: All servers have direct access to all disk and tape arrays through the SAN, once authorization has been established at the network and the data level.
- Storage Consolidation: Since the client, server, and storage units can be scaled separately, and storage units can be shared, fewer units are necessary. This is especially important for expensive, large tape libraries.
- Remote Mirroring and Archiving: Since the SAN links may be up to 10 km. long, disk and tape drives can be remotely located, for disaster recovery.
- LAN-free backup. The servers can move the data between disk and tape arrays over the SAN — so the LAN between server and clients is not impacted by the backups, and is always available.
- Server-free backup. In the ideal case, the disk array and the tape array have enough intelligence to let the servers command 3rd-party transfers, so that, for example, data would flow directly between a disk array and tape library

Figure 1.1
Example of an Enterprise or Service Provider SAN+LAN Topology

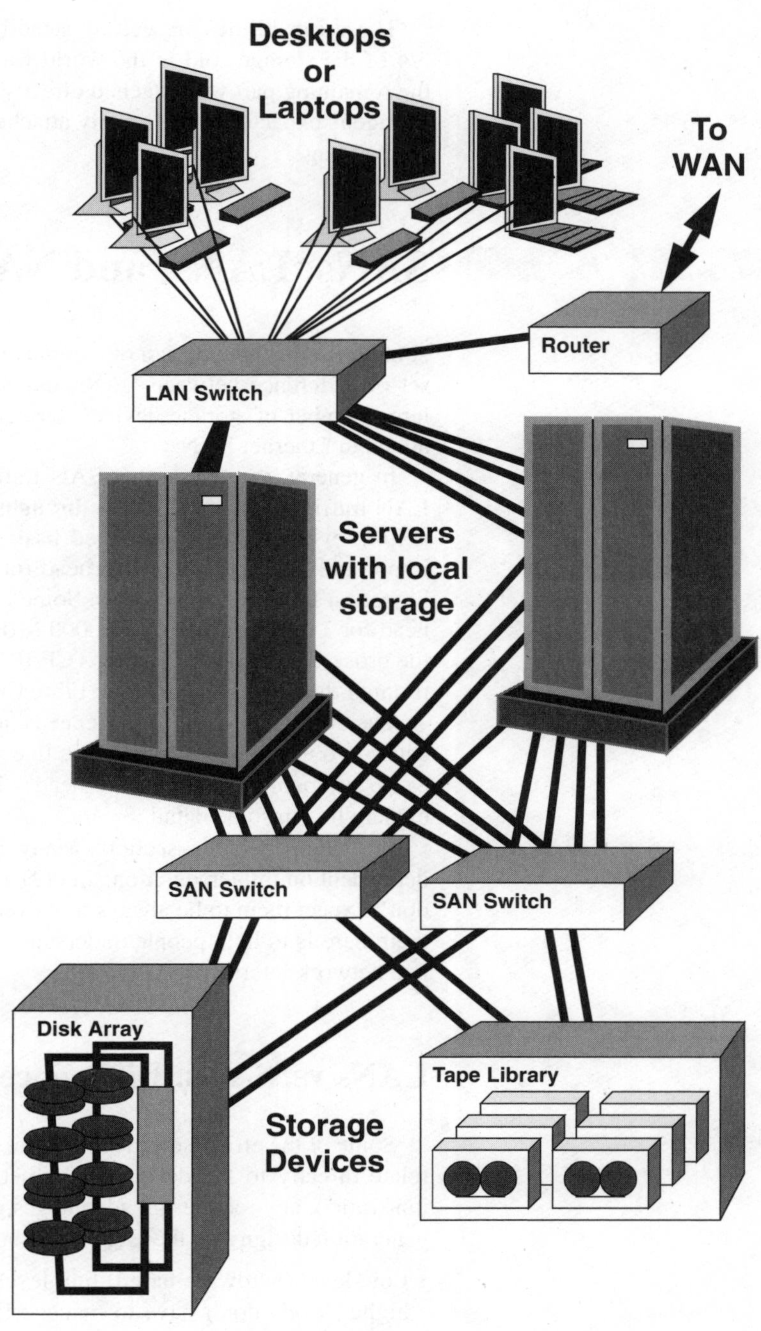

across the SAN, without loading any servers.

These capabilities are getting steadily more important. In 1999, roughly 3/4 of the storage sold in the world was attached directly to servers, while the remaining part was attached directly to the network. In 2003, over 3/4 of storage is expected to be directly attached to the networks, either as SAN or NAS storage.

SANs, LANs, and NAS

A major issue in the design of complex installations such as this involves the set of difference between LANs and SANs, particularly, since there are a large number of storage devices, termed "Network Attached Storage," that attach to Ethernet LANs.

In general, the fact is that SAN traffic is faster and more efficient than LAN traffic. Getting over 80% throughput on SAN links is expected, while getting over 30% on a sustained basis on LAN links is doing well. More importantly, the processor overhead for communications is generally much higher on LANs, than in SANs. Some estimates are that the processor overhead for TCP/IP on a LAN is 1,000 MIPS to receive data at 1 Gb/s, and that the processor overhead running TCP/IP over Ethernet is 30 times higher than running the same data rate over Fibre Channel.

The 30X performance difference is quite amazing — what could possibly cause two networks with the same line speed to use 30X difference in processor protocol-processing overhead? The following sections attempt to explain this in some detail.

A caution on this section. Many of these factors (1) are extremely dependent on implementation, and (2) are changing extremely quickly — so don't expect them to be always true everywhere. The main reason for listing them here is to help people understand how to optimize design of networks and network interfaces.

LANs vs. SANs: Differences in Network Design

Some of the efficiency advantages of Fibre Channel compared to Ethernet relate directly to the design of the network. In an environment of steady innovation, any real design advantages get quickly adopted in all following-generation designs, so these are only short-term advantages.

- Low-level (hardware-based) link-level and end-to-end flow control, so the higher levels don't have to manage flow control and congestion control. High-level flow control and congestion control (e.g., the TCP window mechanism, slow start and congestion avoidance) can require significant overhead, especially on heavily-loaded networks.

- Switch-based transmission (vs. shared medium), so the quality of service for a particular connection can be higher.
- Upper-level protocol information defined in the network-level headers, so low-level hardware can effectively assist higher-level protocol processing.

Again, the network layer for Fibre Channel is not much different than modern Ethernet on a switched fabric (i.e., not shared medium), with link-level backpressure flow control. There are some advantages to the Fibre Channel network vs. Gigabit Ethernet, but not a 30X difference.

LANs vs. SANs: Differences in Protocol Design

The more important advantages in SAN efficiency vs. LAN efficiency and performance relate to the higher levels of protocol design, and have to do with the fact that LANs are, in general, accessed through a TCP/IP (or UDP/IP) protocol stack, where SANs are accessed through a simpler SCSI protocol stack with less overhead on the host processor. This include the following factors.

- Lower-lever error checking. The channels deliver the data to the server intact, or not at all (data corruption, or pulled cable) — so the processors do less checksum calculation or validation of header fields, for example.
- Predictable network performance
  - Ordered transmission — assume no re-ordering of traffic on the network, so the extra overhead associated with checking for correct delivery order, and resource allocation to compensate if you don't have it, are gone.
  - Well-defined network round-trip times, so that the protocol doesn't have to include code to handle the "did the packet get lost, or is it just badly delayed?" problem.
- Request/Response network — the server makes requests to the disk subsystem for reads or writes, so all incoming packets to the server are expected packets. This means:
  - Less header parsing and less handling of special cases, since all packets coming in are expected, and resources for dealing with them have been pre-allocated.
  - Less overhead for flow control — no need to allocate buffer space or do buffer management processing for traffic which may or may not come in.
- Message-based transport: TCP is a sockets stream protocol, where SCSI works in command or data blocks, or messages, with memory space pre-

allocated, so less buffer management, and less data copying, are required in many cases.

- Higher granularity of transfers — Ethernet adapters typically work at the level of Ethernet packets, with all higher-level segmentation and reassembly into IP datagrams, or TCP-level sockets, requires host processor intervention. Fibre Channel adapters typically do reassembly of Frames into Sequences, and deliver the full Sequence to the ULP for processing by the host processor. This means, for example, that there may be fewer processor interrupts, and less context switching.
- Real address operations — SCSI protocols work in the kernel, so there's no switching from user context to kernel context, and real addresses can be used in all the operations, so may be less translation between virtual and physical addresses.

Network-attached Storage (NAS) and Storage Area Networks (SAN)

An area that is closely tied to this difference between LANs and SANs is the difference between NAS and SANs. It is sometimes difficult to be sure of the function difference between the two, partly because they nearly share an acronym, and partly because they both allow networked access to stored data. However, they really are quite different from each other, both in functionality and how they are used.

Part of the difference between Network-attached Storage, and a Storage-Area Network has to do with the network and protocol stack used. Network attached storage emphasizes the network: Ethernet networks and TCP/IP or UDP/IP protocol stacks), where Storage Area Networks use Fibre Channel and a SCSI protocol stack.

The hardware difference is less important than the higher layer differences, however, particularly if both networks operate at nearly the same speed and topology.

A more important key to the difference between NAS and SAN is the distinction in which kind of traffic crosses the network. In NAS, the traffic crossing the network is high-level requests and responses for files, independent of how they are arranged on disks. In SAN, however, the traffic is requests and responses for blocks of data at specific locations on specific disks.

The difference here is that NAS operates above the file system level, where SANs operate below the file system level, at the data block level.

A network-attached storage device is a dedicated file server which holds files, and exports to the clients a picture of a file system. The clients request reads or writes to files, and the network-attached storage device does the file-system work to translate those file requests into operations on disk blocks, then accesses or updates the disk blocks.

A SAN storage device, on the other hand, is much more of a raw, stripped-down storage device. The client or clients do the file system work to translate file access to operations on specific disk blocks, and then send the requests across the network. The storage device does the operations and returns the responses, without any file system work.

This difference in operation, and whether the file system work gets done at the front side or the back side of the network, can make even more of a difference than the difference of whether the traffic goes through a TCP/IP/Ethernet protocol stack, or a SCSI/Fibre Channel protocol stack, since each specific I/O operation may require up to 20,000 processor instructions to complete. Communication overhead can best be minimized by avoiding unnecessary data transfers altogether. Aspects to consider include the following:

- SANs can be much more scalable, since the filesystem work can be distributed among dozens or hundreds of small servers, accessing 1 or 2 large disk arrays. A NAS device would have to do all of the file system processing work itself for all the servers accessing its data, causing a possible bottleneck.
- NAS infrastructure may be cheaper and more easily understood, since a NAS device attaches directly to a standard Ethernet fabric.
- NAS has been around for a long time, since it is essentially a dedicated file server. SANs are newer technology, providing different and better features in many cases.

Often, a combination of the two may be worthwhile: a large network-attached storage device may have many disks inside or behind it, which it may communicate with through a SAN.

It's worth making again the statement about the importance of where the file system work is done. The lowest-overhead communication is communication which is avoided, and avoided communication requires an understanding of what communication is required and what is not. With a SAN, the application requesting the data is running on the same system that's doing the file work, so the policy work of deciding when and where to do disk accesses can be made intelligently to minimize network traffic. With Network Attached Storage, however, the client requesting file access is separate from the NAS device doing the file system work and generating the disk operations, so it's more difficult to make good predictions on which disk accesses will be required and which can be avoided. Data caching may also be easier to optimize using SAN vs. NAS mechanisms.

In sophisticated environments, with complex data management and access requirements, the extra complexity of a SAN based on Fibre Channel can provide a very substantial return on the investment required to learn and build a new and dedicated network infrastructure. Since data is growing tremendously in size and complexity, Storage Area Networking technology has an extremely bright future.

Goals of This Book

In this book, I will try to describe how Fibre Channel works, what strengths and weaknesses it has, and how it fits in with other parts of a modern high-performance computing environment. This is not an easy book — the subject matter is complicated, the treatment is sophisticated, and the discussion goes into more detail than any but a few dedicated readers will actually care to know about the subject. It's necessary, though, to get to this level of detail to achieve what I consider to be the two key goals of this book.

The first goal is to describe the operation of Fibre Channel networks in enough detail that any parts of the specification will make sense. One major characteristic of Fibre Channel is that it tries to solve many different data communications problems within a single architecture. On the negative side, this means that Fibre Channel is quite complicated, with many different options and types of service. On the positive side, this means that Fibre Channel is very flexible and can simultaneously be used for many different types of communications and computer system operations. Much of the work required in implementing Fibre Channel systems is in selecting the parts of the architecture that are best suited to the problem at hand. I will attempt to give a complete picture of all the possible options of a Fibre Channel installation, as well as to show which parts of the architecture are most suitable for usage in particular applications.

The second goal is to help accelerate and improve the development of future networking technologies and architectures. Networking technologies are advancing very rapidly, and as network architects work to integrate these new technologies into new top-to-bottom network architectures, it's helpful to understand at a deep level why existing networks have been designed the way they have. Hopefully, this book will be useful both for driving new technology development and for driving architectures that use those developments while preserving some of the best features of existing networks.

In short, this book is designed to help Fibre Channel network designers and users make best use of the existing technology, and carry further developments in network technology and integrated network architectures well into the future. I hope that this book will be as rewarding to read as it has been to write.

Chapter 2

Overview

Introduction

This chapter provides an overview of the general structure, concepts, organization, and mechanisms of the Fibre Channel protocol. This will provide a background for the detailed discussions of the various parts of the architecture in the following chapters and will give pointers on where to find information about specific parts of the protocol.

A Fibre Channel network is logically made up of one or more bidirectional point-to-point serial data channels, structured for high-performance capability. The basic data rate over the links is just over 1 Gbps, providing >100 MBps data transmission bandwidth, with half-, quarter-, eighth-, double-, and quadruple-speed links defined. Although the Fibre Channel protocol is configured to match the transmission and technological characteristics of single- and multi-mode optical fibers, the physical medium used for transmission can also be copper twisted pair or coaxial cable.

Physically, a Fibre Channel network can be set up as (1) a single point-to-point link between two communication Ports, called "N_Ports," (2) a network of multiple N_Ports, each linked through an "F_Port" into a switching network, called a Fabric, or (3) a ring topology termed an "Arbitrated Loop," allowing multiple N_Port interconnection without switch elements. Each N_Port resides on a hardware entity such as a computer or disk drive, termed a "Node." Nodes incorporating multiple N_Ports can be interconnected in more complex topologies, such as rings of point-to-point links or dual independent redundant Fabrics.

Logically, Fibre Channel is structured as a set of hierarchical functions, as illustrated in Figure 2.1. Interfaces between the levels are defined, but vendors are not limited to specific interfaces between levels if multiple levels are implemented together. A single Fibre Channel Node implementing one or more N_Ports provides a bidirectional link and FC-0 through FC-2 or FC-4 services through each N_Port.

- The FC-0 level describes the physical interface, including transmission media, transmitters and receivers, and their interfaces. The FC-0 level specifies a variety of media and associated drivers and receivers that can operate at various speeds.
- The FC-1 level describes the 8B/10B transmission code that is used to provide DC balance of the transmitted bit stream, to separate transmitted control bytes from data bytes and to simplify bit, byte, and word alignment. In addition, the coding provides a mechanism for detection of some transmission and reception errors.
- The FC-2 level is the signaling protocol level, specifying the rules and mechanisms needed to transfer blocks of data. At the protocol level, the FC-2 level is the most complex level, providing different classes of ser-

Figure 2.1 Fibre Channel structural hierarchy.

Level	Functions		
ULPs	IP · SCSI · IPI-3 · HIPPI · ATM/AAL5 · SBCCS		
FC-4	**Upper Level Protocol Mapping** - Mapping of ULP functions and constructs over Fibre Channel transport service - Policy decisions for use of lower-layer capabilities	Support for one or more FC-4 interfaces on a node	
FC-3	- Common services over multiple N_Ports, e.g., Multicast, Hunt Groups, or striping		
	N_Port		N_Port
FC-2	**Link Service** - Fabric and N_Port Login and Logout - Other Basic and Extended Link Services. Process Login and Logout, determinations of Sequence and Exchange Status, Request Sequence Initiative, Abort Sequences, Echo, Test, end-to-end Credit optimization, etc. **Signaling Protocol** - Frames, Sequences, and Exchanges - N_Ports, F_Ports, and Topologies - Service Classes 1, 2, 3, Intermix, 4, and 6 - Segmentation and reassembly - Flow control, both buffer-to-buffer and end-to-end	• • • One or possibly more N_Ports per Node • • •	
FC-AL	**Arbitrated Loop Functions** - Ordered Sets for loop arbitration, opening and closing communications, enabling/disabling loop Ports - Loop Initialization - AL_PA Physical Address Assignment - Loop Arbitration and Fairness Management	• • •	
FC-1	**Transmission Protocol** - 8B/10B encoding for byte and word alignment, data/special separation, and error minimization through run length minimization and DC balance - Ordered Sets for Frame bounds, low-level flow control, link management - Port Operational State - Error monitoring	• • •	
FC-0	**Physical Interface** - Transmitters and receivers - Link Bandwidth **Media** - Optical or electronic cable plant - Connectors	• • •	

vice, packetization and sequencing, error detection, segmentation and reassembly of transmitted data, and Login services for coordinating communication between Ports with different capabilities.

- The FC-3 level provides a set of services that are common across multiple N_Ports of a Fibre Channel Node. This level is not yet well defined, due to limited necessity for it, but the capability is provided for future expansion of the architecture.
- The FC-4 level provides mapping of Fibre Channel capabilities to preexisting Upper Level Protocols, such as the Internet Protocol (IP) or SCSI (Small Computer Systems Interface), or FICON (Single-Byte Command Code Sets, or ESCON).

FC-0 General Description

The FC-0 level describes the link between two Ports. Essentially, this consists of a pair of either optical fiber or electrical cables along with transmitter and receiver circuitry which work together to convert a stream of bits at one end of the link to a stream of bits at the other end. The FC-0 level describes the various kinds of media allowed, including single-mode and multi-mode optical fibers, as well as coaxial and twisted pair electrical cables for shorter distance links. It describes the transmitters and receivers used for interfacing to the media. It also describes the data rates implemented over the cables. The FC-0 level is designed for maximum flexibility and allows the use of a wide variety of technologies to meet a range of system requirements.

Each fiber is attached to a transmitter of a Port at one end and a receiver of another Port at the other end. The simplest configuration is a bidirectional pair of links, as shown in Figure 2.2. A number of different Ports may be connected through a switched Fabric, and the loop topology allows multiple Ports to be connected together without a routing switch, as shown in Figure 2.3.

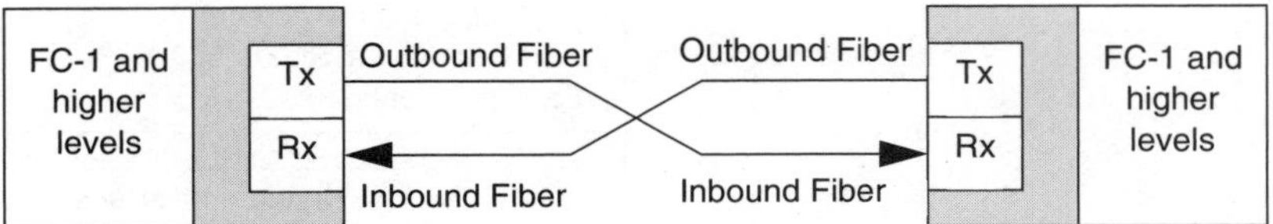

Figure 2.2
FC-0 link.

A multi-link communication path between two N_Ports may be made up of links of different technologies. For example, it may have copper coaxial cable links attached to end Ports for short-distance links, with single-mode

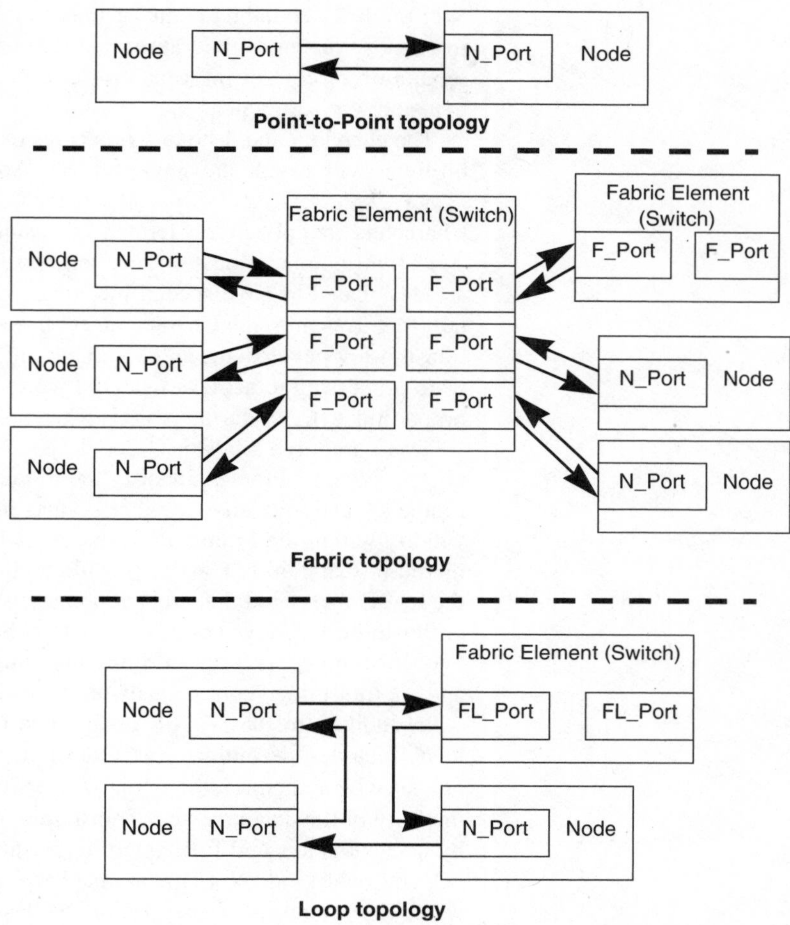

Figure 2.3
Examples of Point-to-point, Fabric, and Arbitrated Loop topologies.

optical fibers for longer-distance links between switches separated by longer distances.

FC-1 General Description

In a Fibre Channel network, information is transmitted using an 8B/10B data encoding. This coding has a number of characteristics which simplify design of inexpensive transmitter and receiver circuitry that can operate at the 10^{-12} bit error rate required. It bounds the maximum run length, assuring that there are never more than 5 identical bits in a row except at synchronization points. It maintains overall DC balance, ensuring that the signals transmitted over the links contain an equal number of 1s and 0s. It minimizes the low-frequency content of the transmitted signals. Also, it allows straightforward separation of control information from the transmitted data, and simplifies byte and word alignment.

The encoding and decoding processes result in the conversion between 8-bit bytes with a separate single-bit "data/special" flag indication and 10-bit "Data Characters" and "Special Characters." Data Characters and Special Characters are collectively termed "Transmission Characters."

Certain combinations of Transmission Characters, called Ordered Sets, are designated to have special meanings. Ordered Sets, which always contain four Transmission Characters, are used to identify Frame boundaries, to transmit low-level status and command information, to enable simple hardware processing to achieve byte and word synchronization, and to maintain proper link activity during periods when no data are being sent.

There are three kinds of Ordered Sets. Frame delimiters mark the beginning and end of Frames, identify the Frame's Class of Service, indicate the Frame's location relative to other Frames in the Sequence, and indicate data validity within the Frame. Primitive Signals include Idles, which are transmitted to maintain link activity while no other data can be transmitted, and the R_RDY Ordered Set, which operates as a low-level acknowledgment for buffer-to-buffer flow control. Primitive Sequences are used in Primitive Sequence protocols for performing link initialization and link-level recovery and are transmitted continuously until a response is received.

In addition to the 8B/10B coding and Ordered Set definition, the FC-1 level includes definitions for "transmitters" and "receivers." These are blocks which monitor the signal traversing the link and determining the integrity of the data received. Transmitter and receiver behavior is specified by a set of states and their interrelationships. These states are divided into "Operational" and "Not Operational" types. FC-1 also specifies monitoring capabilities and special operation modes for transmitters and receivers. Example block diagrams of a transmitter and a receiver are shown in Figure 2.4. The serial and serial/parallel converter sections are part of FC-0, while the FC-1 level contains the 8B/10B coding operations and the multiplexing and demultiplexing between bytes and 4-byte words, as well as the monitoring and error detection functionality.

FC-2 General Description

The FC-2 level is the most complex part of Fibre Channel and includes most of the Fibre Channel-specific constructs, procedures, and operations. The basic parts of the FC-2 level are described in overview in the following sec-

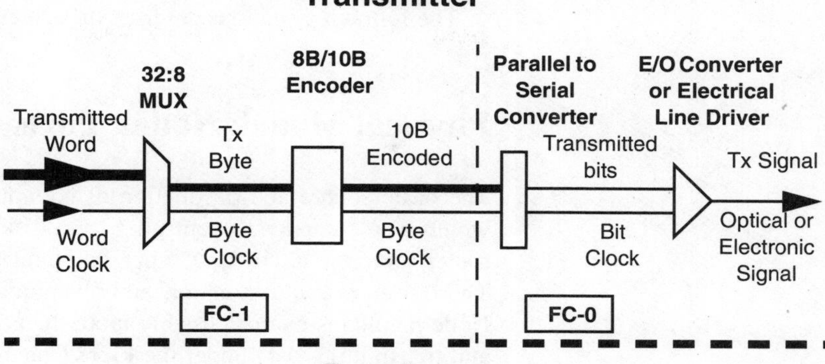

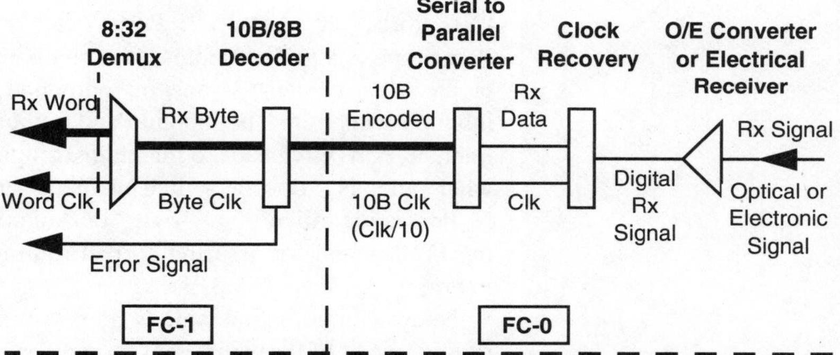

Figure 2.4
Transmitter and receiver FC-1 and FC-0 data flow stages.

tions, with full description left to later chapters. The elements of the FC-2 level include the following:

- Physical Model: Nodes, Ports, and topologies
- Bandwidth and Communication Overhead
- Building blocks and their hierarchy
- Link Control Frames
- General Fabric model
- Flow control
- Classes of service provided by the Fabric and the N_Ports
- Basic and Extended Link Service Commands
- Protocols
- Arbitrated Loop functions

- Segmentation and reassembly
- Error detection and recovery

The following sections describe these elements in more detail.

Physical Model: Nodes, Ports, and Topologies

The basic source and destination of communications under Fibre Channel would be a computer, a controller for a disk drive or array of disk drives, a router, a terminal, or any other equipment engaged in communications. These sources and destinations of transmitted data are termed "Nodes." Each Node maintains one or possibly more than one facility capable of receiving and transmitting data under the Fibre Channel protocol. These facilities are termed "N_Ports." Fibre Channel also defines a number of other types of "Ports," which can transmit and receive Fibre Channel data, including "NL_Ports," "F_Ports," "E_Ports," etc., which are described below. Each Port supports a pair of "fibres" (which may physically be either optical fibers or electrical cables) — one for outbound transmission, and the other for inbound reception. The inbound and outbound fibre pair is termed a "link." Each N_Port only needs to maintain a single pair of fibres, without regard to what other N_Ports or switch elements are present in the network. Each N_Port is identified by a 3-byte "Port identifier," which is used for qualifying Frames and for assuring correct routing of Frames through a loop or a Fabric.

Nodes containing a single N_Port with a fibre pair link can be interconnected in one of three different topologies, shown in Figure 2.3. Each topology supports bidirectional flow between source and destination N_Ports. The three basic types of topologies include:

Point-to-point: The simplest topology directly connecting two N_Ports is termed "Point-to-point," and it has the obvious connectivity as a single link between two N_Ports.

Fabric: More than two N_Ports can be interconnected using a "Fabric," which consists of a network of one or more "switch elements" or "switches." A switch contains two or more facilities for receiving and transmitting data under the protocol, termed "F_Ports." The switches receive data over the F_Ports and, based on the destination N_Port address, route it to the proper F_Port (possibly through another switch, in a multistage network), for delivery to a destination N_Port. Switches are fairly complex units, containing facilities for maintaining routing to all N_Ports on the Fabric, handling flow control, and satisfying the requirements of the different Classes of service supported.

Arbitrated Loop: Multiple N_Ports can also be connected together without benefit of a Fabric by attaching the incoming and outgoing fibers to different Ports to make a loop configuration. A Node Port which incorporates the small amount of extra function required for operation in this topology is termed an "NL_Port." This is a blocking topology — a single NL_Port arbitrates for access to the entire loop and prevents access by any other NL_Ports while it is communicating. However, it provides connectivity between multiple Ports while eliminating the expense of incorporating a switch element.

It is also possible to mix the Fabric and Arbitrated Loop topologies, where a switch Fabric Port can participate on the Loop, and data can go through the switch and around the loop. A Fabric Port capable of operating on a loop is termed an "FL_Port."

Most Fibre Channel functions and operations are topology-independent, although routing of data and control of link access will naturally depend on what other Ports may access a link. A series of "Login" procedures performed after a reset allow an N_Port to determine the topology of the network to which it is connected, as well as other characteristics of the other attached N_Port, NL_Ports, or switch elements. The Login procedures are described further in the "Protocols" section, on page 35 below.

Bandwidth and Communication Overhead

The maximum data transfer bandwidth over a link depends both on physical parameters, such as clock rate and maximum baud rate, and on protocol parameters, such as signaling overhead and control overhead. The data transfer bandwidth can also depend on the communication model, which describes the amount of data being sent in each direction at any particular time.

The primary factor affecting communications bandwidth is the clock rate of data transfer. The base clock rate for data transfer under Fibre Channel is 1.0625 GHz, with 1 bit transmitted every clock cycle. For lower bandwidth, less expensive links, half-, quarter-, and eighth-speed clock rates are defined. Double- and quadruple-speed links have been defined for implementation in the near future as well. The most commonly used data rates will likely be the full-speed and quarter-speed initially, with double- and quadruple-speed components becoming available as the technology and market demand permit.

Figure 2.5 shows a sample communication model, for calculating the achievable data transfer bandwidth over a full speed link. The figure shows a single Fibre Channel Frame, with a payload size of 2048 bytes. To transfer this payload, along with an acknowledgment for data traveling in the reverse

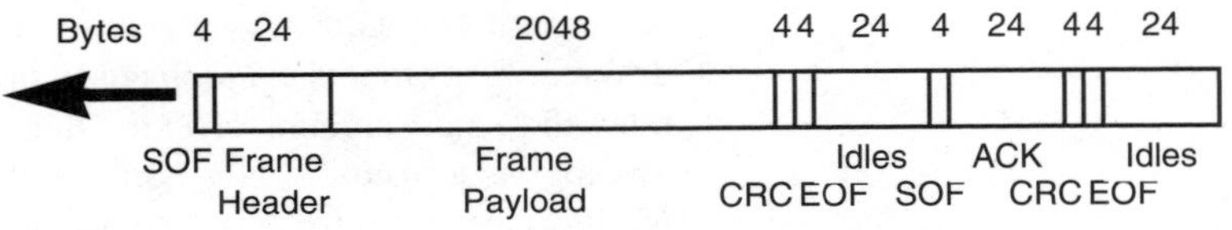

Figure 2.5
Sample Data Frame + ACK Frame transmission, for bandwidth calculation.

direction on a separate fiber for bidirectional traffic, the following overhead elements are required:

SOF: Start of Frame delimiter, for marking the beginning of the Frame (4 bytes),

Frame Header: Frame header, indicating source, destination, sequence number, and other Frame information (24 bytes),

CRC: Cyclic Redundancy Code word, for detecting transmission errors (4 bytes),

EOF: End of Frame delimiter, for marking the end of the Frame (4 bytes),

Idles: Inter-Frame space for error detection, synchronization, and insertion of low-level acknowledgments (24 bytes),

ACK: Acknowledgment for a Frame from the opposite Port, needed for bidirectional transmission (36 bytes), and

Idles: Inter-Frame space between the ACK and the following Frame (24 bytes).

The sum of overhead bytes in this bidirectional transmission case is 120 bytes, yielding an effective data transfer rate of 100.369 MB/s:

$$1.0625[Gbps] \times \frac{2048[payload]}{2168[payload + overhead]} \times \frac{1[byte]}{10[codebits]} = 100.369$$

Thus, the full-speed link provides better than 100 MBps data transport bandwidth, even with signaling overhead and acknowledgments. The achieved bandwidth during unidirectional communication would be slightly higher, since no ACK frame with following Idles would be required. Beyond this, data transfer bandwidth scales directly with transmission clock speed, so that, for example, the data transfer rate over a half-speed link would be 100.369 / 2 = 50.185 MBps.

Building Blocks and Their Hierarchy

The set of building blocks defined in FC-2 are:

Frame: A series of encoded transmission words, marked by Start of Frame and End of Frame delimiters, with Frame Header, Payload, and possibly an optional Header field, used for transferring Upper Level Protocol data

Sequence: A unidirectional series of one or more Frames flowing from the Sequence Initiator to the Sequence Recipient

Exchange: A series of one or more non-concurrent Sequences flowing either unidirectionally from Exchange Originator to the Exchange Responder or bidirectionally, following transfer of Sequence Initiative between Exchange Originator and Responder

Protocol: A set of Frames, which may be sent in one or more Exchanges, transmitted for a specific purpose, such as Fabric or N_Port Login, Aborting Exchanges or Sequences, or determining remote N_Port status

An example of the association of multiple Frames into Sequences and multiple Sequences into Exchanges is shown in Figure 2.6. The figure shows four Sequences, which are associated into two unidirectional and one bidirectional Exchange. Further details on these constructs follow.

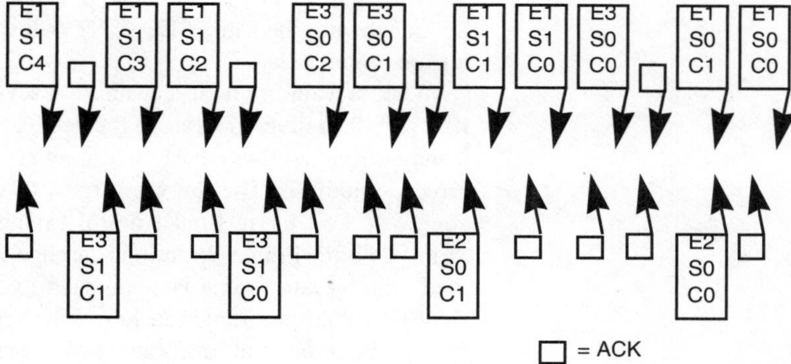

Figure 2.6
Building blocks for the FC-2 Frame / Sequence / Exchange hierarchy.

Frames. Frames contain a Frame header in a common format (see Figure 7.1), and may contain a Frame payload. Frames are broadly categorized under the following classifications:

- Data Frames, including
  - Link Data Frames
  - Device Data Frames
  - Video Data Frames
- Link Control Frames, including

- Acknowledge (ACK) Frames, acknowledging successful reception of 1 (ACK_1), N (ACK_N), or all (ACK_0) Frames of a Sequence
- Link Response ("Busy" (P_BSY, F_BSY) and "Reject" (P_RJT, F_RJT) Frames, indicating unsuccessful reception of a Frame
- Link Command Frames, including only Link Credit Reset (LCR), used for resetting flow control credit values

Frames operate in Fibre Channel as the fundamental block of data transfer. As stated above, each Frame is marked by Start of Frame and End of Frame delimiters. In addition to the transmission error detection capability provided by the 8B/10B code, error detection is provided by a 4-byte CRC value, which is calculated over the Frame Header, optional Header (if included), and payload. The 24-byte Frame Header identifies a Frame uniquely and indicates the processing required for it. The Frame Header includes fields denoting the Frame's source N_Port ID, destination N_Port ID, Sequence ID, Originator and Responder Exchange IDs, routing, Frame count within the Sequence, and control bits.

Every Frame must be part of a Sequence and an Exchange. Within a Sequence, the Frames are uniquely identified by a 2-byte counter field termed SEQ_CNT in the Frame header. No two Frames in the same Sequence with the same SEQ_CNT value can be active at the same time, to ensure uniqueness.

When a Data Frame is transmitted, several different things can happen to it. It may be delivered intact to the destination, it may be delivered corrupted, it may arrive at a busy Port, or it may arrive at a Port which does not know how to handle it. The delivery status of the Frame will be returned to the source N_Port using Link Control Frames if possible, as described in the "Link Control Frames" section, on page 27. A Link Control Frame associated with a Data Frame is sent back to the Data Frame's source from the final Port that the Frame reaches, unless no response is required, or a transmission error prevents accurate knowledge of the Frame Header fields.

Sequences. A Sequence is a set of one or more related Data Frames transmitted unidirectionally from one N_Port to another N_Port, with corresponding Link Control Frames, if applicable, returned in response. The N_Port which transmits a Sequence is referred to as the "Sequence Initiator" and the N_Port which receives the Sequence is referred to as the "Sequence Recipient."

Each Sequence is uniquely specified by a Sequence Identifier (SEQ_ID), which is assigned by the Sequence Initiator. The Sequence Recipient uses the same SEQ_ID value in its response Frames. Each Port operating as Sequence Initiator assigns SEQ_ID values independent of all other Ports, and uniqueness of a SEQ_ID is only assured within the set of Sequences initiated by the same N_Port.

The SEQ_CNT value, which uniquely identifies Frames within a Sequence, is started either at zero in the first Frame of the Sequence or at 1 more than the value in the last Frame of the previous Sequence of the same Exchange. The SEQ_CNT value is incremented by 1 each subsequent Frame. This assures uniqueness of each Frame header active on the network.

The status of each Sequence is tracked, while it is open, using a logical construct called a Sequence Status Block. Normally separate Sequence Status Blocks are maintained internally at the Sequence Initiator and at the Sequence Recipient. A mechanism does exist for one N_Port to read the Sequence Status Block of the opposite N_Port, to assist in recovery operations, and to assure agreement on Sequence state.

There are limits to the maximum number of simultaneous Sequences which an N_Port can support per Class, per Exchange, and over the entire N_Port. These values are established between N_Ports before communication begins through an N_Port Login procedure.

Error recovery is performed on Sequence boundaries, at the discretion of a protocol level higher than FC-2. Dependencies between the different Sequences of an Exchange are indicated by the Exchange Error Policy, as described below.

Exchanges. An Exchange is composed of one or more non-concurrent related Sequences, associated into some higher level operation. An Exchange may be unidirectional, with Frames transmitted from the "Exchange Originator" to the "Exchange Responder," or bidirectional, when the Sequences within the Exchange are initiated by both N_Ports (nonconcurrently). The Exchange Originator, in originating the Exchange, requests the directionality. In either case, the Sequences of the Exchange are nonconcurrent, i.e., each Sequence must be completed before the next is initiated.

Each Exchange is identified by an "Originator Exchange ID," denoted as OX_ID in the Frame Headers, and possibly by a "Responder Exchange ID," denoted as RX_ID. The OX_ID is assigned by the Originator, and is included in the first Frame transmitted. When the Responder returns an acknowledgment or a Sequence in the opposite direction, it may include an RX_ID in the Frame Header to let it uniquely distinguish Frames in the Exchange from other Exchanges. Both the Originator and Responder must be able to uniquely identify Frames based on the OX_ID and RX_ID values, the source and destination N_Port IDs, SEQ_ID, and the SEQ_CNT. The OX_ID and RX_ID fields may be set to the "unassigned" value of x'FFFF' if the other fields can uniquely identify Frames. If an OX_ID or RX_ID is assigned, all subsequent Frames of the Sequence, including both Data and Link Control Frames, must contain the Exchange ID(s) assigned.

An Originator may initiate multiple concurrent Exchanges, even to the same destination N_Port, as long as each uses a unique OX_ID. Exchanges may not cross between multiple N_Ports, even multiple N_Ports on a single Node.

Large-scale systems may support up to thousands of potential Exchanges, across several N_Ports, even if only a few Exchanges (e.g., tens) may be active at any one time within an N_Port. In these cases, Exchange resources may be locally allocated within the N_Port on an "as needed" basis. An "Association Header" construct, transmitted as an optional header of a Data Frame, provides a means for an N_Port to invalidate and reassign an X_ID (OX_ID or RX_ID) during an Exchange. An X_ID may be invalidated when the associated resources in the N_Port for the Exchange are not needed for a period of time. This could happen, for example, when a file subsystem is disconnecting from the link while it loads its cache with the requested data. When resources within the N_Port are subsequently required, the Association Header is used to locate the "suspended" Exchange, and an X_ID is reassigned to the Exchange so that operation can resume. X_ID support and requirements are established between N_Ports before communication begins through an N_Port Login procedure.

Fibre Channel defines four different Exchange Error Policies. Error policies describe the behavior following an error, and the relationship between Sequences within the same Exchange. The four Exchange Error policies include:

Abort, discard multiple Sequences: Sequences are interdependent and must be delivered to an upper level in the order transmitted. An error in one Frame will cause that Frame's Sequence and all later Sequences in the Exchange to be undeliverable.

Abort, discard a single Sequence: Sequences are not interdependent. Sequences may be delivered to an upper level in the order that they are received complete, and an error in one Sequence does not cause rejection of subsequent Sequences.

Process with infinite buffering: Deliverability of Sequences does not depend on all the Frames of the Sequence being intact. This policy is intended for applications such as video data where retransmission is unnecessary (and possibly detrimental). As long as the first and last Frame of the Sequence are received, the Sequence can be delivered to the upper level.

Discard multiple Sequences with immediate retransmission: This is a special case of the "Abort, discard multiple Sequences" Exchange Error Policy, where the Sequence Recipient can use a Link Control Frame to request that a corrupted Sequence be retransmitted immediately. This Exchange Error Policy can only apply to Class 1 transmission.

The Error Policy is determined at the beginning of the Exchange by the Exchange Originator and cannot change during the Exchange. There is no dependency between different Exchanges on error recovery, except that errors serious enough to disturb the basic integrity of the link will affect all active Exchanges simultaneously.

The status of each Exchange is tracked, while it is open, using a logical construct called a Exchange Status Block. Normally separate Exchange Status Blocks are maintained internally at the Exchange Originator and at the Exchange Responder. A mechanism does exist for one N_Port to read the Exchange Status Block of the opposite N_Port of an Exchange, to assist in recovery operations, and to assure agreement on Exchange status. These Exchange Status Blocks maintain connection to the Sequence Status Blocks for all Sequences in the Exchange while the Exchange is open.

Link Control Frames

Link Control Frames are used to indicate successful or unsuccessful reception of each Data Frame. Link Control Frames are only used for Class 1 and Class 2 Frames — all link control for Class 3 Frames is handled above the Fibre Channel level. Every Data Frame should generate a returning Link Control Frame (although a single ACK_N or ACK_0 can cover more than one Data Frame). If a P_BSY or F_BSY is returned, the Frame may be retransmitted, up to some limited and vendor-specific number of times. If a P_RJT or F_RJT is returned, or if no Link Control Frame is returned, recovery processing happens at the Sequence level or higher; there is no facility for retransmitting individual Frames following an error.

General Fabric Model

The Fabric, or switching network, if present, is not directly part of the FC-2 level, since it operates separately from the N_Ports. However, the constructs it operates on are at the same level, so they are included in the FC-2 discussion.

The primary function of the Fabric is to receive Frames from source N_Ports and route them to their correct destination N_Ports. To facilitate this, each N_Port which is physically attached through a link to the Fabric is characterized by a 3-byte "N_Port Identifier" value. The N_Port Identifier values of all N_Ports attached to the Fabric are uniquely defined in the Fabric's address space. Every Frame header contains S_ID and D_ID fields containing the source and destination N_Port identifier values, respectively, which are used for routing.

To support these functions, a Fabric Element or switch is assumed to provide a set of "F_Ports," which interface over the links with the N_Ports, plus a "Connection-based" and/or "Connectionless" Frame routing functionality. An F_Port is a entity which handles FC-0, FC-1, and FC-2 functions up to the Frame level to transfer data between attached N_Ports. A Connection-based router, or Sub-Fabric, routes Frames between Fabric Ports through Class 1 Dedicated Connections, assuring priority and non-interference from any other network traffic. A Connectionless router, or Sub-Fabric, routes Frames between Fabric Ports on a Frame-by-Frame basis, allowing multiplexing at Frame boundaries.

Implementation of a Connection-based Sub-Fabric is incorporated for Class 1, Class 4, and Class 6 service, while a Connectionless Sub-Fabric is incorporated for supporting Class 2 and 3 service. Although the term "Sub-Fabric" implies that separate networks are used for the two types of routing, this is not necessary. An implementation may support the functionality of Connection-based and Connectionless Sub-Fabrics either through separate internal hardware or through priority scheduling and routing management operations in a single internal set of hardware. Internal design of a switch element is largely implementation-dependent, as long as the priority and bandwidth requirements are met.

Fabric Ports. A switch element contains a minimum of two Fabric Ports. There are several different types of Fabric Ports, of which the most important are F_Ports. F_Ports are attached to N_Ports and can transmit and receive Frames, Ordered Sets, and other information in Fibre Channel format. An F_Port may or may not verify the validity of Frames as they pass through the Fabric. Frames are routed to their proper destination N_Port and intervening F_Port based on the destination N_Port identifier (D_ID). The mechanism used for doing this is implementation dependent, although address translation and routing mechanisms within the Fabric are being addressed in current Fibre Channel development work.

In addition to F_Ports, which attach directly to N_Ports in a switched Fabric topology, several other types of Fabric Ports are defined. In a multi-layer network, switches are connected to other switches through "E_Ports" (Expansion Ports), which may use standard media, interface, and signaling protocols or may use other implementation-dependent protocols. A Fabric Port that incorporates the extra Port states, operations, and Ordered Set recognition to allow it to connect to an Arbitrated Loop, as shown in Figure 2.3, is termed an "FL_Port." A "G_Port" has the capability to operate as either an E_Port or an F_Port, depending on how it is connected, and a "GL_Port" can operate as an F_Port, as an E_Port, or as an FL_Port. Since implementation of these types of Ports is implementation-dependent, the discussion in this book will concentrate on F_Ports, with clear requirements for extension to other types of Fabric Ports.

Each F_Port may contain receive buffers for storing Frames as they pass through the Fabric. The size of these buffers may be different for Frames in different Classes of service. The maximum Frame size capabilities of the Fabric for the various Classes of service are indicated for the attached N_Ports during the "Fabric Login" procedure, as the N_Ports are determining network characteristics.

Connection-Based Routing. The Connection-based Sub-Fabric function provides support for Dedicated Connections between F_Ports and the N_Ports attached to these F_Ports for Class 1, Class 4, or Class 6 service. Such Dedicated Connections may be either bidirectional or unidirectional and may support the full transmission rate concurrently in each direction, or some lower transmission rate. Class 1 Dedicated Connection is described here. Class 4 and Class 6 are straightforward modifications of Class 1, and are described in the "Classes of Service" section, on page 31.

On receiving a Class 1 connect-request Frame from an N_Port, the Fabric begins establishing a Dedicated Connection to the destination N_Port through the connection-based Sub-Fabric. The Dedicated Connection is pending until the connect-request is forwarded to the destination N_Port. If the destination N_Port can accept the Dedicated Connection, it returns an acknowledgment. In passing the acknowledgment back to the source N_Port, the Fabric finishes establishing the Dedicated Connection. The exact mechanisms used by the Fabric to establish the Connection are vendor-dependent. If either the Fabric or the destination Port are unable to establish a Dedicated Connection, they return a "BSY" (busy) or "RJT" (reject) Frame with a reason code to the source N_Port, explaining the reason for not establishing the Connection.

Once the Dedicated Connection is established, it appears to the two communicating N_Ports as if a dedicated circuit has been established between them. Delivery of Class 1 Frames between the two N_Ports cannot be degraded by Fabric traffic between other N_Ports or by attempts by other N_Ports to communicate with either of the two. All flow control is managed using end-to-end flow control between the two communicating N_Ports.

A Dedicated Connection is retained until either a removal request is received from one of the two N_Ports or an exception condition occurs which causes the Fabric to remove the Connection.

A Class 1 N_Port and the Fabric may support "stacked connect-requests." This function allows an N_Port to simultaneously request multiple Dedicated Connections to multiple destinations and allows the Fabric to service them in any order. This allows the Fabric to queue connect-requests and to establish the Connections as the destination N_Ports become available.

While the N_Port is connected to one destination, the Fabric can begin processing another connect-request to minimize the connect latency. If stacked connect-requests are not supported, connect-requests received by the Fabric for either N_Port in a Dedicated Connection will be replied to with a "BSY" (busy) indication to the requesting N_Port, regardless of Intermix support.

If a Class 2 Frame destined to one of the N_Ports established in a Dedicated Connection is received, and the Fabric or the destination N_Port doesn't support Intermix, the Class 2 Frame may be busied and the transmitting N_Port is notified. In the case of a Class 3 Frame, the Frame is discarded and no notification is sent. The destination F_Port may be able to hold the Frame for a period of time before discarding the Frame or returning a busy Link Response. If Intermix is supported and the Fabric receives a Class 2 or Class 3 Frame destined to one of the N_Ports established in a Dedicated Connection, the Fabric may allow delivery with or without a delay, as long as the delivery does not interfere with the transmission and reception of Class 1 Frames.

Class 4 Dedicated Connections are similar to Class 1 connections, but they allow each connection to occupy a fraction of the source and destination N_Port link bandwidths, to allow finer control on the granularity of Quality of Service guarantees for transmission across the Fabric. The connect-request for a Class 4 dedicated connection specifies the requested bandwidth, and maximum end-to-end latency, for connection, in each direction, and the acceptance of connection by the Fabric commits it to honor those Quality of Service parameters during the life of the connection.

Class 6 is a Uni-Directional Dedicated Connection service allowing an acknowledged multicast connection, which is useful for efficient data replication in systems providing high availability. In Class 6 service, each Frame transmitted by the source of the Dedicated Connection is replicated by the Fabric and delivered to each of a set of destination N_Ports. The destination N_Ports then return acknowledgements indicating correct and complete delivery of the Frames, and the Fabric aggregates the acknowledgments into a single response which is returned to the source N_Port.

Connectionless Routing. A Connectionless Sub-Fabric is characterized by the absence of Dedicated Connections. The connectionless Sub-Fabric multiplexes Frames at Frame boundaries between multiple source and destination N_Ports through their attached F_Ports.

In a multiplexed environment, with contention of Frames for F_Port resources, flow control for connectionless routing is more complex than in the Dedicated Connection circuit-switched transmission. For this reason, flow control is handled at a finer granularity, with buffer-to-buffer flow control across each link. Also, a Fabric will typically implement internal buffering to temporarily store Frames that encounter exit Port contention until the congestion eases. Any flow control errors that cause overflow of the buffering mechanisms may cause loss of Frames. Loss of a Frame can clearly be extremely detrimental to data communications in some cases and it will be avoided at the Fabric level if at all possible.

In Class 2, the Fabric will notify the source N_Port with a "BSY" (busy) or a "RJT" (reject) indication if the Frame can't be delivered, with a code explaining the reason. The source N_Port is not notified of non-delivery of a Class 3 Frame, since error recovery is handled at a higher level.

Classes of Service

Fibre Channel currently defines five Classes of service, which can be used for transmitting different types of traffic with different delivery requirements. The Classes of service are not mandatory, in that a Fabric or N_Port may not support all Classes. The Classes of service are not topology-dependent. However, topology will affect performance under the different Classes, e.g., performance in a Point-to-point topology will be affected much less by the choice of Class of service than in a Fabric topology.

The five Classes of service are as follows. Class 1 service is intended to duplicate the functions of a dedicated channel or circuit-switched network, guaranteeing dedicated high-speed bandwidth between N_Port pairs for a defined period. Class 2 service is intended to duplicate the functions of a packet-switching network, allowing multiple Nodes to share links by multiplexing data as required. Class 3 service operates as Class 2 service without acknowledgments, allowing Fibre Channel transport with greater flexibility and efficiency than the other Classes under a ULP which does its own flow control, error detection, and recovery. In addition to these three, Fibre Channel Ports and switches may support Intermix, which combines the advantages of Class 1 with Class 2 and 3 service by allowing Class 2 and 3 Frames to be intermixed with Class 1 Frames during Class 1 Dedicated Connections. Class 4 service allows the Fabric to provide quality of service guarantees for bandwidth and latency over a fractional portion of a link bandwidth. Class 6 service operates as an acknowledged multicast, with unidirectional transmission from 1 source to multiple destinations at full channel bandwidth.

Class 1 Service: Dedicated Connection. Class 1 is a service which establishes Dedicated Connections between N_Ports through the Fabric, if available. A Class 1 Dedicated Connection is established by the transmission of a "Class 1 connect-request" Frame, which sets up the Connection and may or may not contain any message data. Once established, a Dedicated Connection is retained and guaranteed by the Fabric and the destination N_Port until the Connection is removed by some means. This service guarantees maximum transmission bandwidth between the two N_Ports during the established Connection. The Fabric, if present, delivers Frames to the destination N_Port in the same order that they are transmitted by the source N_Port. Flow control and error recovery are handled between the communicating N_Ports, with no Fabric intervention under normal operation.

Management of Class 1 Dedicated Connections is independent of Exchange origination and termination. An Exchange may be performed within one Class 1 Connection or may be continued across multiple Class 1 Connections.

Class 2 Service: Multiplex. Class 2 is a connectionless service with the Fabric, if present, multiplexing Frames at Frame boundaries. Multiplexing is supported from a single source to multiple destinations and to a single destination from multiple sources. The Fabric may not necessarily guarantee delivery of Data Frames or acknowledgments in the same sequential order in which they were transmitted by the source or destination N_Port. In the absence of link errors, the Fabric guarantees notification of delivery or failure to deliver.

Class 3 Service: Datagram. Class 3 is a connectionless service with the Fabric, if present, multiplexing Frames at Frame boundaries. Class 3 supports only unacknowledged delivery, where the destination N_Port sends no acknowledgment of successful or unsuccessful Frame delivery. Any acknowledgment of Class 3 service is up to and determined by the ULP utilizing Fibre Channel for data transport. The transmitter sends Class 3 Data Frames in sequential order within a given Sequence, but the Fabric may not necessarily guarantee the order of delivery. In Class 3, the Fabric is expected to make a best effort to deliver the Frame to the intended destination but may discard Frames without notification under high-traffic or error conditions. When a Class 3 Frame is corrupted or discarded, any error recovery or notification is performed at the ULP level. Class 3 can also be used for an unacknowledged multicast service, where the destination ID of the Frames specifies a pre-arranged multicast group ID, and the Frames are replicated without modification and delivered to every N_Port in the group.

Intermix. A significant problem with Class 1 as described above is that if the source N_Port has no Class 1 data ready for transfer during a Dedicated Connection, the N_Port's transmission bandwidth is unused, even if there might be Class 2 or 3 Frames which could be sent. Similarly, the destination

N_Port's available bandwidth is unused, even if the Fabric might have received Frames which could be delivered to it.

Intermix is an option of Class 1 service which solves this efficiency problem by allowing interleaving of Class 2 and Class 3 Frames during an established Class 1 Dedicated Connection. In addition to the possible efficiency improvement described, this function may also provide a mechanism for a sender to transmit high-priority Class 2 or Class 3 messages without the overhead required in tearing down an already-established Class 1 Dedicated Connection.

Support for Intermix is optional, as is support for all other Classes of service. This support is indicated during the Login period, when the N_Ports, and Fabric, if present, are determining the network configuration. Both N_Ports in a Dedicated Connection as well as the Fabric, if present, must support Intermix, for it to be used.

Fabric support for Intermix requires that the full Class 1 bandwidth during a Dedicated Connection be available, if necessary — insertion of Class 2 or 3 Frames cannot delay delivery of Class 1 Frames. In practice, this means that the Fabric must implement Intermix to the destination N_Port either by waiting for unused bandwidth or by inserting Intermixed Frames "in between" Class 1 Frames, removing Idle transmission words between Class 1 Frames to make up the bandwidth used for the Intermixed Class 2 or 3 Frame. If a Class 1 Frame is generated during transmission of a Class 2 or Class 3 Frame, the Class 2 or Class 3 Frame should be terminated with an End of Frame marker indicating that it is invalid, so that the Class 1 Frame can be transmitted immediately.

Class 4. A different, but no less significant problem with Class 1 is that it only allows Dedicated Connection from a single source to a single destination, at the full channel bandwidth. In many applications, it is useful to allocate a fraction of the resources between the N_Ports to be used, so that the remaining portion can be allocated to other connections. In Class 4, a bidirectional circuit is established, with one "Virtual Circuit" (VC) in each direction, with negotiated Quality of Service guarantees on bandwidth and latency for transmission in each direction's VC. A source or destination N_Port may support up to 254 simultaneous Class 4 circuits, with a portion of its link bandwidth dedicated to each one. Class 4 does not specify how data is to be multiplexed between the different VCs, or how it is to be implemented in the Fabrics — these functions are determined by the implementation of the Fabric supporting Class 4 traffic.

Class 6. A primary application area for Fibre Channel technology is in enterprise-class data centers or Internet Service Providers, supporting high-reliability data storage and transport. In these application areas, data replication is a very common requirement, and a high load on the SAN. Class 6 is intended to provide additional efficiency in data transport, by allowing data to be replicated by the Fabric without modification and delivered to each destination N_Port in a multicast group. Class 6 differs from Class 3 multicast in that the full channel bandwidth is guaranteed, and that the destination N_Ports each generate responses, which are collected by the Fabric and delivered to the source N_Port as a single aggregated response Frame.

Basic and Extended Link Service Commands

Beyond the Frames used for transferring data, a number of Frames, Sequences, and Exchanges are used by the Fibre Channel protocol itself, for initializing communications, overseeing the transmission, allowing status notification, and so on. These types of functions are termed "Link Services," and two types of Link Service operations are defined.

"Basic Link Service commands" are implemented as single Frame messages that transfer between N_Ports to handle high priority disruptive operations. These include an Abort Sequence (ABTS) request Frame, which may be used to determine the status of and possibly to abort currently existing Sequences and/or Exchange for error recovery. Aborting (and possibly retransmitting) a Sequence or Exchange is the main method of recovering from Frame- and Sequence-level errors. Acceptance or Rejection of the Abort Sequence (ABTS) command is indicated by return of either a Basic Accept (BA_ACC) or a Basic Reject (BA_RJT) reply. A Remove Connection (RMC) request allows a Class 1 Dedicated Connection to be disruptively terminated, terminating any currently active Sequences. A No Operation (NOP) command contains no data but can implement a number of control functions, such as initiating Class 1 Dedicated Connections, transferring Sequence Initiative, and performing normal Sequence termination, through settings in the Frame Header and the Frame delimiters.

"Extended Link Service commands" implement more complex operations, generally through establishment of a completely new Exchange. These include establishment of initial operating parameters and Fabric or topology configuration through the Fabric Login (FLOGI) and N_Port Login (PLOGI) commands, and the Logout (LOGO) command. The Abort Exchange (ABTX) command allows a currently existing Exchange to be terminated through transmission of the ABTX in a separate Exchange. Several commands can request the status of a particular Connection, Sequence, or Exchange or can read timeout values and link error status from a remote Port, and one command allows for requesting the Sequence Initiative within an already existing Exchange. Several commands are defined to be used as part of a protocol to establish the best end-to-end credit value between two

Ports. A number of Extended Link Service commands are defined to manage Login, Logout, and Login state management for "Processes." Implementation of the Process Login and related functions allows targeting of communication to one of multiple independent entities behind a single N_Port. This allows for a multiplexing of operations from multiple Processes, or "images" over a single N_Port, increasing hardware usage efficiency. A set of Extended Link Service commands allow management of Alias_IDs, which allows a single N_Port or group of N_Ports to be known to other N_Ports and by the Fabric by a different ID, allowing different handling of traffic delivered to the same physical destination Port. Finally, a set of Extended Link Service commands allow reporting or querying of the state or the capabilities of a Port in the Fabric.

Arbitrated Loop Functions

The management of the Arbitrated Loop topology requires some extra operations and communications beyond those required for the point-to-point and Fabric topologies. These include new definitions for Primitive Sequences and Primitive Signals for initialization and arbitration on the loop, an additional initialization scheme for determining addresses on the loop, and an extra state machine controlling access to the loop and transmission and monitoring capabilities.

Protocols

Protocols are interchanges of specific sets of data for performing certain defined functions. These include operations to manage the operating environment, transfer data, and do handshaking for specific low-level management functions. Fibre Channel defines the following protocols:

Primitive Sequence protocols: Primitive Sequence protocols are based on single-word Primitive Sequence Ordered Sets and do low-level handshaking and synchronization for the Link Failure, Link Initialization, Link Reset, and Online to Offline protocols.

Arbitrated Loop Initialization protocol: In an Arbitrated Loop topology, the assignment of the 127 possible loop address to different Ports attached on the loop is carried out through the transmission of a set of Sequences around the loop, alternately collecting and broadcasting mappings of addresses to Nodes.

Fabric Login protocol: In the Fabric Login protocol, the N_Port interchanges Sequences with the Fabric, if present, to determine the service parameters determining the operating environment. This specifies parameters such as flow control buffer credit, support for different Classes of service, and support for various optional Fibre Channel services. The equivalent of this procedure can be carried out through an "implicit Login" mechanism, whereby an external agent such as a system administrator or preloaded initialization program notifies a Port of what type of environment it is attached to. There is no explicit Fabric Logout since the Fabric has no significant resources dedicated to an N_Port which could be made available. Transmission of the OLS and NOS Primitive Sequences cause an implicit Fabric Logout, requiring a Fabric re-Login before any further communication can occur.

N_Port Login protocol: The N_Port Login protocol performs the same function with a particular destination N_Port that the Fabric Login protocol performs with the Fabric.

N_Port Logout protocol: An N_Port may request removal of its service parameters from another Port by performing an N_Port Logout protocol. This request may be used to free up resources at the other N_Port.

Segmentation and Reassembly

Segmentation and reassembly are the FC-2 functions provided to subdivide application data to be transferred into Payloads, embed each Payload in an individual Frame, transfer these Frames over the link(s), and reassemble the application data at the receiving end. Within each Sequence, there may be multiple "Information Categories." The Information Categories serve as markers to separate different blocks of data within a Sequence that may be handled differently at the receiver.

The mapping of application data to Upper Level Protocols (ULPs) is outside the scope of Fibre Channel. ULPs maintain the status of application data transferred. The ULPs at the sending end specify to the FC-2 layer:

- blocks or sub-blocks to be transferred within a Sequence,
- Information Category for each block or sub-block,
- a Relative Offset space starting from zero, representing a ULP-defined origin, for each Information Category, and
- an Initial Relative Offset for each block or sub-block to be transferred.

The "Relative Offset" relationship between the blocks to be transferred in multiple Sequences is defined by an upper level and is transparent to FC-2. Relative Offset is a field transmitted in Data Frame Header used to indicate the displacement of the first data byte of the Frame's Payload into an Information Category block or collection of blocks at the sending end. Relative Offset is not a required function in a Fibre Channel implementation. If Relative Offset is not supported, SEQ_CNT is used to perform the segmentation and reassembly. Since Frame sizes are variable, Frames without Relative Offset cannot be placed into their correct receive block locations before all Frames with lower SEQ_CNT values have been received and placed.

The Sequence Recipient indicates during Login its capability to support Continuously Increasing or Random Relative Offset. If only the former is supported, each Information Category transferred within a Sequence is treated as a block by upper levels. If Random Relative Offset is supported, an Information Category may be specified as sub-blocks by upper levels and the sub-blocks may be transmitted in a random order.

Data Compression

Another function included in Fibre Channel is the capability for data compression, for increasing the effective bandwidth of data transmission. ULP data may be compressed on a per Information Category basis within a Sequence, using the Adaptive Lossles Data Compress Lempel Ziv-1 algorithm. When the compression and decompression engines can operate at link speed or greater, the effective rate of data transmission can be multiplied by the inverse of the compression ratio.

Error Detection and Recovery

In general, detected errors fall into two broad categories: Frame errors and link-level errors. Frame errors result from missing or corrupted Frames. Corrupted Frames are discarded and the resulting error is detected and possibly recovered at the Sequence level. At the Sequence level, a missing Frame is detected at the Recipient due to one or more missing SEQ_CNT values and at the Initiator by a missing or timed-out acknowledgment. Once a Frame error is detected, the Sequence may either be discarded or be retransmitted, depending on the Exchange Error Policy for the Sequence's Exchange. If one of the discard Exchange Error policies is used, the Sequence is aborted at the Sequence level once an error is detected. Sequence errors may also cause Exchange errors which may also cause the Exchange to be aborted. When a retransmission Exchange Error policy is used, error recovery may be performed on the failing Sequence or Exchange with the involvement of the sending ULP. Other properly performing Sequences are unaffected.

Link-level errors result from errors detected at a lower level of granularity than Frames, where the basic signal characteristics are in question. Link-level errors include such errors as Loss of Signal, Loss of Synchronization, and link timeout errors which indicate no Frame activity at all. Recovery from link-level errors is accomplished by transmission and reception of Primitive Sequences in one of the Primitive Sequence protocols. Recovery at the link level disturbs normal Frame flow and may introduce Sequence errors which must be resolved following link level recovery.

The recovery of errors may be described by the following hierarchy, from least to most disruptive:

1. Abort Sequence: Recovery through transmitting Frames of the Abort Sequence protocol;
2. Abort Exchange: Recovery through transmitting Frames of the Abort Exchange protocol;
3. Link Reset: Recovery from link errors such as Sequence timeout for all active Sequences, E_D_TOV timeout without reception of an R_RDY Primitive Signal, or buffer-to-buffer overrun;
4. Link Initialization: Recovery from serious link errors such that a Port needs to go offline or halt bit transmission;
5. Link Failure: Recovery from very serious link errors such as loss of signal, loss of synchronization, or timeout during a Primitive Sequence protocol.

The first two protocols require transmission of Extended Link Service commands between N_Ports. The last three protocols are Primitive Sequence protocols operating at the link level. They require interchange of more fundamental constructs, termed Primitive Sequences, to allow interlocked, clean bring-up when a Port (N_Port or F_Port) may not know the status of the opposite Port on the link.

FC-3 General Description

The FC-3 level is intended to provide a framing protocol and other services that manage operations over multiple N_Ports on a single Node. This level is under development, since the full requirements for operation over multiple N_Ports on a Node have not become clear. A example function would be "striping," where data could be simultaneously transmitted through multiple N_Ports to increase the effective bandwidth.

A number of FC-3-related functions have been described in the FC-PH-2 and FC-PH-3 updates. These include (1) broadcast to all N_Ports attached to the Fabric, (2) Alias_ID values, for addressing a subset of the Ports by a single alias, (3) multi-cast, for a restricted broadcast to the Ports in an alias group, (4) Hunt groups, for letting any member of a group handle requests directed to the alias group.

FC-4 General Description

The FC-4 level defines mappings of Fibre Channel constructs to ULPs. There are currently defined mappings to a number of significant channel, peripheral interface, and network protocols, including:

- SCSI (Small Computer Systems Interface)
- IPI-3 (Intelligent Peripheral Interface-3)
- HIPPI (High Performance Parallel Interface)
- IP (the Internet Protocol) — IEEE 802.2 (TCP/IP) data
- ATM/AAL5 (ATM adaptation layer for computer data)
- SBCCS (Single Byte Command Code Set) or ESCON/FICON/SBCON.

The general picture is of a mapping between messages in the ULP to be transported by the Fibre Channel levels. Each message is termed an "Information Unit," and is mapped as a Fibre Channel Sequence. The FC-4 mapping for each ULP describes what Information Category is used for each Information Unit, and how Information Unit Sequences are associated into Exchanges.

The following sections give general overviews of the FC-4 ULP mapping over Fibre Channel for the IP, SCSI, and FICON protocols, which are three of the most important communication and I/O protocols for high-performance modern computers.

IP over Fibre Channel

Establishment of IP communications with a remote Node over Fibre Channel is accomplished by establishing an Exchange. Each Exchange established for IP is unidirectional. If a pair of Nodes wish to interchange IP packets, a separate Exchange must be established for each direction. This improves bidirectional performance, since Sequences are non-concurrent under each Exchange, while IP allows concurrent bidirectional communication.

A set of IP packets to be transmitted is handled at the Fibre Channel level as a Sequence. The maximum transmission unit, or maximum IP packet size, is 65,280 (x'FF00') bytes, to allow an IP packet to fit in a 64-kbyte buffer with up to 255 bytes of overhead.

IP traffic over Fibre Channel can use any of the Classes of service, but in an networked environment, Class 2 most closely matches the characteristics expected by the IP protocol.

The Exchange Error Policy used by default is "Abort, discard a single Sequence," so that on a Frame error, the Sequence is discarded with no retransmission, and subsequent Sequences are not affected. The IP and TCP levels will handle data retransmission, if required, transparent to the Fibre Channel levels, and will handle ordering of Sequences. Some implementations may specify that ordering and retransmission on error be handled at the Fibre Channel level by using different Abort Sequence Condition policies.

An Address Resolution Protocol (ARP) server must be implemented to provide mapping between 4-byte IP addresses and 3-byte Fibre Channel address identifiers. Generally, this ARP server will be implemented at the Fabric level and will be addressed using the address identifier x'FF FFFC.'

SCSI over Fibre Channel

The general picture is of the Fibre Channel levels acting as a data transport mechanism for transmitting control blocks and data blocks in the SCSI format. A Fibre Channel N_Port can operate as a SCSI source or target, generating or accepting and servicing SCSI commands received over the Fibre Channel link. The Fibre Channel Fabric topology is more flexible than the SCSI bus topology, since multiple operations can occur simultaneously. Most SCSI implementation will be over an Arbitrated Loop topology, for minimal cost in connecting multiple Ports.

Each SCSI-3 operation is mapped over Fibre Channel as a bidirectional Exchange. A SCSI-3 operation requires several Sequences. A read command, for example, requires (1) a command from the source to the target, (2) possibly a message from the target to the source indicating that it's ready for the transfer, (3) a "data phase" set of data flowing from the target to the source, and (4) a status Sequence, indicating the completion status of the command. Under Fibre Channel, each of these messages of the SCSI-3 operation is a Sequence of the bidirectional Exchange.

Multiple disk drives or other SCSI targets or initiators can be handled behind a single N_Port through a mechanism called the "Entity Address." The Entity Address allows commands, data, and responses to be routed to or from the correct SCSI target/initiator behind the N_Port. The SCSI operating environment is established through a procedure called "Process Login," which determines operating environment such as usage of certain non-required parameters.

FICON, or ESCON over Fibre Channel

ESCON is the standard mechanism for attaching storage control units on S/390 mainframe systems. ESCON over Fibre Channe is conceptually quite similar to SCSI over Fibre Channel, with a set of command and data Information Units transmitted as payloads of Fibre Channel Sequences. The FICON "Channel Control Words" (CCWs), control blocks containing the I/O requests, are different and more sophisticated than the SCSI command and data blocks, to accomodate the different format of data storage on these systems, and the higher throughput and reliability requirements.

ESCON channels were the first storage networking infrastructure, since they allowed multiple host systems to access the same storage control units across long-distance, switched fabrics. The mapping to Fibre Channel lets these systems operate through ANSI standard Fibre Channel switch Fabrics, while preserving the unique features and functions of the host systems and storage control units.

Chapter 3

Initialization and Data Transfer

Introduction

To give the general idea of how communication operates under the Fibre Channel protocol, we will start with a fairly detailed example of how an initially unpowered computer and disk drive device connected over a Fibre Channel link can begin interchanging data. This kind of example gives an understanding of how the various elements of Fibre Channel are used, particularly as a function of time.

The example we will use is for a very simple configuration, consisting of a host computer and a single disk drive, connected in a two-port Arbitrated Loop topology as shown in Figure 3.1. The figure looks similar to the Point-to-point topology shown in Figure 2.3, but since both Ports in the network incorporate the NL_Port functionality for operation on an Arbitrated Loop, they will operate as NL_Ports when connected directly. It will be clear from the example which Port operations are specific to the Arbitrated Loop topology, and which operations are applicable to any topology.

In this example we will refer to the host computer as the "computer," the disk drive attached over an FC-AL interface as the "drive," and their respective NL_Ports as the "computer NL_Port" and "drive NL_Port." Any time we refer to a generic N_Port or NL_Port, we are referring to computer's NL_Port. A similar set of operations and states occurs simultaneously on the drive's NL_Port as well, although we won't always emphasize it.

Figure 3.1 Physical configuration of a host-to-disk Fibre Channel Arbitrated Loop connection.

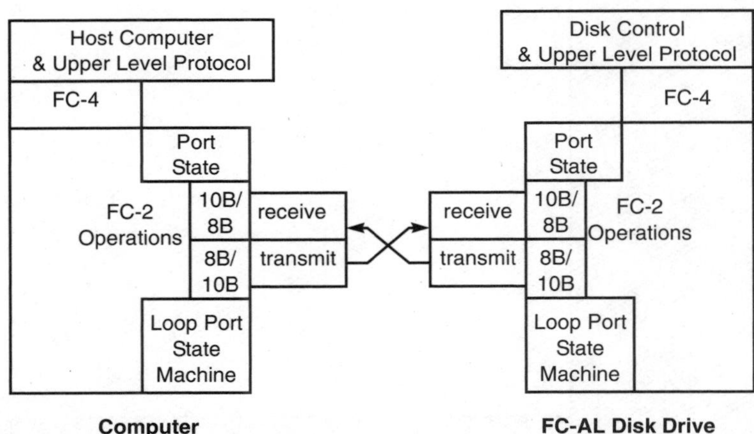

The example starts from the time when both machines are physically connected and about to be powered on.

Link Initialization

When the system is powered on, several things begin happening. Many are generic to all computer hardware and software and unrelated to Fibre Channel, such as the boot of the computer's operating system and the power-on self test of the CMOS chips in the computer adapter and drive logic. We will describe here only the operations involved in the Fibre Channel-specific parts of the system.

When the power is turned on, neither transmitter is sending any data, and the receivers are not detecting any incoming signal. The receivers at the computer and the drive detect this "Loss of Signal," and go into a Port state of "Link Failure 2." Being in the Link Failure state, they are unable to field any requests for data transmission by the FC-1 and higher levels. In the following discussion, equivalent operations occur simultaneously in both directions, enabling transmission in both directions between the drive and the computer.

As shown in Figure 6.3, a Port, whether an N_Port, NL_Port, or F_Port, which is in the "Link Failure 2" state will continuously transmit the NOS, or "Not Operational," Ordered Set. The FC-1 layer on the computer NL_Port will therefore drive its transmitter with the 40-bit stream for the Not Operational Ordered Set continuously. This bit stream is converted into a continuous signal of light and darkness traveling through the optical fiber. As this signal arrives at the receiver, a clock recovery circuit detects that the incoming signal is being modulated at the correct operating frequency, and synchronizes its clock to the frequency and phase of the incoming data stream. This provides bit synchronization between the transmitter and receiver.

The received bit stream is passed to a deserialization circuit, which continuously searches for the "comma" character, a series of bits with either the pattern b'001 1111' or b'110 0000.' When this series of bits is detected, the byte and word synchronization circuitry marks the first bit as a byte and word boundary. Every tenth bit following this comma character is the beginning of an encoded byte, and every 40th bit following is the beginning of an encoded word. Each byte and word boundary bit should also be the beginning of another comma, until the link begins transmitting real data.

Having established byte and word synchronization, the circuitry passes each 40-bit transmission word into a 10B/8B decoder, which decodes each 10-bit code and demultiplexes the output into words. The received bit stream is therefore transformed into a stream of received transmission words (at one-fortieth the rate), with "Data" or "Special" indications on each character. All transmitted words are sent through a corresponding 8B/10B encoder.

At this point, the Port is receiving decoded words but is still in the "Link Failure 2" state. The transitions between this state and the Active state, where the Port can send and receive data, are shown in Figure 3.2. A set of circuitry examines the incoming words, searching for Ordered Sets. When it receives three consecutive Ordered Sets of the same kind, it will transition to a different Port state. The next state depends on the current state and the Ordered Set received.

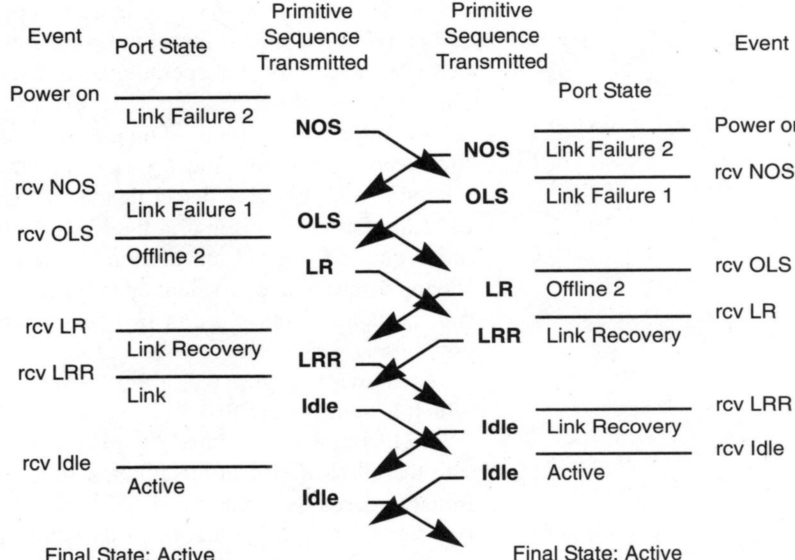

Figure 3.2 Link initialization through Ordered Set Transmissions.

As shown in Figure 6.3, when a state is in the Link Failure 2 Port state and receives a NOS Ordered Set, it will transition to the Link Failure 1 state and start transmitting the "OLS" (Offline) Ordered Set. A Port in the Link Failure 1 state that receives an OLS Ordered Set transitions to the Offline 2 state. Further transitions are followed, with a fully interlocked handshaking protocol between a half-dozen different Port states, until the Ports on both sides of the link are in the Active state, continuously transmitting Idle Ordered Sets. At this stage the transmitters and receivers are constantly sending and receiving the Idle Ordered Set, so the bit, byte, and word synchronization functions are constantly operating to keep the links synchronized and error-free in both directions.

At this point, the link is active in both directions, transmitting Idles, and is ready to begin fielding transmission requests from the FC-2 and higher levels in both Ports.

Loop Initialization

When the Ports have all made the transition to the Active state, it is possible to begin procedures to initialize the Loop, assigning Arbitrated Loop Physical Addresses (AL_PA values) to each Port. If the two NL_Ports in this example were N_Ports (i.e., they didn't incorporate Arbitrated Loop functionality), processing would skip to the steps described in the "Fabric and N_Port Login" section, on page 49.

When the Loop Port state machine (LPSM) has detected that the Ports are in the Active state, without loop initialization, the Ports continuously transmit a "Loop Initialization Primitive, no valid AL_PA" (LIP) Ordered Sequence. Receipt of three consecutive Ordered Sets of this type causes the LPSM to transition into the OPEN-INIT state, triggering the Ports go through the procedure to select a temporary loop master and determine AL_PA values.

After transitioning into the OPEN-INIT state, the NL_Port transmits an LISM Frame and monitors the receiver for an LISM Frame received. This Frame consists of a stream of transmission characters in the form of a normal Data Frame, as shown in Figure 7.1. The Payload of each LISM Frame, shown in Figure 16.2, contains the transmitting NL_Port's Port_Name. Each Port_Name is an 8-byte value, guaranteed to be unique worldwide.

At this point, each NL_Port is transmitting an LISM Frame and monitoring the incoming fiber for a similar Frame received. All the time they are doing anything other than transmitting Frames, they will transmit the Idle Ordered Set. The Frames are transmitted using Class 3 rules, so there are no Link Control Frames traveling in the opposite direction. When each NL_Port receives the first LISM Frame from the other Port, it compares the Port_Name contained in the Frame to its own Port_Name to decide which has the higher priority. If the received Port_Name has higher priority, it will forward the received LISM on. Otherwise, it will transmit its own LISM again. If the Port_Name in the received LISM is its own, it becomes the loop master.

Suppose that the computer's NL_Port has the higher priority. The drive's NL_Port, seeing the higher priority LISM arriving, will forward it around the loop. The computer's NL_Port, seeing a lower-priority LISM arriving, will discard it and transmit an LISM with its own Port_Name. Very soon, the computer's NL_Port will receive an LISM containing its own Port_Name and will know that it is the loop master for the purposes of Loop initialization.

As loop master, it is the computer NL_Port's responsibility to initiate the four loop initialization Frames (LIFA, LIPA, LIHA, and LISA) to allow the NL_Ports to determine AL_PA addresses. The procedure for assigning AL_PA values provides four different ways to have addresses assigned. The

NL_Ports may either (1) remember an address assigned by an attached Fabric (LIFA), (2) remember an address assigned by some other means (LIPA), (3) have some hard-wired preference for an address (LIHA), or (4) take an address left over after all other higher priority assignment methods have reserved their own AL_PA values (LISA).

The LIFA, LIPA, LIHA, and LISA Frames are used to collect AL_PA values for each of these four assignment methods. They are each passed around the loop, with each NL_Port reserving an AL_PA value by setting a bit in one of the Frames. The Payload contains a field slightly larger than 127 bits, to contain bit masks for the 127 valid AL_PA values on the network.

At this point, assume that the computer's NL_Port remembers that the last time it was powered on it had an AL_PA value of x'01' and that the drive's NL_Port also has a hard-wired preference for the AL_PA value x'01.'

The computer's NL_Port transmits a LIFA Frame, a blank Payload, since it was not assigned an address by a Fabric. The NL_Port receives the Frame and passes it back to the NL_Port unmodified, since it had no Fabric-assigned AL_PA value either. After receiving the Frame back, the computer's NL_Port transmits a LIPA Frame with the bit corresponding to the AL_PA value x'01' set to 1, to express its request for that value. The drive's NL_Port passes this Frame on unmodified too, since it has no previously assigned address.

The NL_Port then transmits the LIHA Frame, with the bit for x'01' still set to 1, since this is a cumulative address collection. The drive's NL_Port receives the LIHA and would set the bit for x'01' to 1, but it sees that this value is already taken, so it passes on the Frame unmodified. The computer's NL_Port then transmits an identical LISA Frame, with the bit for AL_PA x'01' set to 1. The drive's NL_Port receives the Frame, sees that the bit for x'01' is taken, so it sets the next available AL_PA, which is x'02.' The computer's NL_Port receives the LISA Frame back with the 2 bits set for AL_PA values x'01' and x'02,' so it knows that it is on a two-port loop.

On this small network, it is known where the AL_PA values physically live, but the LIFAs through LISA Frames do not inherently have any spatial information, and it's useful to know where the different AL_PA values are located around the network. The LIRP and LILP Frames are used for this. The computer's NL_Port sets an offset count to x'01,' sets the first byte in the 128-byte AL_PA position map field to its own AL_PA value, and passes it on. The drive's NL_Port receives the LIRP Frame, increments the count to x'02,' then fills in the second byte with its own AL_PA value, and forwards it. When the computer's NL_Port receives the LIRP Frame, it forwards it on as a LILP Frame, reporting the Address map to every other Port on the loop.

At this point, both NL_Ports have valid AL_PA address maps for the entire loop, and although the drive's NL_Port doesn't have the preferred hard-wired AL_PA value, it can still be addressed by every other NL_Port on the loop.

On an Arbitrated Loop, we would normally have to worry about the arbitration mechanism for determining which Ports have access to the Loop resources at any particular time. We would also have to consider the method a Port uses to release the Loop resources once it is finished communicating. In this case, though, since we have only two Ports, we don't have to worry about other Ports needing access to the loop. Therefore, we can arbitrate for the Loop once and be done with it.

Arbitration for the loop operates similarly to the operation for the LISM arbitration to determine the loop master. Each Port monitors the incoming fiber for an ARBx Ordered Set and either forwards it on or, if it wants loop access and has higher priority, replaces the ARBx received with its own ARBx (x indicates the AL_PA value).

In the example, both NL_Ports try to arbitrate for the Loop simultaneously. Both transmit an ARBx Ordered Set, and monitor the incoming fiber for other ARBx Ordered Sets to decide whether to forward or replace any incoming ARBx Ordered Sets having lower priority with their own. The computer's NL_Port is using AL_PA x'01,' so it transmits an ARB(01), and receives the ARB(02) from the other NL_Port. The x'02' value is lower priority than x'01,' so the computer's NL_Port discards the ARB(02) and replaces it with ARB(01). The NL_Port at the drive is forwarding the ARB(01) as higher priority, so as soon as the computer's NL_Port receives its own ARB, it knows that it owns the loop. Therefore, it goes into the OPEN state, as shown in Figure 16.4. It then sends the OPNyx Primitive Signal with y = x'02' and x='01' (OPN(02,01)), so that the drive's NL_Port transitions to the OPENED state.

In the OPEN and OPENED state, with no other L_Ports arbitrating for access, the two Ports are engaged in the equivalent of a Point-to-point topology, and further communication is carried out independently of the fact that both Ports have the capability of operating as L_ports. We will not discuss Arbitrated Loop anymore in this section, except to state that if there were other L_Ports on the loop, Frame transmission would be identical to a Point-to-point topology, except that in the space between Frames arbitration signals may be transmitted.

Fabric and N_Port Login

Having become active, an N_Port's next step is to determine the operating environment that it is attached to. This determination of operating environment is termed "Login" and involves two steps: Fabric Login and N_Port Login. The first is to determine what kind of Port is attached on the other end of the link, since the Port could either have been wired up into a Point-to-point topology with an N_Port or to an F_Port of a Fabric. This is termed "Fabric Login," since the attached Port could be part of a Fabric. Fabric Login is also the mechanism that would have been used to assign the Port an N_Port Identifier, if a Fabric was present. The second step is to determine the operating characteristics of the other N_Ports on the network that the Port may wish to communicate with. This is termed "N_Port Login," and has to happen separately for every pair of N_Ports that wish to communicate with each other.

Fabric and N_Port Login are carried out by a Port by transmitting a Sequence containing a test set of operating characteristics, or "Service Parameters," and receiving a Sequence containing a similar or modified set of Service Parameters in return. The combination of the two Sequences constitutes a negotiation, where the Sequence transmitted contains the Service Parameters that the transmitter would like to use, and the Sequence returned contains the Service Parameters that the two Ports can agree on. It's a very quick bartering system, where the transmitted Sequence indicates a requested set of Service Parameters and the returned Sequence indicates the Service Parameters agreed-upon. Further communication uses only the agreed-upon parameters. The formats of the FLOGI (Fabric Login), PLOGI (N_Port Login), and ACC (Login Accept) Sequences are shown in Figure 9.3.

The first Login operation is Fabric Login, which (1) establishes the Address Identifier the Port will use and (2) establishes the buffer-to-buffer Credit to use for flow control with the attached Port. To initiate the procedure, the computer's NL_Port initiates a Sequence, within an Exchange, containing the Payload shown in Figure 9.3. The Payload of this Sequence is actually less important than the S_ID and D_ID header fields. Remember that, although both NL_Ports in our example have AL_PA values assigned now, they do not yet have the 3-byte Address Identifiers assigned, which means they will accept Frames addressed to any D_ID values.

The computer's NL_Port transmits an FLOGI Frame with the S_ID value set to x'00 0001,' meaning that the least-significant byte has been assigned during the AL_PA Login, and the most-significant 2 bytes have not been assigned yet. The D_ID value is set to x'FF FFFE,' indicating that is directed to the Fabric. If a Fabric Port were present on the Loop, the computer's NL_Port would expect the Fabric to assign the top 2 bytes to be the same as the F_Port's top 2 bytes. In this case, there is no Fabric, so the attached NL_Port will return an ACC with swapped S_ID and D_ID values and a flag in the Payload indicating that the attached Port is an N_Port, instead of an F_Port.

At this point, all NL_Ports on the loop have established unique Address Identifiers, with values set to x'00 00 AL_PA.'

Following transmission of the FLOGI, the computer's NL_Port sends a PLOGI Frame to address x'00 0002' (the other Port's Address Identifier) to establish operating characteristics between the two communicating N_Ports. As described earlier, the Service Parameters operating between N_Ports are established during this N_Port Login. These include factors such as level of Fibre Channel supported, buffer-to-buffer Credit, buffer-to-buffer receive data field size for Class 2, Class 3, and connect-request Frames; total concurrent Sequences, Relative Offset Support, Error_Detect_Timeout values. It also establishes per-Class Service Parameters such as support for each of the three Classes of service, support for Intermix, and ACK Form usage as Sequence Recipient and Initiator. The Payloads of the PLOGI and ACC also contain the Ports' worldwide unique Port_Name and Node_Name values.

At the end of N_Port Login, the two NL_Ports know enough to communicate with each other, using commonly supported Service Parameters, and are both ready to support FC-4 requests for transmission of data in both directions.

Initial Sequence Transmission

The ULP interfaces with the FC-2 level through an FC-4 mapping level. The FC-4 specifies what Information Units can be sent by the Fibre Channel levels. The Information Units are messages, such as control, data, or status, for the ULP, that are transported as Sequences by the Fibre Channel levels.

Suppose that the ULP, SCSI in this case, has requested that an I/O operation be initiated to write a block of data to the drive. The full I/O operation requires four Sequences: a Sequence containing the SCSI command to the drive, a Sequence from the drive requesting the data, a Sequence to the drive containing the data, and a Sequence from the drive containing a status response. These four nonconcurrent Sequences are grouped into an Exchange, which describes a single I/O operation in this case.

The FC-4 level requests the FC-2 level to build and maintain data structures to keep track of the Exchange status and the Sequence status for each Sequence required. It then supplies a command to the FC-2 level requesting transmission of the FCP_CMND SCSI command. The command carries indications that the Sequence is the first of a new Exchange, that the Sequence Initiative should be transferred, and carries the service options are to be used for the Sequence transmission.

There are a number of service options specified for Sequence transmission, which we will not discuss in detail here. Two very general options to specify for the transmission are which Class of service to use for the Sequence and what Exchange Error Policy to use. In the current case, it does not matter much which Class of service we choose. With no Fabric, and only two Ports on the loop, using Class 1 would not provide any advantage in eliminating interference from other Ports, or any penalty in time to set up a Dedicated Connection. Class 3 operation would not provide acknowledgment of successful delivery, but this function is handled at the SCSI protocol level, through return of the SCSI response Information Unit so there is no reason not to use Class 3 transmission. (In actual SCSI operation, Class 3 is used for simplicity of implementation.) For this example, however, we will assume Class 2 transmission, since it illustrates use of acknowledgments and both the buffer-to-buffer and end-to-end flow control mechanisms.

Another option is the choice of Exchange Error Policy. There are four different Exchange Error Policies, the use of which depends on the relationship between different Sequences and on the ordered delivery requirements of the ULP. In this case, ordered delivery of Sequences is important, but the decision of whether a corrupted Sequence must be retransmitted is made at the ULP level, so the appropriate Exchange Error Policy is "Abort, discard multiple Sequences." This ensures that if an error occurs in one Sequence, all following Sequences in the Exchange will be discarded.

On receiving the request, the FC-2 level builds one or more Frames, of the format shown in Figure 7.1. In this case we will assume the Sequence contains one Frame. Here is how the Frame Header fields are filled in.

The R_CTL field is filled with the code b'0000 0110.' The first 4 bits specify that it is an FC-4 Device_Data Frame, containing an "unsolicited command." The D_ID and S_ID fields contain the destination and source Address Identifiers, which were established as x'00 0002' and x'00 0001.' The TYPE field contains x'0000 1000,' showing that the Frame is for SCSI-FCP, the Fibre Channel Protocol FC-4 mapping for SCSI. We will skip the F_CTL field for now, and describe it later. The SEQ_ID and OX_ID are set to values entirely of the NL_Port's choosing — in practice they will likely be indexes into arrays of Exchange and Sequence Status Blocks. This would allow the NL_Port, for example, to later use the SEQ_ID field value to do a fast lookup into a data structure which tracks Sequence status. The DF_CTL field is zero, since no optional headers are included between the Frame Header and the Payload. The Frame is the first of a new Exchange, so the SEQ_CNT field, which sequentially numbers Frames, is reset to x'0000.' The RX_ID field is set to the "unassigned" value of x'FFFF,' for possible later assignment by the Exchange Responder, which is the drive's NL_Port. Finally, there is no Relative Offset of the Frame contents into a buffer space, so the Parameter field is set to binary zeros.

The F_CTL field contains an array of 24 bits, which must all be set as described in Figure 7.7. This is where most of the information describing the Frame is placed. In this case, the bits could be set to b'0010 1001 0000 0000 0000 0010,' indicating that [23–22] the Exchange Originator and Sequence initiator sent the Frame (i.e., it is not a response Sequence or an ACK); [21–20] it is the first and not the last Sequence of an Exchange; [19] it is the last

Frame of the Sequence; [18] it is not the last Frame of a Class 1 connection; [16] the Sequence Initiative is transferred, so the other Port will send the next Sequence in the Exchange; [13–12] the Sequence Recipient decides what form of ACK to use; [9–8] it is not a retransmitted Sequence or part of a unidirectional Class 1 Dedicated Connection; [7–6] there is no indication of when the next Sequence will be transmitted (meaningless, since the Sequence Initiative is being transferred anyway); [5–4] the Exchange Error Policy requested is "Abort, discard multiple Sequences"; [3] there is no Relative Offset in the Parameter field; and [1–0] the Frame Payload has 2 filler bytes, to round out the contents to a word boundary (the useful data is the payload is 2 bytes less than a multiple of 4 bytes long. The other F_CTL bits are ignored for this type of Frame, either because they relate to functions unsupported by the system because of the settings of other F_CTL bits or because the bits are reserved.

The F_CTL field does not indicate which Class of service is being used for the Frame. The Class of service is indicated in the Start of Frame delimiter. This is the first and last Frame of a Class 2 Sequence, so the Frame delimiters are SOFi2 and EOFt, as shown in Figure 6.1.

By the way, all of these settings of Frame Header fields and Frame delimiters were set similarly in the FLOGI, PLOGI, and ACC Sequences that were sent earlier for Fabric and N_Port Login. We ignored them earlier to help focus the discussion on Login.

Once the FC-2 level knows what Frame delimiters, Frame Header fields, and Frame Payload to transmit, it can start passing words down to a small set of circuitry which calculates a 4-byte Cyclic Redundancy Code (CRC) value over the full Frame (except the Frame delimiters) and then appends the CRC value calculated to the end of the Frame. A similar CRC value will be calculated at the destination end on the Frame received. If the two CRC values do not match, or there is any invalid 10-bit code, then the receiver know that there has been a transmission error and knows not to receive the Frame.

The stream of transmission words is then passed to the FC-1 level for word-to-byte multiplexing and 8B/10B encoding and then to the FC-0 level for conversion into optical (or electronic) pulses over the cables.

At the destination Node, the received stream of optical or electronic pulses is digitized, decoded, demultiplexed, and parsed out to tell what type of Frame was transmitted. Recognizing the SCSI code in the TYPE field, the FC-2 level passes the received data up to the SCSI FC-4 level, along with whatever other control information was derived from the F_CTL bit settings to let the SCSI protocol level know how to process the data received.

In parallel with this, the FC-2 level does the processing necessary to track transmission and Sequence-level protocol handling at the Fibre Channel level. Since the Frame was transmitted in Class 2, both buffer-to-buffer and end-to-end Credit mechanisms are used, meaning that the Frame recipient must do two forms of flow control. First, when resources are ready to receive another Frame, it transmits the R_RDY Ordered Set over the link, indicating readiness for a new Frame. Second, it transmits an ACK Frame, which looks like a Data Frame without any data and with Sequence Context = 1 to show it was sent by the Sequence Recipient.

Since the received Sequence was the start of a newly-created Exchange, the drive's NL_Port must build an Exchange Status Block and a Sequence Status block, to be used for tracking status. The amount of information stored is implementation-dependent, but at least the Exchange must be marked as Open and Active, since it is ongoing, and the Sequence must be marked as Not Open and not Active, since the last Frame was received and successfully delivered to the FC-4 and upper level for processing.

The Exchange responder (the drive's Port, in this case) will assign an RX_ID to the Exchange to help it separate Sequences from different Exchanges. As with the OX_ID at the Exchange Originator, the value used is probably the index into the array of Exchange Status Blocks, to help the NL_Port locate Exchange information. The OX_ID is only valid on the Originator, and the RX_ID is only valid on the Responder, and they may have no relation with each other. The Originator does, however, have to know what the RX_ID value is for the Exchange so it can include the value in any later Frames sent. Therefore, the Responder must have put the assigned RX_ID value into the ACK Frame mentioned a few paragraphs ago. Once the Sequence Initiator receives the ACK, both Ports know both the OX_ID and RX_ID for the Exchange and can use them in all later Frames of the Exchange.

Since it was the last Frame of the Sequence, and it was delivered successfully to the FC-4 level, the FC-2 level is done and will wait for either a request from an FC-4 level or a new piece of information to come in over the receiver.

Completion of an Exchange

The SCSI ULP will parse the received Payload to understand that it is a SCSI write command. It therefore allocates whatever resources are necessary to be able to receive the data, then sends an indication to the FC-4 level to initiate transmission of another Sequence in the Exchange, using the Payload of a FCP_XFER_RDY command. This will tell the SCSI protocol level on the computer side of the link that the drive is ready for the data to come. The FXP_XFER_READY command is packaged into a Fibre Channel Frame, with header fields and Frame delimiters set to describe the Sequence and Frame, and the Frame is returned to the computer's NL_Port. All Frames of this new Sequence use the OX_ID and RX_ID values determined before in all the Frame Headers, to identify the Frames as part of the same Exchange.

The computer's NL_Port receives the Frame, parses it, passes it up to the SCSI protocol level, and builds a Sequence Status Block for the received Sequence, indicating that it was received and delivered. The SCSI FC-4 level then requests that the data be packaged up into a Fibre Channel Sequence for transmission to the drive NL_Port. Depending on the amount of data, the Sequence may have to be transmitted as multiple Frames, with different SEQ_CNT values. This is a new Sequence for the same Exchange, so the SEQ_CNT numbering for the Sequence can either restart with SEQ_CNT = x'0000,' or continue on from the last number used with SEQ_CNT = x'0001,' which is a little better.

When the drive's NL_Port passes the received Sequence data up to the SCSI level on the drive and determines that it has been successfully written, it is ready to transmit the last Sequence of the Exchange, indicating that the Exchange is finished. This last Sequence contains the Payload of an FCP_RSP message, to notify the computer's SCSI level that the data was written successfully. Since the drive's SCSI level knows that it is the last Sequence of the I/O operation or Exchange, it tells the FC-2 level to finish the Exchange. The FC-2 level sets the Last_Sequence bit (F_CTL bit 20) to 1 in the last data Frame of the Sequence transmitted. It updates the Exchange Status Block and Sequence Status Block when it receives the ACK back for the final Frame to show that all Sequences have been transmitted, received, and delivered to their corresponding ULP entities. The FC-2 level on the computer's NL_Port does the same thing, and both sides are finished with the operation.

Both NL_Ports go back to transmitting a continuous stream of Idles, waiting for more work to do.

Protocol Not Covered by the Example

The example here has used most of the main pieces of the Fibre Channel architecture, and full understanding of all the steps of the example will give a very good understanding of most of the operations of data communication over Fibre Channel. However, there are some significant parts of the protocol that were not included, and rather than just ignoring them, it is worthwhile to give a general idea of what was not included.

Error Detection and Recovery

The main subject left out was error recovery processing, since no errors occurred. This can be a complex subject. In any complete data communications protocol that hopes to assure reliable (error-free) communication, much more work must be spent designing the procedures for detecting and recovering from errors than in designing the procedures for normal operation. More things can go wrong than can go right.

Fortunately for understanding the Fibre Channel architecture, the Fibre Channel levels are targeted as the lowest levels of a layered communications architecture, so not much error recovery can be done. This is because, in any complete communication protocol, error recovery must be done at the highest layers, since (1) only the highest layers know what the actual data delivery requirements are, and (2) any error recovery at a lower layer is useless when errors can occur between layers. In this case, this means that most error recovery must be done by the ULP processing.

The function of the lower levels in a communication architecture are therefore primarily to detect unambiguously when errors have occurred, to be able to notify the upper levels what happened. Most of these functions incorporated at the Fibre Channel level for doing this detection have already been described: the unique SEQ_CNT values for identifying Frames; the SEQ_ID, OX_ID, and RX_ID values for uniquely associating Frames together; the 8B/10B transmission coding and CRC error detection code for detecting transmission errors; and the Port state functionality for handling situations when the incoming signal is too degraded to derive binary signals from.

The only other major mechanism for error recovery is a mechanism that unambiguously notifies the Initiator of a Sequence of what happened at the receiver end. This is necessary because of the basic transmitter side indeterminism: If a signal of some type is sent out, and an expected reply does not come back, the source does not know whether (1) the reply was corrupted, in which case the destination received the signal, or (2) the signal was corrupted, in which case the destination did not receive it. Fibre Channel provides several mechanisms for letting a source Port ask a destination Port about the delivery status of a Sequence or set of Sequences. These include an "Abort Sequence" protocol (for determining what Sequences need to be aborted), a "Read Sequence Status" request, and a "Read Exchange Status" request. Once the delivery status has been unambiguously determined, the Fibre Channel levels can notify the ULP of the true situation so that it can decide what data needs to be retransmitted.

Class 1 and Fabric Operation

The second main part of the missing architecture from the example was Class 1 transmission, with management of Class 1 Dedicated Connections and their relationship to Sequence transmission in other Classes of service. This is a sufficiently complex subject that its description covers most of Chapter 10 and all of Chapter 14.

The last main part of the architecture not discussed was operation over a Fabric of switch elements. Fortunately, many details of Fabric management can be hidden from the N_Ports. To a large extent, the N_Ports can simply manage the link to the attached F_Port, and trust the Fabric to route the Frames it receives to the destination N_Port identified in the Address Identifier. For example, there is no time when an N_Port needs to have any knowledge of the detailed configuration of a Fabric topology, such as how many switches are implemented, how they are connected to each other, and what N_Ports are attached to what switches. This makes the Fabric's switch elements fairly complex, since they must track this information, but it greatly simplifies operation of the N_Ports. All transmission is based, from the N_Ports' view, on the 3-byte Address Identifiers, rather than on any knowledge of how the Ports with those Address Identifiers are attached to the network.

Other Protocol Operations

Beyond this, there are a number of details, but the major parts of the protocol have been covered. There are a number of other Extended Link Service Commands other than the FLOGI and PLOGI commands for doing control and status related functions at the Fibre Channel level. There are several different timeout mechanisms used in error detection for deciding whether an expected event will occur or not. If something doesn't happen within a particular timeout period, it's assumed that it will not ever happen. There are several different lengths of timeout periods for detecting the occurrence of different kinds of events. There are different Exchange Error Policies, which can be used to provide different rules determining the effect a transmission error will actually have on delivery of the corrupted Sequence and of later Sequences. Also, there are some details on how flow control works over links and between N_Ports that have not been covered in much detail. All of these elements will all be covered in detail in later chapters.

Once these elements of the protocol are understood, it should be straightforward to go through the ANSI standards documents to understand the details of how Fibre Channel really works.

Chapter 4

FC-0: Physical Interface

Introduction

The FC-0 level is the physical link layer in the Fibre Channel architecture. Although its characteristics are well suited to optical fiber implementation, it is designed for maximum flexibility and allows use of several electrical technologies as well. Data transfer interfaces are defined for optical transmission over both single- and multi-mode fibers, with both lasers and LEDs, at both long (1300–1360 nm) and short (770–850 nm) wavelengths, and for electrical transmission over video coax, miniature coax, and shielded twisted pair cables. Transmission speeds are defined at 1.0625 Gbps, at half-, quarter-, and eighth-speed data rates, and recently at double- (2.125 Gbps) and quadruple-speed rates as well. This variety of possible technologies makes for a large range of customer and manufacturer choices regarding transmission medium, transmission distance, and price.

General Characteristics

The FC-0 level converts 10-bit transmission characters at a transmitting Port into serialized transmission signals that traverse a "cable plant" to a receiving Port. A cable plant consists of optical fibers or electrical cables with associated connectors and/or splices for optical fibers and grounding mechanisms for electrical cables. The FC-0 level at the receiving Port digitizes and deserializes the received signal into retimed 10-bit characters to be passed to the receiving Port's FC-1 level.

The transmission signal is binary, with a two-state transmission signal. A digital b'1' is encoded relative to b'0' as either more optical power (optical fiber) or a higher voltage on the center conductor (coaxial cable) or higher voltage on the conductor labeled '+' (twisted pair cable). Bits flow at either 132.821, 265.625, and 531.25 Mbps or 1.062.5, 2.125, and 4.25 Gbps, with clock frequency tolerances of 100 parts per million.

Each FC-0 connection is a point-to-point link, with any branching, switching, or broadcast functions handled at the higher levels of the protocol. This greatly simplifies transmitter and receiver design for high-speed operation relative to schemes with multiple receivers or multiple transmitters on a shared medium, such as buses.

The interface to the FC-0 level from the FC-1 level is specified in terms of 10-bit encoded binary transmission characters, and it hides the physical implementation of the data transmission from the FC-1 and higher levels. In principle, this allows implementation of FC-1 and higher levels completely independent (aside from bit rate) of whether transmission occurs over optical fibers, coaxial cable, twisted pair cable, or any other medium that may be developed in the future. The 10-way serialization and deserialization allows the FC-0/FC-1 interface to operate at a much lower clock speed than the link data rate.

Transmitter and Receiver

An FC-0 implementation must contain one transmitter, for the outgoing fiber, and one receiver, for the incoming fiber. In the Point-to-point and Fabric topologies, the remote ends of the two fibers will be connected to the same remote Port, but in the Arbitrated Loop topology, they are connected to different Ports around the Loop. The FC-0 level has digital interfaces to the FC-1 level and analog interfaces to the transmission medium, and it must convert between the two formats to perform the interface.

The receiver must always be operational, but the transmitter, which may need to be turned off and on, has four states of operation. It may be (1) enabled (capable of transmitting bits passed down by FC-1), (2) not enabled (turned off, or transmitting b'0' continuously), (3) transitioning (warming up or cooling down, depending on the medium and the implementation), or (4) failed (unable to operate). Laser transmitters typically have internal circuitry for detecting a failed state, but this state may not be detectable on other types of transmitters.

The receiver's function is to convert the incoming signal into retimed characters and to present the characters along with a synchronized byte clock to the FC-1 level. Generation of the synchronized byte clock requires some kind of clock recovery circuit such as a phase locked loop, for bit synchronization, and a small state machine for detecting the "comma" bit stream, which occurs at the beginning of Ordered Sets, for byte and word synchronization. The receiver has no states. When powered on, it is always attempting to convert the received signal into valid data bits to pass to the FC-1 level. It may take some time (up to 2500 bit periods) for bit synchronization to be acquired following initial receipt of a new bit stream.

Note that the internal processing clock (which is generally the transmission clock) is generated at the local Port, while the receive clock is derived from the received data, which was generated using a different Port's transmission clock. They will generally not, therefore, have exactly the same frequency (within tolerances) or the same phase. Compensation for this frequency difference is handled at the FC-1 level, through an asynchronous FIFO buffer and discarding of Idle Ordered Sets, to synchronize received data with internal processing clocks.

Intentional Transmission of Invalid Code

There are times when it is useful for an FC-1 level to intentionally transmit invalid data, such as data with bad disparity or with code points outside of the valid code set, to check that error detection or error recovery operations are working. This is legal, but the proper functioning of the FC-0 level in correctly converting digital data into a transmission signal and back to digital data depends on the transmitted data having certain characteristics. These include ensuring that the transmitted data has a maximum run length of 5 bits, that the number of 1s in any 10 bits must be between four and six, and that the number of 1s in any 1,000 bits must be between 495 and 505.

If these rules are violated by the transmitting FC-1 level, the FC-0 level (particularly the receiver) may have difficulty in properly transmitting and receiving a signal that can be converted into binary data for delivery to the receiving FC-1 level. The final result of this may be a higher bit error rate in transmission than can be tolerated. This limitation on transmission of invalid data illustrates the dependency of the transmitter/receiver pair on the transmission characteristics of the binary data stream and the 8B/10B code functionality.

FC-0 Nomenclature and Technology Options

The FC-0 specification provides for a large variety of technology options, allowing flexibility in transmission speed, distance, transmission medium and transmitter/receiver technology. Each link is described using a nomenclature described in Figure 4.1.

Not all of the possible combinations of media, transmission speeds, and distances are actually specified or used. Twisted pair cable, for example, will not operate at more than the full gigabit speed, currently, and it would be a waste to specify single-mode optical fiber for an eighth-speed link. Figure 4.2 shows the various optical and electronic options currently specified. The general trend is for the optical links to be geared toward the higher speeds and longer distances, with electrical links being used for shorter distance, particularly inside enclosures, since they can be expected to have lower cost. Also, there is a growing convergence on the higher speeds (gigabit and 2 gigabit), with, for example, 1/2 and 1/8 speed links being practically unused.

Figure 4.1
Nomenclature for describing FC-0 cable plant options.

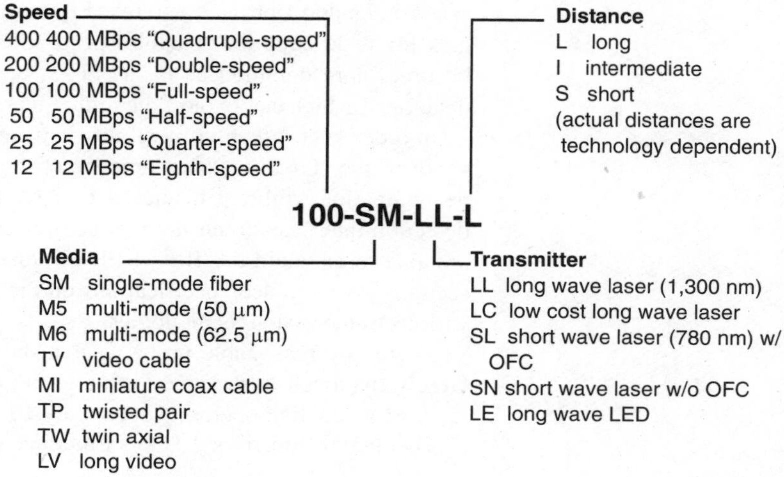

Figure 4.2
Defined cable plant technology options.

Optical			
Single-mode (9 µm)	**Multi-mode (62.5 µm)**	**Multi-mode (50 µm)**	**Multi-mode (50 µm) w/o OFC**
400-SM-LL-L	100-M6-SL-I*	200-M5-SL-I	400-M5-SN-I
200-SM-LL-L	50-M6-SL-I*	100-M5-SL-I	200-M5-SN-I
100-SM-LL-L	25-M6-SL-I*	50-M5-SL-I	100-M5-SN-I
100-SM-LL-I	25-M6-LE-I*	25-M5-SL-I	
50-SM-LL-I	12-M6-LE-I	25-M5-LE-I*	
25-SM-LL-L		12-M5-LE-I*	

Electrical				
Long Video	**Video Coax**	**Twin Axial**	**Shielded Twisted Pair**	**Miniature Coax**
100-LV-EL-S	100-TV-EL-S	100-TW-EL-S	100-TP-EL-S	100-MI-EL-S
50-LV-EL-S	50-TV-EL-S	50-TW-EL-S	50-TP-EL-S	50-MI-EL-S
25-LV-EL-S	25-TV-EL-S	25-TW-EL-S	25-TP-EL-S	25-MI-EL-S
12-LV-EL-S	12-TV-EL-S	12-TW-EL-S	12-TP-EL-S	12-MI-EL-S

Clearly there is a wide range of specified options, and not all of them will be used equally frequently. The Fibre Channel industry will likely agree on providing widespread support for only a few of the technology options. As of this writing, it appears that the most generally used technology options will be full-speeds, and double-rate, with SL and LL technologies for applications which exit computer enclosures, and twisted pair or coaxial cable for short-distance links which remain within an enclosure. The half- and eighth-speed links don't provide enough differentiation from competing technologies for wide adoption. Double-speed and faster links can be expected to become more common as FC-1, FC-2, upper levels, and applications are developed which can exploit the higher bandwidth.

In order to ensure interoperability between components made by different vendors, the signaling characteristics of the transmitters and receivers must be compatible. Fibre Channel places requirements on the signaling and detection interfaces to ensure that components adhering to the requirements are able to maintain the 10^{-12} BER (bit error rate) objective. The following sections give detailed specifications on the requirements established for the various transmission technologies.

There is a reasonable probably that the quadruple-speed option will be largely bypassed in practice (although it's specified and clearly works) in favor of a new link operating at roughly 10 Gb/s, to leverage the link techology being developed for 10 Gb/s Ethernet. Watch this space.

Long-Wavelength Laser Single-Mode Link

Links using 9-μm core diameter single-mode optical fiber and 1.3-μm lasers are the highest-performance class of link technology, both in transmission speed and distance. The detailed technical requirements for single-mode links are shown in Figure 4.3. In general, these requirements are relatively conservative and easy to meet using current technology through the gigabit and 2 gigabit link speed. Operations at higher rates are mainly limited by the cost of driver and receiver electronics, rather than by optical parameters.

All laser optical links are restricted to have a maximum amount of optical power in the fiber, to prevent danger of eye damage. Since the cables may be placed in non-controlled environments such as offices, they must conform to Class 1 laser safety operational specifications. (Class 1 laser safety is unrelated to Class 1 service with Dedicated Connections.) Class 1 laser safety means that the system cannot be capable of causing eye damage under any condition, even if a curious user puts the fiber directly up to his or her eye to see if the laser is on. For the relatively short lengths used for Fibre Channel data communications, it is relatively easy to design receivers at 1.3 μm that can operate with Class 1 laser transmitters to the required BER. At the shorter wavelengths, where detectors are less sensitive and eyes are more sensitive, this is more difficult, so special procedures to ensure safety may be required, as described in the "Open Fiber Control Safety System for SW Laser Links" section, on page 70.

Figure 4.3
Specifications for single-mode optical fiber links.

Parameter	Units	400-SM-LL-L	200-SM-LL-L	100-SM-LL-L	100-SM-LL-I	50-SM-LL-L	25-SM-LL-L	25-SM-LL-I
FC-0 and Cable Plant								
data rate	MBps	400	200	100	100	50	25	25
nominal bit rate/1062.5	Mbaud	4	2	1	1	1/2	1/4	1/4
operating range (typical)	m	2-2k	2-2k	2-10k	2-2k	2-10k	2-10k	2-2k
loss budget	dB	6	6	14	6	14	14	6
dispersion related penalty	dB	1	1	1	1	1	1	1
reflection related penalty	dB	1	1	1	1	1	1	1
cable plant dispersion	ps/nm·km	12	12	35	12	60	60	12
Transmitter (S)								
min. center wavelength	nm	1270	1270	1270	1270	1270	1270	1270
max. center wavelength	nm	1355	1355	1355	1355	1355	1355	1355
RMS spectral width	nm (max.)	6	6	3	6	3	6	30
launched power, max.	dBm (ave.)	–3	–3	–3	–3	–3	–3	–3
launched power, min.	dBm (ave.)	–12	–12	–9	–12	0	–9	–12
extinction ratio	dB (min.)	9	9	9	9	3	6	6
RIN_{12} (maximum)	dB/Hz	–116	–116	–116	–116	–114	–112	–112
eye opening @ $BER=10^{-12}$	% (min.)	57	57	57	57	61	63	63
deterministic jitter	% (p-p)	20	20	20	20	20	20	20
Receiver (R)								
received power, min.	dBm (ave.)	–20	–20	–25	–20	–25	–25	–20
received power, max.	dBm (ave.)	–3	–3	-3	–3	–3	–3	–3
optical path power penalty	dB (max.)	2	2	2	2	2	2	2
return loss of receiver	dB (min.)	12	12	12	12	12	12	12

Cable Plant

The single-mode optical fiber cable plant is generally insensitive to the data rate, so any installed part of the cable plant may be used at any data rate, provided dispersion and loss budget penalty specifications are met. For all links, the loss budget is calculated by taking the difference between minimum transmitter output power and receiver sensitivity and subtracting 2 dB for the sum of dispersion- and reflection-related penalties. In some cases the dispersion limit induced by the fiber's modal bandwidth limitation may be a more stringent limit than the loss budget limit.

Perhaps the most difficult optical fiber cable parameter to measure is optical return loss. This describes the ratio between the transmitted optical power and the power reflected back into the transmitter. Reflected power can cause transmitter instabilities and inter-symbol interference. The specified return loss for the entire cable plant with a receiver connected is 12 dB, which is equivalent to a single glass-air interface at the destination receiver. The return loss for each splice or connector must be higher than 26 dB.

Optical fiber links use a duplex SC connector as shown in Figure 4.4. A matching receptacle is implemented at the transmitter and receiver. The keys at the top of the connector distinguish between transmitter (outgoing) and receiver (incoming) fibers. The difference between narrow and wide keys also distinguishes between connectors for single-mode and multi-mode optical fibers, which is important since coupling a multi-mode fiber connector into a single-mode receptacle could cause serious laser safety problems.

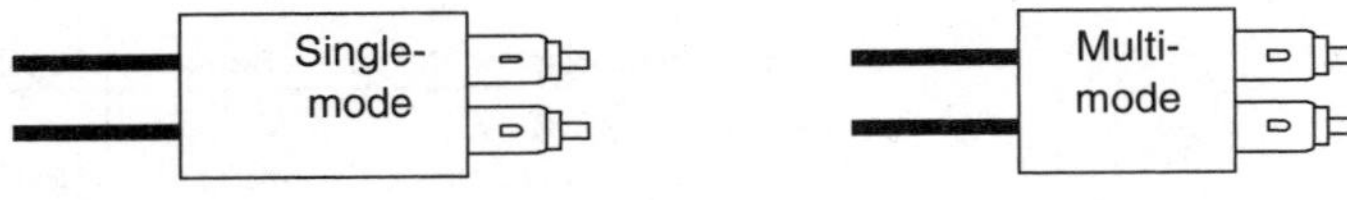

Figure 4.4
Duplex SC connector for single- and multi-mode optical links.

Transmitter

The transmitter's required center wavelength and root-mean-square (RMS) spectral width are specified to bound the effect of spectral dispersion in the optical fiber. The maximum and minimum optical power requirements are specified to balance the necessity of assuring adequate receiver optical power for signal detection with laser safety and receiver overload limitations. The extinction ratio value describes the ratio between the average optical energy in a 1 and a 0, measured under fully modulated conditions in the presence of worst-case reflections. The RIN (Relative Intensity Noise) value describes the amount of intensity noise in the signal, which affects for example the amount of optical power in two different transmitted 1s.

The eye opening specification describes the eye diagram mask governing the allowable pulse shape at the output of the transmitter. These characteristics include rise time, fall time, pulse overshoot, undershoot, and ringing, all of which are controlled to prevent excessive degradation of the receiver sensitivity.

The parameters specifying the mask of the transmitter eye diagram are shown in Figure 4.5. As shown in the diagram, the optical waveform must not overshoot by more than 30% or undershoot on negative transitions by more than 20%, and the eye opening must be larger than the mask shown. The waveforms shown in the diagram are not intended to resemble actual optical waveforms but are intended to show the type of eye diagram mask violations that can occur. For measuring the mask of the eye diagram, a standardized reference receiver must be used, with a specified transfer function. This is not intended to represent the performance of a functioning optical receiver but will provide a uniform measurement condition.

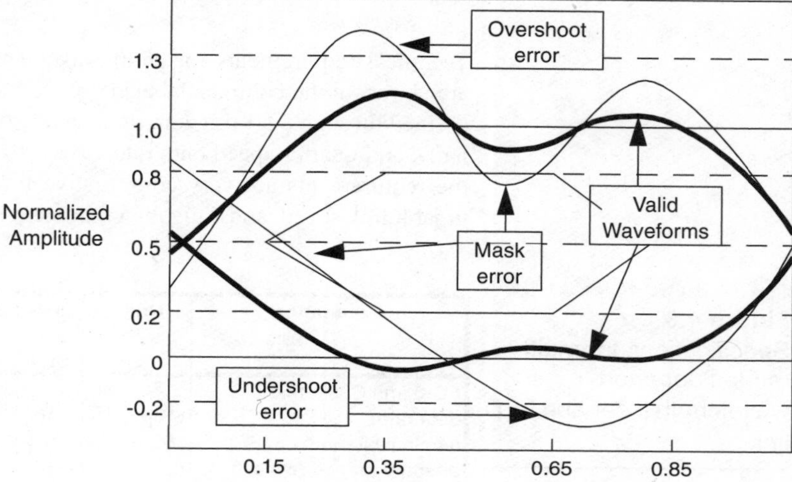

Figure 4.5
Eye diagram mask for the transmitter output.

Receiver

A receiver complying with the Fibre Channel specifications will operate with a BER of 10^{-12} over the lifetime and temperature range of operation when driven by signals of the correct power and eye diagram as shown in Figure 4.5. The receiver characteristics specified are minimum received power (receiver sensitivity), maximum received power (overload level), return loss (reflectance), and allowable optical path power penalty from combined effects of dispersion, reflections, and jitter. The power penalty accounts for the signal degradation along the optical path from all non-static effects. For example, in the case of the 100-SM-LL-L class the difference between the minimum transmitter output and receiver sensitivity is (–9 dBm) – (–25 dBm) = 16 dBm; 2 dB are allocated to the optical path power penalty and the remaining 14 dB are allocated to the cable plant.

Short-Wavelength Laser Multi-Mode Fiber Links

Technical requirements for short-wavelength (SW) laser multi-mode links are shown in the columns labeled "XXX-M5-SL_I" in Figure 4.6. Links are defined through a 50-μm core multi-mode optical fiber for the double-, full-, half-, and quarter-speed data rates. As with the long wavelength laser links, the requirements are very conservative for current technology through the gigabit link speed, and 2 gigabit equipment is becoming readily available.

Figure 4.6
Specifications for multi-mode fiber short wavelength laser and LED links.

Parameter	Units	200-M5-SL-I	100-M5-SL-I	50-M5-SL-I	25-M5-SL-I	25-M6-LE-I	25_M5_LE-I
FC-0 and Cable Plant							
data rate	MBps	400	200	100	100	25	25
nominal bit rate/1062.5	Mbaud	2	1	1/2	1/4	1/4	1/4
operating range (typical)	m	2-300	2-500	2-1k	2-2k	2-1.5k	2-1.5k
fiber core diameter	μm	50	50	50	50	62.5	50
loss budget	dB	6	6	8	12	6	5.5
multi-mode fiber bandwidth							
@ 850 nm	MHz·km	500	500	160	500	160	500
@ 1,300 nm	MHz·km	500	500	500	500	500	500
numerical aperture	(unitless)	0.20	0.20	0.20	0.20	0.275	0.275
Transmitter (S)							
type		laser	laser	laser	laser	LED	LED
min. center wavelength	nm	770	770	770	770	1280	1280
max. center wavelength	nm	850	850	850	850	1380	1380
RMS spectral width	nm(max.)	4	4	4	4	(see	(see
FWHM spectral width	nm(max.)					text)	text)
launched power, max.	dBm(ave)	1.3	1.3	1.3	0	−14	−17
launched power, min.	dBm(ave)	−7	−7	−7	−5	−20	−23.5
extinction ratio	dB (min.)	6	6	6	6	9	9
RIN_{12} (max.)	dB/Hz	−116	−114	−114	−112		
eye opening @BER=10^{-12}	% (min.)	57	61	61	63		
deterministic jitter	% (p-p)	20	20	20	20	16	16
random jitter	% (p-p)					9	9
optical rise/fall time	ns (max.)					2/2.2	2/2.2
Receiver (R)							
received power, min.	dBm(ave)	−13	−13	−15	−17	−26	−29
received power, max.	dBm(ave)	1.3	1.3	1.3	0	−14	−14
return loss of receiver	dB (min.)	12	12	12	12		
deterministic jitter	% (p-p)					19	19
random jitter	% (p-p)					9	9
optical rise/fall time	ns (max.)					2.5	2.5

Cable Plant

The cable specified for SW laser links is multi-mode optical fiber with a 50-μm core diameter. The 50-μm core fiber gives greater modal bandwidth and a lower dispersion penalty than the 62.5-μm fiber used for LED links. The return loss must be 12 dB for the entire cable with receiver connected and 20 dB at each connector or splice. In some cases the limit induced by the fiber's modal dispersion may be more stringent than the loss budget limit.

Alternate Cable Plant Usage. Fibre Channel specifies several alternate cable plants which may be desirable to use in some cases. Some examples of usage are the need for extended distances or operation in locations where standard cables are already installed. These fiber types have not been fully studied and details for their use are not provided directly. However, given the conservative operating characteristics, they may be expected to operate correctly for short distances. These alternate cable plants use the ordinary nomenclature. The five alternate cable plants, with their operating distances, are (1) 100-M6-SL-I: 2 m to 175 m, (2) 50-M6-SL-I: 2 m to 350 m, (3) 25-M6-SL-I: 2 m to 700M, (4) 25-M5-LE-I: 2 m to 1.5km, (5) 12-M5-LE-L: 2 m to 1.5 km. The BER performance of these alternate cable plants (particularly the LED links) might have to be verified on a per-installation basis.

Transmitter

Most transmitter requirements are specified for reasons equivalent to the reasons for long-wavelength laser links. For SW multi-mode links, dispersive effects are much stronger than for long-wavelength single-mode links. The most difficult problems in building these links are minimizing the effects of modal dispersion, mode-selective loss, and reflection-induced pulse distortion. These effects can be minimized by tailoring laser characteristics and laser/fiber coupling parameters to minimize reflection effects and by taking care to avoid mode-selective loss, particularly at locations in the fiber near the transmitter. Since pulse distortion increases with increasing transmission distance, mode-selective loss occurring near a receiver will have much less of an effect than near a transmitter.

The maximum and minimum of the allowed range of average transmitter power coupled into the fiber are worst-case values to account for manufacturing variances, drift due to temperature variations, aging effects, and operation within the specified minimum value of the extinction ratio. The minimum value of the extinction ratio is the minimum acceptable value of the ratio between the average optical energies in a 1 and a 0, measured under fully modulated conditions in the presence of worst-case reflections.

Receiver

The characteristics and sensitivities of optical receivers for SW transmission are essentially similar to those for 1.3-µm transmission. However, some differences arise in accounting for optical path penalties when considering multi-mode versus single-mode cable plants.

The optical path power penalty (in decibels) typically accounts for the total degradation along the optical path resulting from reflections, jitter, and from the combined effects of dynamic effects such as inter-symbol interference, mode-partition noise, and laser chirp. However, it is a common practice for multi-mode fiber data links to include this optical path power penalty into the actual link budget specifications of both the power budget (producing amplitude degradation) and the timing budget (producing pulse spreading degradation). Therefore, the SW laser data links have no specified optical path power penalty in Figure 4.6. The related link degradations of both amplitude and pulse spreading (primarily modal dispersion) are already accounted for in the power budget and time budget.

Open Fiber Control Safety System for SW Laser Links

As stated in the "Long-Wavelength Laser Single-Mode Link" section, on page 64, it is a requirement of Fibre Channel transmission systems that they meet Class 1 laser safety standards, i.e., a user must be able unplug any connector and look directly into any fiber without fear of eye damage. For long-wavelength (and LED) systems, this is not difficult — the BER objectives can be met with transmitter output power levels well below Class 1 specifications. However, for the SW systems, it was originally difficult to provide enough optical power for adequate detection without exceeding laser safety standards. The "Open Fibre Control safety system," or OFC safety system, is a mechanism for circumventing this problem. The OFC system assures that when any link is broken, the transmitter goes into a disabled state, and only eye-safe amounts of light are transmitted, to test connectivity, until the link is reconnected.

The OFC algorithm operates on the principle that each end of an optical link will only transmit optical signal power if it is receiving signal power. If an optical connection is broken, then the receiver of the broken connection will turn off its transmitter, with the result that both transmitters over the broken link will turn off their lasers within less than 1 round-trip time over the link. A state machine then controls each side turning on the laser for short, eye-safe, periods, to test connectivity of the link.

Since original development of the Fibre Channel technology, however, further transceiver development has allowed construction of SW transceivers that are able to operate at Class 1 Laser Safety power levels under normal operating conditions without Open Fibre Control state capability. Most transceivers are currently built to Class 1 Laser Safety without OFC capability, and only have protection circuitry for turning off the transmitters when they are operated outside of normal operating conditions (for example, when a power supply voltage exceeds the design specification).

LED Multi-Mode Fiber Links

Technical requirements for LED multi-mode links are shown in the last three columns of Figure 4.6. Most parameters are interpreted similarly to the parameters for laser links. The optical return loss is not specified for LED links, since LEDs are much less sensitive to reflections than lasers are, but it is expected that the cables will be similar to those for laser links.

Two parameters which do arise in LED links, which don't arise in laser links, are the related parameters of pulse rise and fall times and the allowable spectral width of the transmitter LED. They are related since LEDs in general have much wider and less "clean" spectra than lasers, as well as slower rise and fall times. These parameters are related since the rise/fall times at the receivers depend on both the transmitter rise/fall times and the degradation induced by modal and chromatic dispersion during transmission.

The requirement is for a 2.5-ns rise/fall time at the receiver for quarter-speed links after transmission through 1.5 km of 62.5 μm multi-mode fiber. This limits the LED FWHM spectral bandwidth to between 100 and 200 nm, depending on the transmitter's rise/fall time and its deviation in center wavelength from the fiber's chromatic dispersion minimum.

75-Ohm Coaxial Cable Electrical Links

For shorter distance and more inexpensive links than the optical fiber media described, Fibre Channel allows signaling over electrical links as well. There are three different kinds of cables used, including shielded twisted pair

(STP), which is described in the "150-Ohm Shielded Twisted Pair Electrical Links" section, on page 74, a miniature style coaxial cable, and video coaxial cable. Specifications for the various classes of electrical cable links are shown in Figure 4.7.

Figure 4.7 Specifications for electrical cable links.

Parameter	Units	100-TV-EL-S 100-MI-EL-S 100-TP-EL-S	50-TV-EL-S 50-MI-EL-S 50-TP-EL-S	25-TV-EL-S 25-MI-EL-S 25-TP-EL-S
FC-0 and Cable Plant				
data rate	MBps	100	50	25
bit rate/1062.5	Mbaud	1	1/2	1/4
operating distance				
video coax	m	0–25	0–50	0–75
mini-coax	m	0–10	0–15	0–25
twisted pair	m	<1	<1	0–50
S_{11} reflected power @ 0.1–1.0 bit rate				
video coax	dB	–15	–15	–15
mini-coax	dB	–7	7	–7
twisted pair	dB			–12
attenuation @ 0.5 bit rate				
video coax	dB/m	0.288	0.167	0.096
mini-coax	dB/m	0.62	0.46	0.31
twisted pair	dB/m			0.138
loss per connector				
video coax	dB	0.25	0.25	0.25
mini-coax	dB	0.5	0.5	0.25
twisted pair	dB	0.25	0.25	0.25
Transmitter (S)				
type		ECL/PECL	ECL/PECL	ECL/PECL
output voltage (p-p)				
maximum	mV	1600	1600	1600
minimum	mV	600	600	600
deterministic jitter	%(p-p)	10	10	10
random jitter	%(p-p)	12	12	8
20–80% rise/fall time	ns, max.	0.4	0.6	1.2
Receiver (R)				
min. data sensitivity	mv (p-p)	200	200	200
max. input voltage	mv (p-p)	1600	1600	1600
S_{11}@(0.1–1.0 bit rate)	dB	–17	–17	–17
min. connector return loss (0.3MHz–1GHz)				
video coax	dB	20	20	20
mini-coax	dB	15	15	15
twisted pair	dB	-----	-----	12

Both of the coaxial cable types are rated at 75 Ω nominal impedance, and are electrically and physically interoperable with conversion connectors, with minor impact on link length capability. Both transmitter and receiver are assumed to share a common ground return or ground plane to minimize safety and interference concerns caused by the voltage differences between equipment grounds.

The video coaxial cable links use either type RG 6/U or type RG 59/U coaxial cable. The mini-coaxial cable links use type RG 179B/U coaxial cable. The cable and connector electrical characteristics for both types are shown in Figure 4.7. Connector return loss is limited to 20 dB in both types to minimize distortions on the transmitted signal. No more than four miniature coaxial connectors can be present from the transmitter to the receiver.

A specific design goal of the specifications for the coaxial cable plants is to allow interoperability with restricted lengths of miniature and video style coaxial cable. For example, as long as the miniature coaxial cable is less than 2 m in length at each end of two cabinets, the full 50 m capability of the video style coax should be achievable for cabinet interconnections at 50 MBps. Especially at transmission rates of 531 Mbps and higher, particular attention must be given to the transition between miniature and video-style coaxial cable.

The output drivers are assumed to have Emitter Coupled Logic (ECL) output levels. The output drivers should have peak-to-peak levels as given in Figure 4.8 when measured across a 75 Ω resistor at the transmitter output. The normalized mask of the eye diagram at the transmitter output is shown in Figure 4.8. The rise/fall time requirements in the eye mask are somewhat less stringent for the full speed link than for the fractional speed links.

The electrical receivers may be expected to have an equalization network, particularly at the higher data rates, to compensate for frequency-dependent signal distortion induced in transmission over the cable link. Such a network is not required if the BER objectives can be met without it. Such an equalizer would have greater attenuation and phase delay at lower frequencies than at higher frequencies, to compensate for cable characteristics, and would have a high return loss to minimize signal reflections.

Figure 4.8
Transmitter eye diagram mask for coaxial cable links.

Eye Diagram Mask			
Rate	x1	X2	y
half, quarter, eighth	0.15	0.35	0.30
full	0.15	0.30	0.30

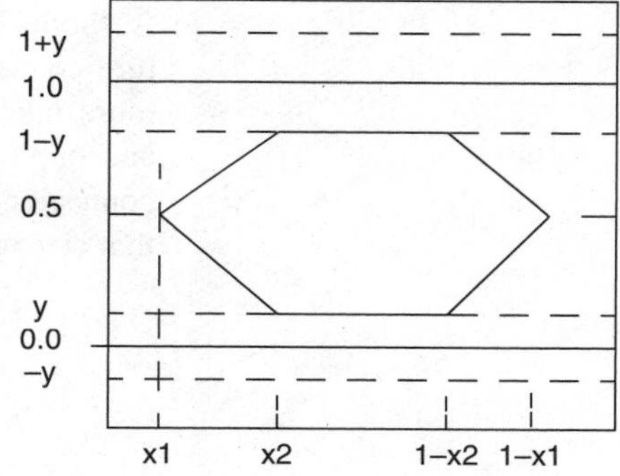

The cables are expected to be grounded only at the transmitter end, and the shield at the receiver end and the center conductor at both ends are expected to be ac coupled to the circuitry through transformer or capacitive coupling. In order to simplify connection of transmitters to receivers and vice versa, Fibre Channel specifies that the video coax cable has a male BNC connector on the transmitter end and a male TNC connector on the receiver end, both with the 75 Ω characteristic impedance of the line. The miniature coaxial cable uses standard mini-coaxial cable connectors on both ends. These cables are intended for use inside cabinets or enclosures, with infrequent disconnection and reconnection, so correct polarization is easier to maintain.

150-Ohm Shielded Twisted Pair Electrical Links

Specification for shielded twisted pair (STP) electrical cable links are shown in Figure 4.7, including operation at up to full-speed rate. This data link is based on 150 Ω STP cable, which contains one twisted pair for signals traveling in each direction. The two conductors are connected to the differential transmitter outputs and receiver inputs, and the shield is grounded on both ends.

The output drivers are assumed to have ECL output levels, with eye mask as shown in Figure 4.8 when measured with both differential outputs grounded through 75 Ω at the transmitter output. The receiver end of the cable is terminated to 150 Ω through an equalization network to compensate for cable response and to improve the signal-to-noise ratio at the data recovery circuit input.

There are several different types of connectors appropriate for these links. A "Style-1" connector for balanced cables is a 9-pin shielded D-subminiature connector. A "Style-2" connector (also termed "HSSDC") provides a more integrated plug with simpler mechanical characteristics for plugging and unplugging cables. There is also specified an integral intra-enclosure connector, termed the "SCA-2," for backplane or rackmount applications, that can also carry signals as well as power for numerous devices.

Chapter 5

FC-1: 8B/10B, Transmitters, and Receivers

Introduction

This chapter covers the design of the 8B/10B transmission code used for Fibre Channel transmission, and the transmitter and receiver states.

8B/10B Transmission Code Overview

When data transmitted at high speed over either an electrical or optical link, there are several reasons why it can be important to encode the data being sent. First, the "transmission characteristics" of the transmitted bits can have a significant effect on the error rate and achievable transmission bandwidth with a given technology. These transmission characteristics include such factors as the ratio of 1s to 0s within a window of transmitted bits and the maximum number of clock periods between transitions. Second, the ability of a receiver to perform clock recovery or derive bit synchronization can be improved greatly with data encoding. Third, data encoding can also greatly increase the possibility of detecting and even correcting single- or multiple-bit transmission or reception errors. Fourth, it helps in distinguishing bits transmitted as data from bits transmitted as control. Fifth, encoding can allow simple detection of byte and word boundaries for higher-level synchronization. These five functions, (1) improving transmission characteristics, (2) enabling bit-level clock recovery, (3) improving error detection, (4) separation of data bits from control bits, and (5) derivation of byte and word synchronization, are all accomplished by the 8B/10B transmission coding used under Fibre Channel.

In Fibre Channel, all information transmitted over the links is encoded 8 bits at a time into a 10-bit "Transmission Character" and then sent serially bit by bit. Bits received over a fiber are collected ten bits at a time, and those Transmission Characters that are used for data, called Data Characters, are decoded into the corresponding 8-bit codes. The 10-bit Transmission Code supports all 256 8-bit combinations. Some of the remaining 10-bit Transmission Characters are used as Special Characters, which are distinguishable from Data Characters and are used for control.

To improve error detection, the code recognizes the idea of a "Running Disparity," which is the disparity between the number of 1s and 0s transmitted. Every data byte to be transmitted has two corresponding 10-bit Transmission Characters (which may be identical), one with either five or six 1s and the other with either five or four 1s. The sender keeps track of the Running Disparity and selects the Transmission Character that keeps the Running Disparity as close to zero as possible. This way, some transmission errors which convert a Transmission Character into another valid Transmission character can still be detected as errors, since they affect the Running Disparity.

The 8B/10B transmission encoding guarantees that sufficient transitions are present in the serial bit stream to make clock recovery possible at the receiver. It also greatly increases the likelihood of detecting any single- or multiple-bit errors that may occur during transmission and reception of information. In addition, the code also provides a "comma" feature, a particular 7-bit string that can only occur at the beginning of the first byte of a control word, allowing simple byte and word synchronization.

When the Fibre Channel architecture was being designed, the 4B/5B code used for FDDI was also considered. Its main disadvantage is the lack of DC balance — under transmission of certain data streams, a transmitter may send more 1s than 0s or vice versa. Receivers must then build dc-coupled level-restoring circuitry, which negatively affects receiver sensitivity and dynamic range, and makes the transmission link more sensitive to low-frequency noise. Lack of DC balance also causes data-dependent heating of lasers and LEDs. All these factors make it more difficult to build low-error-rate hardware that operates at high speeds using 4B/5B coding than 8B/10B coding.

Notation Conventions

FC-1 uses a special letter notation for describing information bits and control variables. There is also a prescribed notation for naming valid Transmission Characters. This notation is not used outside the context of the 8B/10B encoding FC-1 level.

An unencoded FC-1 information byte is composed of 8 information bits, denoted A through H, and the control variable Z. This information is encoded into bits denoted a, b, c, d, e, i, f, g, h, and j of the 10-bit Transmission Character. The correspondence between letter and number naming conventions is as follows:

```
                          data            control
FC-2 bit notation:  7 6 5 4 3 2 1 0  variable
FC-1 unencoded      | | | | | | | |      |
   bit notation:    H G F E D C B A      Z
FC-1 encoded
   bit notation: a b c d e i f g h j
```

Each Transmission Character is given a name using the following convention: Zxx.y, where Z is the control variable of the unencoded FC-1 information byte, xx is the decimal value of the binary number composed of the bits E, D, C, B, and A of the unencoded FC-1 information byte in that order, and y is the decimal value of the binary number composed of the bits H, G, and F of the unencoded FC-1 information byte, in that order. The value of Z indicates whether the Transmission Character is a Data Character (Z = D) or a Special Character (Z = K). A sample conversion from FC-2 byte notation to FC-1 Transmission Character notation is shown in Figure 5.1 for conversion of Control x'BC' to K28.5. Similarly, Data x'60' = b'0110 0000' = D0.3, Data x'8F' = b'1000 1111' = D15.4, and Control x'FB' = b'1111 1011' = K27.7.

Figure 5.1
Sample conversion of an FC-2 byte to FC-1 Transmission Character notation.

```
FC-2 byte notation:            x'BC'        -- Special Code

FC-2 bit notation:       7 6 5 4 3 2 1 0     Control
                         1 0 1 1 1 1 0 0        K

FC-1 unencoded bit notation:
                         H G F E D C B A        Z
                         1 0 1 1 1 1 0 0        K

reordered to conform with Zxx.y naming convention:
                         Z E D C B A H G F
                         K 1 1 1 0 0 1 0 1

FC-1 Transmission Character name:
                         K 28.5
```

Each of the 256 Dxx.y data bytes has two corresponding Transmission Characters, but most Kxx.y bytes do not result in valid Transmission Characters. Only 12 Kxx.y bytes have corresponding Transmission Characters defined in the 8B/10B code, and only the K28.5 Special Character is used in Fibre Channel transmission.

Character Encoding and Decoding

Conversion of data bytes or special bytes to Transmission Characters is carried out on the basis of the 5- and 3-bit groupings, using separate 5B/6B and 3B/4B encoding and decoding sub-blocks. This division of coding work into smaller blocks lets it be carried out more efficiently in CMOS hardware than a full 8-bit single-shot coding could be. The code conversions used are shown in Figure 5.2. The characters D11.7, D13.7, and D14.7 are coded as exceptions to these rules, since following the unmodified coding with a positive Current Running Disparity would allow a stream of five adjacent 0s.

Within the 8B/10B transmission code the Transmission Characters are labeled a, b, c, d, e, i, f, g, h, and j. The sub-blocks HGF and EDCBA of the unencoded byte are converted into sub-blocks fghj and abcdei, respectively, of the coded Transmission Characters. Every sub-block is assigned one of two possibly identical encoded sub-blocks, dependent on the current Running Disparity.

Figure 5.2
5B/6B and 3B/4B coding conversions.

5B/6B coding for Data Characters		
Unencoded EDCBA	Current RD- abcdei	Current RD+ abcdei
D0:00000	100111	011000
D1:00001	011101	100010
D2:00010	101101	010010
D3:00011	110001	110001
D4:00100	110101	001010
D5:00101	101001	101001
D6:00110	011001	011001
D7:00111	111000	000111
D8:01000	111001	000110
D9:01001	100101	100101
D10:01010	010101	010101
D11:01011	110100	110100
D12:01100	001101	001101
D13:01101	101100	101100
D14:01110	011100	011100
D15:01111	010111	101000
D16:10000	011011	100100
D17:10001	100011	100011
D18:10010	010011	010011
D19:10011	110010	110010
D20:10100	001011	001011
D21:10101	101010	101010
D22:10110	011010	011010
D23:10111	111010	000101
D24:11000	110011	001100
D25:11001	100110	100110
D26:11010	010110	010110
D27:11011	110110	001001
D28:11100	001110	001110
D29:11101	101110	010001
D30:11110	011110	100001
D31:11111	101011	010100

3B/4B coding for Data Characters		
Unencoded HGF	Current RD- fghj	Current RD+ fghj
--.0: 000	1011	0100
--.1: 001	1001	1001
--.2: 010	0101	0101
--.3: 011	1100	0011
--.4: 100	1101	0010
--.5: 101	1010	1010
--.6: 110	0110	0110
--.7: 111	1110/0111	0001

5B/6B coding for Special Characters		
Unencoded EDCBA	Current RD - abcdei	Current RD + abcdei
K28:11100	001111	110000
K23:10111	111010	000101
K27:11011	110110	001001
K29:11101	101110	010001
K30:11110	011110	100001

3B/4B coding for Special Characters		
Unencoded HGF	Current RD - fghj	Current RD + fghj
--.0: 000	1011	0100
--.1: 001	0110	1001
--.2: 010	1010	0101
--.3: 011	1100	0011
--.4: 100	1101	0010
--.5: 101	0101	1010
--.6: 110	1001	0110
--.7: 111	0111	1000

After powering on or exiting the diagnostic mode, the transmitter assumes the negative value for its initial Running Disparity. The Current Running Disparity is calculated following transmission of each sub-block. The Running Disparity at the beginning of the 6-bit sub-block is the Running Disparity at the end of the last Transmission Character. Running Disparity at the beginning of the 4-bit sub-block is the Running Disparity at the end of the 6-bit sub-block. Running Disparity at the end of the Transmission Character is the Running Disparity at the end of the 4-bit sub-block.

The rules for calculating Running Disparity for sub-blocks are as follows:

- Running Disparity at the end of any sub-block is positive if (1) the encoded sub-block contains more 1s than 0s, (2) if the 6-bit sub-block is b'00 0111,' or (3) if the 4-bit sub-block is b'0011.'

- Running Disparity at the end of any sub-block is negative if (1) the encoded sub-block contains more 0 than 1 bits, (2) if the 6-bit sub-block is b'11 1000,' or (3) if the 4-bit sub-block is b'1100.'
- Otherwise, Running Disparity at the end of the sub-block is the same as at the beginning of the sub-block.

In order to limit the run length of 0 or 1 bits between sub-blocks, the 8B/10B transmission code rules specify that sub-blocks encoded as b'00 0111' or b'0011' are generated only when Running Disparity at the beginning of the sub-block is positive; thus, Running Disparity at the end of these sub-blocks will also be positive. Likewise, sub-blocks containing b'11 1000' or b'1100' are generated only when the Running Disparity at the beginning of the sub-block is negative; thus Running Disparity at the end of these sub-blocks will also be negative.

The full set of valid special characters in the 8B/10B code is shown in Figure 5.3. The valid Data Characters are shown in Figures 5.4, 5.5, and 5.6.

Figure 5.3
Full 8B/10B code: Valid Special Characters.

Special Characters

Char Name	#	Current RD- abcdei	fghj	Current RD+ abcdei	fghj
K28.0	1C	001111	0100	110000	1011
K28.1	3C	<u>001111</u>	<u>1</u>001	<u>110000</u>	<u>0</u>110
K28.2	5C	001111	0101	110000	1010
K28.3	7C	001111	0011	110000	1100
K28.4	9C	001111	0010	110000	1101
K28.5	BC	<u>001111</u>	<u>1</u>010	<u>110000</u>	<u>0</u>101
K28.6	DC	001111	0110	110000	1001
K28.7	FC	<u>001111</u>	<u>1</u>000	<u>110000</u>	<u>0</u>111
K23.7	F7	111010	1000	000101	0111
K27.7	FB	110110	1000	001001	0111
K29.7	FD	101110	1000	010001	0111
K30.7	FE	011110	1000	100001	0111

Notes:
12 Special Characters are defined, but only K28.5 is actually used in Fibre Channel transmission.

The comma series b'001 1111' or b'110 0000' does not appear in any data character or combination of data characters

Transmission, Reception, and Error Detection

When the FC-1 level receives a byte to be transmitted from the FC-2 level, it uses (1) the byte to be transmitted, (2) the Data versus Special Byte indicator, and (3) the current Running Disparity, to determine the encoded trans-

Figure 5.4 Full 8B/10B code: Valid Data Characters, 1 of 3.

Char Name	#	Current RD- abcdei	Current RD- fghj	Current RD+ abcdei	Current RD+ fghj
Characters x'00' to x'1F'					
D0.0	00	100111	0100	011000	1011
D1.0	01	011011	0100	100010	1011
D2.0	02	101101	0100	010010	1011
D3.0	03	110001	1011	110001	0100
D4.0	04	110101	0100	001010	1011
D5.0	05	101001	1011	101001	0100
D6.0	06	011001	1011	011001	0100
D7.0	07	111000	1011	000111	0100
D8.0	08	111001	0100	000110	1011
D9.0	09	100101	1011	100101	0100
D10.0	0A	010101	1011	010101	0100
D11.0	0B	110100	1011	110100	0100
D12.0	0C	001101	1011	001101	0100
D13.0	0D	101100	1011	101100	0100
D14.0	0E	011100	1011	011100	0100
D15.0	0F	010111	0100	101000	1011
D16.0	10	011011	0100	100100	1011
D17.0	11	100011	1011	100011	0100
D18.0	12	010011	1011	010011	0100
D19.0	13	110010	1011	110010	0100
D20.0	14	001011	1011	001011	0100
D21.0	15	101010	1011	101010	0100
D22.0	16	011010	1011	011010	0100
D23.0	17	111010	0100	000101	1011
D24.0	18	110011	0100	001100	1011
D25.0	19	100110	1011	100110	0100
D26.0	1A	010110	1011	010110	0100
D27.0	1B	110110	0100	001001	1011
D28.0	1C	001110	1011	001110	0100
D29.0	1D	101110	0100	010001	1011
D30.0	1E	011110	0100	100001	1011
D31.0	1F	101011	0100	010100	1011
D0.1	20	100111	1001	011000	1001
D1.1	21	011101	1001	100010	1001
D2.1	22	101101	1001	010010	1001
D3.1	23	110001	1001	110001	1001
D4.1	24	110101	1001	001010	1001
D5.1	25	101001	1001	101001	1001
D6.1	26	011001	1001	011001	1001
D7.1	27	111000	1001	000111	1001
D8.1	28	111001	1001	000110	1001
D9.1	29	100101	1001	100101	1001
D10.1	2A	010101	1001	010101	1001
D11.1	2B	110100	1001	110100	1001
D12.1	2C	001101	1001	001101	1001
D13.1	2D	101100	1001	101100	1001
D14.1	2E	011100	1001	011100	1001
D15.1	2F	010111	1001	101000	1001

Char Name	#	Current RD- abcdei	Current RD- fghj	Current RD+ abcdei	Current RD+ fghj
Characters x'30' to x'5F'					
D16.1	30	011011	1001	100100	1001
D17.1	31	100011	1001	100011	1001
D18.1	32	010011	1001	010011	1001
D19.1	33	110010	1001	110010	1001
D20.1	34	001011	1001	001011	1001
D21.1	35	101010	1001	101010	1001
D22.1	36	011010	1001	011010	1001
D23.1	37	111010	1001	000101	1001
D24.1	38	110011	1001	001100	1001
D25.1	39	100110	1001	100110	1001
D26.1	3A	010110	1001	010110	1001
D27.1	3B	110110	1001	001001	1001
D28.1	3C	001110	1001	001110	1001
D29.1	3D	101110	1001	010001	1001
D30.1	3E	011110	1001	100001	1001
D31.1	3F	101011	1001	010100	1001
D0.2	40	100111	0101	011000	0101
D1.2	41	011101	0101	100010	0101
D2.2	42	101101	0101	010010	0101
D3.2	43	110001	0101	110001	0101
D4.2	44	110101	0101	001010	0101
D5.2	45	101001	0101	101001	0101
D6.2	46	011001	0101	011001	0101
D7.2	47	111000	0101	000111	0101
D8.2	48	111001	0101	000110	0101
D9.2	49	100101	0101	100101	0101
D10.2	4A	010101	0101	010101	0101
D11.2	4B	110100	0101	110100	0101
D12.2	4C	001101	0101	001101	0101
D13.2	4D	101100	0101	101100	0101
D14.2	4E	011100	0101	011100	0101
D15.2	4F	010111	0101	101000	0101
D16.2	50	011011	0101	100100	0101
D17.2	51	100011	0101	100011	0101
D18.2	52	010011	0101	010011	0101
D19.2	53	110010	0101	110010	0101
D20.2	54	001011	0101	001011	0101
D21.2	55	101010	0101	101010	0101
D22.2	56	011010	0101	011010	0101
D23.2	57	111010	0101	000101	0101
D24.2	58	110011	0101	001100	0101
D25.2	59	100110	0101	100110	0101
D26.2	5A	010110	0101	010110	0101
D27.2	5B	110110	0101	001001	0101
D28.2	5C	001110	0101	001110	0101
D29.2	5D	101110	0101	010001	0101
D30.2	5E	011110	0101	100001	0101
D31.2	5F	101011	0101	010100	0101

Figure 5.5
Full 8B/10B code: Valid Data Characters, 2 of 3.

Characters x'60' to x'8F'

Char Name	#	Current RD- abcdei	fghj	Current RD+ abcdei	fghj
D0.3	60	100111	0011	011000	1100
D1.3	61	011101	0011	100010	1100
D2.3	62	101101	0011	010010	1100
D3.3	63	110001	1100	110001	0011
D4.3	64	110101	0011	001010	1100
D5.3	65	101001	1100	101001	0011
D6.3	66	011001	1100	011001	0011
D7.3	67	111000	1100	000111	0011
D8.3	68	111001	0011	000110	1100
D9.3	69	100101	1100	100101	0011
D10.3	6A	010101	1100	010101	0011
D11.3	6B	110100	1100	110100	0011
D12.3	6C	001101	1100	001101	0011
D13.3	6D	101100	1100	101100	0011
D14.3	6E	011100	1100	011100	0011
D15.3	6F	010111	0011	101000	1100
D16.3	70	011011	0011	100100	1100
D17.3	71	100011	1100	100011	0011
D18.3	72	010011	1100	010011	0011
D19.3	73	110010	1100	110010	0011
D20.3	74	001011	1100	001011	0011
D21.3	75	101010	1100	101010	0011
D22.3	76	011010	1100	011010	0011
D23.3	77	111010	0011	000101	1100
D24.3	78	110011	0011	001100	1100
D25.3	79	100110	1100	100110	0011
D26.3	7A	010110	1100	010110	0011
D27.3	7B	110110	0011	001001	1100
D28.3	7C	001110	1100	001110	0011
D29.3	7D	101110	0011	010001	1100
D30.3	7E	011110	0011	100001	1100
D31.3	7F	101011	0011	010100	1100
D0.4	80	100111	0010	011000	1101
D1.4	81	011101	0010	100010	1101
D2.4	82	101101	0010	010010	1101
D3.4	83	110001	1101	110001	0010
D4.4	84	110101	0010	001010	1101
D5.4	85	101001	1101	101001	0010
D6.4	86	011001	1101	011001	0010
D7.4	87	111000	1101	000111	0010
D8.4	88	111001	0010	000110	1101
D9.4	89	100101	1101	100101	0010
D10.4	8A	010101	1101	010101	0010
D11.4	8B	110100	1101	110100	0010
D12.4	8C	001101	1101	001101	0010
D13.4	8D	101100	1101	101100	0010
D14.4	8E	011100	1101	011100	0010
D15.4	8F	010111	0010	101000	1101

Characters x'90' to x'BF'

Char Name	#	Current RD- abcdei	fghj	Current RD+ abcdei	fghj
D16.4	90	011011	0010	100100	1101
D17.4	91	100011	1101	100011	0010
D18.4	92	010011	1101	010011	0011
D19.4	93	110010	1101	110010	0010
D20.4	94	001011	1101	001011	0010
D21.4	95	101010	1101	101010	0010
D22.4	96	011010	1101	011010	0010
D23.4	97	111010	0010	000101	1101
D24.4	98	110011	0010	001100	1101
D25.4	99	100110	1101	100110	0010
D26.4	9A	010110	1101	010110	0010
D27.4	9B	110110	0010	001001	1101
D28.4	9C	001110	1101	001110	0010
D29.4	9D	101110	0010	010001	1101
D30.4	9E	011110	0010	100001	1101
D31.4	9F	101011	0010	010100	1101
D0.5	A0	100111	1010	011000	1010
D1.5	A1	011101	1010	100010	1010
D2.5	A2	101101	1010	010010	1010
D3.5	A3	110001	1010	110001	1010
D4.5	A4	110101	1010	001010	1010
D5.5	A5	101001	1010	101001	1010
D6.5	A6	011001	1010	011001	1010
D7.5	A7	111000	1010	000111	1010
D8.5	A8	111001	1010	000110	1010
D9.5	A9	100101	1010	100101	1010
D10.5	AA	010101	1010	010101	1010
D11.5	AB	110100	1010	110100	1010
D12.5	AC	001101	1010	001101	1010
D13.5	AD	101100	1010	101100	1010
D14.5	AE	011100	1010	011100	1010
D15.5	AF	010111	1010	101000	1010
D16.5	B0	011011	1010	100100	1010
D17.5	B1	100011	1010	100011	1010
D18.5	B2	010011	1010	010011	1010
D19.5	B3	110010	1010	110010	1010
D20.5	B4	001011	1010	001011	1010
D21.5	B5	101010	1010	101010	1010
D22.5	B6	010101	1010	011010	1010
D23.5	B7	111010	1010	000101	1010
D24.5	B8	110011	1010	001100	1010
D25.5	B9	100110	1010	100110	1010
D26.5	BA	010110	1010	010110	1010
D27.5	BB	110110	1010	001001	1010
D28.5	BC	001110	1010	001110	1010
D29.5	BD	101110	1010	010001	1010
D30.5	BE	011110	1010	100001	1010
D31.5	BF	101011	1010	010100	1010

Figure 5.6
Full 8B/10B code: Valid Data Characters, 3 of 3.

Characters x'C0' to x'DF'

Char Name	#	Current RD- abcdei fghj	Current RD+ abcdei fghj
D0.6	C0	100111 0110	011000 0110
D1.6	C1	011101 0110	100010 0110
D2.6	C2	101101 0110	010010 0110
D3.6	C3	110001 0110	110001 0110
D4.6	C4	110101 0110	001010 0110
D5.6	C5	101001 0110	101001 0110
D6.6	C6	011001 0110	011001 0110
D7.6	C7	111000 0110	000111 0110
D8.6	C8	111001 0110	000110 0110
D9.6	C9	100101 0110	100101 0110
D10.6	CA	010101 0110	010101 0110
D11.6	CB	110100 0110	110100 0110
D12.6	CC	001101 0110	001101 0110
D13.6	CD	101100 0110	101100 0110
D14.6	CE	011100 0110	011100 0110
D15.6	CF	010111 0110	101000 0110
D16.6	D0	011011 0110	100100 0110
D17.6	D1	100011 0110	100011 0110
D18.6	D2	010011 0110	010011 0110
D19.6	D3	110010 0110	110010 0110
D20.6	D4	001011 0110	001011 0110
D21.6	D5	101010 0110	101010 0110
D22.6	D6	011010 0110	011010 0110
D23.6	D7	111010 0110	000101 0110
D24.6	D8	110011 0110	001100 0110
D25.6	D9	100110 0110	100110 0110
D26.6	DA	010110 0110	010110 0110
D27.6	DB	110110 0110	001001 0110
D28.6	DC	001110 0110	001110 0110
D29.6	DD	101110 0110	010001 0110
D30.6	DE	011110 0110	100001 0110
D31.6	DF	101011 0110	010100 0110

Characters x'E0' to x'FF'

Char Name	#	Current RD- abcdei fghj	Current RD+ abcdei fghj
D0.7	E0	110111 0001	011000 1110
D1.7	E1	011101 0001	100010 1110
D2.7	E2	101101 0001	010010 1110
D3.7	E3	110001 1110	110001 0001
D4.7	E4	110101 0001	001010 1110
D5.7	E5	101001 1110	101001 0001
D6.7	E6	011001 1110	011001 0001
D7.7	E7	111000 1110	000111 0001
D8.7	E8	111001 0001	000110 1110
D9.7	E9	100101 1110	100101 0001
D10.7	EA	010101 1110	010101 0001
D11.7	EB	110100 1110	110100 1000
D12.7	EC	001101 1110	001101 0001
D13.7	ED	101100 1110	101100 1000
D14.7	EE	011100 1110	011100 1000
D15.7	EF	010111 0001	101000 1110
D16.7	F0	011011 0001	100100 1110
D17.7	F1	100011 0111	100011 0001
D18.7	F2	010011 0111	010011 0001
D19.7	F3	110010 1110	110010 0001
D20.7	F4	001011 0111	001011 0001
D21.7	F5	101010 1110	011010 0001
D22.7	F6	011010 1110	011010 0001
D23.7	F7	111010 0001	000101 1110
D24.7	F8	110011 0001	001100 1110
D25.7	F9	100110 1110	100110 0001
D26.7	FA	010110 1110	010110 0001
D27.7	FB	110110 0001	001001 1110
D28.7	FC	001110 1110	001110 0001
D29.7	FD	101110 0001	010001 1110
D30.7	FE	011110 0001	100001 1110
D31.7	FF	101011 0001	010100 1110

mission character to transmit. These factors determine the coded bit to be transmitted. A sample block diagram of a logical structure used for this function is shown in Figure 5.7.

Over the serial link, bit a is transmitted first, followed by bits b, c, d, e, i, f, g, h, and j, in that (non-alphabetical) order. That is to say, the 6- and 4-bit sub-blocks are transmitted in what would correspond to a least-significant-bit to most-significant-bit order.

The bytes in the word are transmitted in the most-significant to least-significant order, to correspond with usage in the FC-2 level. The transmission order of all bits in a word is shown in Figure 5.8. Ordered Sets are transmitted sequentially beginning with the Special Character and proceeding Data Character by Data Character left to right within the definition of the Ordered Set until all characters of the Ordered Set are transmitted.

When bits arrive at the receiver, they are converted by deserialization into 10-bit codes. Byte and word synchronization at the FC-0 level are recovered

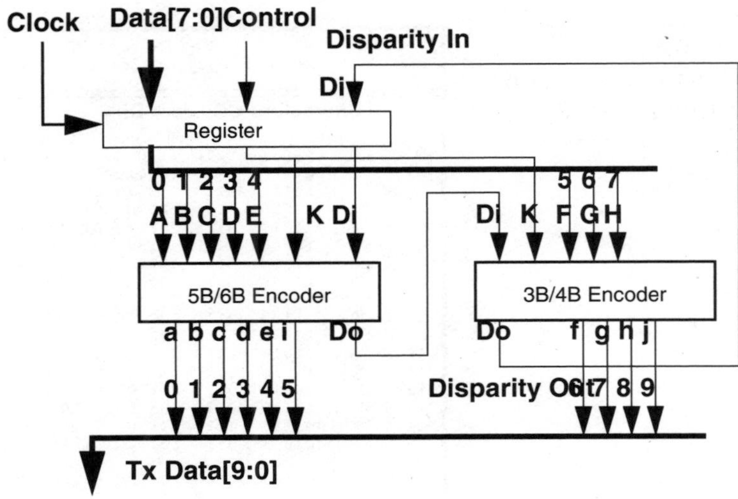

Figure 5.7 Example logical structure of an 8B/10B encoder.

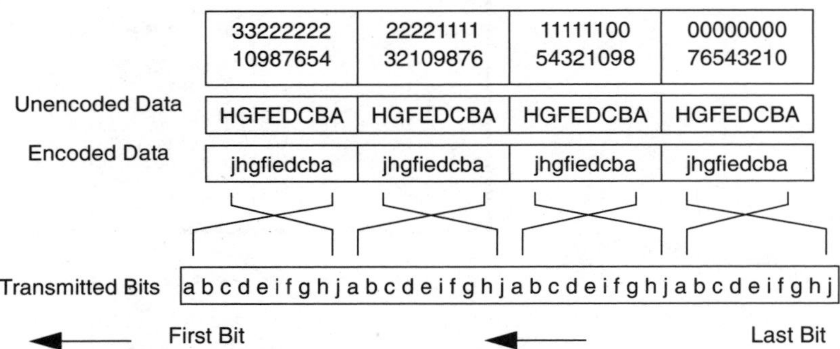

Figure 5.8 Transmission order of bits and bytes in a word.

by detection of the K28.5 character, which contains the "comma" 7-bit string (x'001 1111' or b'110 0000'). This 7-bit string cannot appear in any other valid Transmission characters or combination of Characters. Since these bits only appear at the beginning of the K28.5 character, and the K28.5 character only appears in Ordered Sets, which are aligned with transmission word boundaries, a receiver can derive byte and word synchronization easily by searching for this bit string. Any time the comma string is detected, the first bit is aligned to the first bit of the first byte of a transmission word. Once bit, byte, and word synchronization are derived, the receiver can begin receiving and decoding transmission characters.

For decoding, each 10-bit character received is matched on a sub-block basis against the possible Transmission Character sub-blocks. A sample block diagram of how a decoder could possibly be implemented is shown in

Figure 5.9. If matches are found for both sub-blocks, the Character is converted into its decoded byte and passed up to the FC-2 level. If no match is found, then the Transmission Character is considered invalid and a code violation is detected and reported to the Port. Independent of the Transmission Character's validity, the received Transmission Character is used to calculate a new value of Running Disparity. This new value is used as the receiver's Current Running Disparity for the next received Transmission Character.

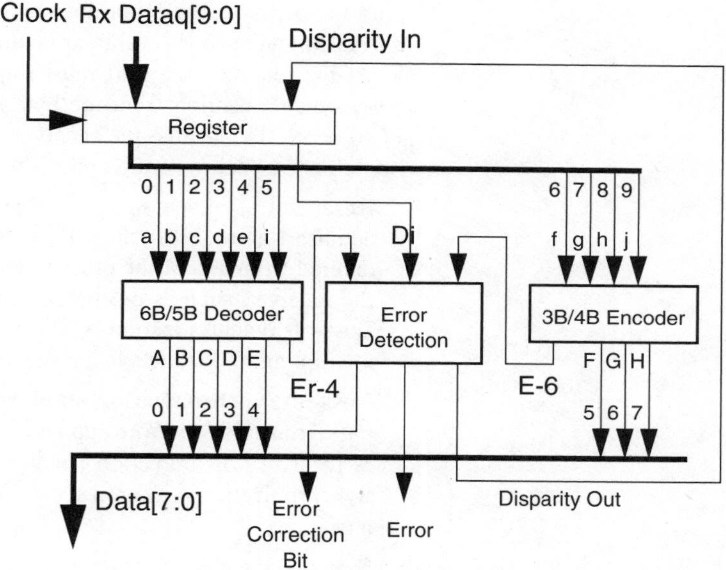

Figure 5.9
Example logical structure of a 10B/8B decoder.

Detection of a code violation does not necessarily indicate that the Transmission Character in which the Code Violation was detected is in error — the error may have occurred in an earlier Character. When a single- or multi-bit error converts a valid code point into a different valid code point, no error is detected in that character. However, if the relative number of 1s and 0s is changed, a later character may be falsely flagged as corrupted when the error causes an invalid Running Disparity to be detected.

Transmitter and Receiver States

The interface between the FC-0 and FC-1 levels at a receiver involves synchronization at the character or byte level, generated in FC-0 using comma detection circuitry. Interface between FC-1 and FC-2 is on a word-by-word basis, so the FC-1 level must provide procedures to transmit words passed down from the FC-2 level and to generate word-level synchronization for data passed up to the FC-2 level.

A transmitter is always either attempting to send Transmission Words, or it is not. There are several reasons for not transmitting a signal, including (1) the transmitter may not be enabled, as when the transmitter is initializing, (2) a laser safety condition such as an open fiber may exist, or (3) the transmitter may have failed or broken in some way. The transmitter maintains a four-state state machine, incorporating states for each of the above conditions, plus an operational and working state, in which the transmitter is attempting to transmit an encoded bit stream onto the fiber.

At the receiver, the Port must implement procedures to generate word-level synchronization from the byte-level synchronization passed from the FC-0 level. To do this, the receiver contains a small state machine which implements synchronization procedures, containing the following states:

Reset state, which indicates that the FC-1 level is not attempting to acquire synchronization or pass data up to the FC-2 level. This state is entered from any of the other states when a reset condition is imposed on the receiver, such as during power-on. Exit from the reset state is to the Loss-of-Synchronization state, when the receiver is ready to begin attempting to acquire synchronization.

Loss-of-Synchronization state, which indicates that the receiver cannot pass Transmission Words up and is attempting to acquire synchronization. A receiver will transition from the Loss-of-Synchronization state to the Synchronization Acquired state after detecting three valid Ordered Sets in a row.

Synchronization Acquired, which indicates that the receiver is passing Transmission Words to the higher levels for interpretation. It contains five sub-states termed (0) "No, (1) "first, (2) "second, (3) "third, and (4) "fourth invalid Transmission Word detected," respectively. The state machine will transition from a lower-numbered sub-state to a higher numbered sub-state if either of two consecutive Transmission Words is invalid. The state machine will transition from a higher-numbered sub-state to a lower numbered sub-state if both Transmission Words are valid. If continuously valid Transmission Words are detected, the receiver will stay in the "no invalid Transmission Words detected" state. If an invalid Transmission Words is detected while in the "Fourth Transmission Word detected" state, the receiver assumes all synchronization has been lost and enters the Loss-of-Synchronization state.

Some comments about the description above. First, an "Invalid Transmission Word" is a word which doesn't match the requirements of an Ordered Set, as described in the "Ordered Sets" section, on page 91. This includes words containing a code violation in one of the characters, a Special Character in either the second, third, or fourth character position, or an incorrect

Beginning Running Disparity, such as a Start-of-Frame delimiter with positive Running Disparity. Second, transitions depend on detection of Ordered Sets since the procedure is only expected to be used when the receiver is following one of the Primitive Sequence Protocols described in the "Primitive Sequence Protocols" section, on page 104, where Ordered Sets are transmitted continuously. Third, the transitions between sub-states in the Synchronization Acquired state may seem more complicated than required, but they are implemented in order to make the receiver less sensitive to transitive (soft) errors. Fourth, some receivers may implement a function to detect a loss of input signal, such as a light level detector in a photodiode receiver. If such a function is implemented, the receiver can transition directly from the Synchronization Acquired to the Loss-of-Synchronization state on detection of a loss of signal.

Loss of synchronization is a serious error condition, encountered when it is absolutely clear that the receiver is not synchronized to Transmission Word boundaries. The complex state machine assures that the transitions between Loss-of-Synchronization and Synchronization Acquired states do not occur easily, such as after a single-bit or small multi-bit transmission error.

When the receiver is in the Synchronization Acquired state, it can pass Transmission Words up to higher levels along with a word clock and indication of Transmission Word validity.

Chapter 6

Ordered Sets

Introduction

In any serial communications protocol, the only thing the receiver sees is a long stream of signals denoting transmitted bits. Some of these bits are included in transmitted data, while some are part of control information, such as is used for marking the ends of Frames or for notifying the receiver that the transmitter is shutting down. The protocol must specify a way of distinguishing between data and the various types of control information.

In Fibre Channel, these functions are carried out using "Ordered Sets." These are four-character Transmission Words that all start with the K28.5 Special Character. K28.5, as described in the "Transmission, Reception, and Error Detection" section, on page 80, are distinguished by a series of 7 bits, termed a "comma," which will not occur in any valid Data Transmission Character or combination of Transmission Characters. Besides maintaining byte and word synchronization, detection of the comma notifies the receiver that the transmitter is sending control information, rather than data information.

The three Transmission Characters following the K28.5 character in the 4-byte word indicate what control information is being sent. Currently, there are defined the following types of Ordered Sets:

- Frame delimiters mark the boundaries of Frames and describe Frame contents.
  - Start of Frame delimiters, including **SOFc1**, **SOFi1**, **SOFn1**, **SOFi2**, **SOFn2**, **SOFi3**, **SOFn3**, **SOFf, SOFc4, SOFi4,** and **SOFn4** mark the start of various types of Frames.
  - End of Frame delimiters, including **EOFn**, **EOFt**, **EOFdt**, **EOFa**, **EOFdti**, **EOFni**, **EOFrt,** and **EOFrti** mark the ends of Frames and describe the Frame contents.
- Primitive Signals are used to signal events.
  - **Idle** is used for maintaining link activity when there is no data to send.
  - Receiver_Ready (**R_RDY**) is used for buffer-to-buffer flow control.
  - **ARBx**, **ARB(val)**, **OPNyx**, **OPNyy**, **OPNfr**, **OPNyr**, **CLS**, and **MRKtx**, for procedures specific to operation on an Arbitrated Loop, such as arbitrating for access, opening and closing communications, and synchronizing.
- Primitive Sequences are used to signal Port states.
  - Not_Operational (**NOS**) indicates that the transmitter is not operational.
  - Offline (**OLS**) indicates that the transmitter is offline.
  - Link_Reset (**LR**) resets a link after a link error.

- Link_Reset_Response (**LRR**) is used for handshaking on resetting a link after a link error.
- Seven versions of Loop Initialization (**LIP**), Loop port enable (**LPEyx**, **LPEfx**), and Loop port bypass (**LPByx**, **LPBfx**) Primitive Sequences are in an Arbitrated Loop topology.

The Primitive Signals and Primitive Sequences used exclusively in the Arbitrated Loop topology are described in detail in Chapter 16. The others are described later in this chapter.

This chapter also describes the states that a Port can be in. These states allow a Port to come up and go down without confusing the Port on the other end of the link. A Port and its attached Port can control transitions between active, reset, and failure Port states using a protocol of exchanged Primitive Sequences.

Ordered Sets

Start of Frame Delimiters

The **SOF** delimiter is an Ordered Set that immediately precedes the Frame content. The **SOF** delimiter is always transmitted on a word boundary. There are eleven different **SOF** Ordered Sets defined for the Fabric and for N_Port Sequence control. The bit encodings for the **SOF** delimiters are shown in Figure 6.1.

SOFc1: "SOF Connect Class 1" is used to request a Class 1 Dedicated Connection. The delimiter may also identify the start of the first Sequence, functioning as an implicit SOFi1.

SOFi1: The first Sequence of a Dedicated Connection is initiated with SOFc1. All subsequent Sequences within that Dedicated Connection are delimited with "SOF Initiate Class 1."

SOFn1: "SOF Normal Class 1" is used for all Frames except the first Frame of a Sequence for Class 1 service.

SOFi2: "SOF Initiate Class 2" is used on the first Frame to initiate a Sequence for Class 2 service.

SOFn2: "SOF Normal Class 2" is used for all Frames except the first Frame of a Sequence for Class 2 service.

SOFi3: "SOF Initiate Class 3" is used on the first Frame to initiate a Sequence for Class 3 service.

Figure 6.1
Frame Delimiter Ordered Sets.

Frame Delimiters			
Delimiter Function SOF = Start of Frame, EOF = End of Frame	**Abbrev.**	**Beginning RD**	**Ordered Set**
SOF Connect Class 1	SOFc1	Negative	K28.5 D21.5 D23.0 D23.0
SOF Initiate Class 1 SOF Normal Class 1	SOFi1 SOFn1	Negative Negative	K28.5 D21.5 D23.2 D23.2 K28.5 D21.5 D23.1 D23.1
SOF Initiate Class 2 SOF Normal Class 2	SOFi2 SOFn2	Negative Negative	K28.5 D21.5 D21.2 D21.2 K28.5 D21.5 D21.1 D21.1
SOF Initiate Class 3 SOF Normal Class 3 (Also SOFil, for AL initialization)	SOFi3 SOFn3	Negative Negative	K28.5 D21.5 D22.2 D22.2 K28.5 D21.5 D22.1 D22.1
SOF Fabric	SOFf	Negative	K28.5 D21.5 D24.2 D24.2
SOF Activate Class 4	SOFc4	Negative	K28.5 D21.5 D25.0 D25.0
SOF Initiate Class 4	SOFi4	Negative	K28.5 D21.5 D25.2 D25.2
SOF Normal Class 4	SOFn4	Negative	K28.5 D21.5 D25.1 D25.1
EOF Normal	EOFn	Negative Positive	K28.5 D21.4 D21.3 D21.3 K28.5 D21.5 D21.3 D21.3
EOF Terminate	EOFt	Negative Positive	K28.5 D21.4 D21.4 D21.4 K28.5 D21.5 D21.4 D21.4
EOF Disconnect-Terminate (Class 1) EOF Deactivate-Terminate (Class 4)	EOFdt	Negative Positive	K28.5 D21.4 D21.7 D21.7 K28.5 D21.5 D21.7 D21.7
EOF Remove-Terminate (Class 4)	EOFrt	Negative Positive	K28.5 D21.4 D25.4 D25.4 K28.5 D21.5 D25.4 D25.4
EOF Disconnect-Terminate-Invalid (Class 1) EOF Deactivate-Terminate-Invalid (Class 4)	EOFdti	Negative Positive	K28.5 D10.4 D21.4 D21.4 K28.5 D10.5 D21.4 D21.4
EOF Remove-Terminate-Invalid (Class 4)	EOFrti	Negative Positive	K28.5 D10.4 D25.4 D25.4 K28.5 D10.5 D25.4 D25.4
EOF Normal-Invalid	EOFni	Negative Positive	K28.5 D10.4 D21.6 D21.6 K28.5 D10.5 D21.6 D21.6
EOF Abort	EOFa	Negative Positive	K28.5 D21.4 D21.6 D21.6 K28.5 D21.5 D21.6 D21.6

SOFn3: "SOF Normal Class 3" is used for all Frames except the first Frame of a Sequence for Class 3 service.

SOFf: A Fabric_Frame may be used by the Fabric for intra-Fabric communication, i.e., communications between switches in a multi-switch Fabric. A Fabric_Frame is composed of a "SOF Fabric" delimiter, a Frame Header and content, and an **EOFn** or **EOFt** delimiter. Generation and handling of Fabric_Frames are described further in Chapter 17. If a Fabric_Frame, indicated by **SOFf**, is received by an N_Port, the Fabric_Frame is discarded and ignored.

SOFc4: "SOF Circuit Activate Class 4" is used to move a Class 4 circuit from the Dormant to the Live state.

SOFi4: "SOF Initiate Class 4" is used on the first Frame to initiate a Sequence for Class 4 Service.

SOFn4: "SOF Normal Class 4" is used for all Frames except the first Frame of a Sequence for Class 4 Service.

End-of-Frame Delimiter

The **EOF** delimiter is an Ordered Set that designates the end of the Frame content. It follows the CRC field and is immediately followed by six or more Idle Primitive Signals at the N_Port transmitter end and by two or more at the receiver end, after traversing the network. Figure 6.1 shows the eight defined **EOF** delimiters.

There are three categories of **EOF** delimiters, of which the first contains four types, the second contains three types, and the third contains a single type.

The first category of delimiter indicates that the Frame is valid from the sender's perspective and potentially valid from the receiver's perspective.

EOFn: "EOF Normal" is used to identify the end of a valid Data Frame or ACK, BSY, or RJT Link Control Frame which does not terminate a Sequence.

EOFt: "EOF Terminate" indicates that the Sequence associated with the SEQ_ID is complete. It is transmitted in the ACK, BSY, or RJT Link Control Frame to the last Data Frame of a Class 1 or Class 2 Sequence.

EOFdt: "EOF Disconnect-Terminate" or "EOF Deactivate Terminate" is transmitted in the ACK, BSY, or RJT to the last Data Frame of a Class 1 or Class 4 Sequence, to remove a Dedicated Connection through the Fabric. The Frame Header End_Connection bit (F_CTL bit 18) indicates the Sequence Initiator's request to remove the Connection. If any Sequences in the Connection are still open, they will be abnormally terminated and may require Sequence recovery on a ULP protocol-dependent basis.

EOFrt: "EOF Remove Terminate" is transmitted in the ACK, BSY, or RJT to the last Data Frame of a Class 4 Sequence to remove the entire Class 4 circuit. Handling of open Sequences is the same as with **EOFdt**.

The second category of **EOF** delimiter indicates that the Frame content is invalid. This occurs when, for example, an intermediate (non-destination)

Port examines the Frame content and detects a code violation or CRC error. This type of delimiter can therefore only be generated on a Fabric topology, where there are intermediate Ports that can examine Frame contents.

Support for generation of these **EOF** delimiters is actually optional, since intermediate (i.e., non-destination) Ports may just discard invalid Frames, rather than forwarding them. Further, it is preferable for an intermediate Port to discard rather than forward invalid Frames, since there are a number of difficulties with forwarding Frames that may have invalid destination or source ID fields. These **EOF** delimiters are defined for use in Fabrics which use "cut-through" or "wormhole" routing, where the head of the Frame is forwarded before the end of the Frame, with the CRC field, is received. These Fabrics cannot discard the entire Frame, so must mark it invalid in the **EOF** delimiter.

If an invalid Frame is received by an intermediate Port, where the Fabric decodes the Frame and detects that one or more transmission characters cannot be decoded, it may continue routing the Frame. In this case, the Fabric can replace the invalid characters with valid characters without generating a new CRC, and replace the **EOFn**, **EOFt**, or **EOFdt** delimiter with either a **EOFni** or **EOFdti** delimiter. Both **EOFni** and **EOFdti** delimiters are defined because of the importance of the facility for removing Class 1 Dedicated Connections and terminating any open Sequences, independent of Frame content in the invalid Frame. A similar delimiter for Class 4, **EOFrti**, performs the same function for Class 4 circuits.

The intermediate Port that replaces the valid Frame delimiter with an invalid one should report the replacement and keep count of the error. The destination N_Port will report reception of the invalid Frame, but intermediate Ports that receive Frames with **EOFni** or **EOFdti** delimiters should just forward the Frames.

EOFni: "EOF Invalid" replaces a recognized **EOFn** or **EOFt** delimiter, indicating that the Frame content is invalid. The destination N_Port will not send any response Frame, and may or may not use the Frame data.

EOFdti: "EOF Disconnect-Terminate-Invalid" replaces a recognized **EOFdt** delimiter on a Frame with invalid Frame content. **EOFdti** removes a Class 1 Dedicated Connection through a Fabric, if present, and deactivaes a Class 4 Virtual Circuit. It also indicates that all Class 1 or Class 4 open Sequences associated with the connected N_Port are abnormally terminated and may require Sequence recovery on a ULP protocol-dependent basis. The destination N_Port will remove the Connection, will not send any response Frame, and may or may not use the Frame data under very restricted conditions, with knowledge that all the Frame Header fields might be corrupted.

EOFrti: "EOF Remove-Terminate-Invalid" terminates the Sequence and removes the Class 4 circuit, marking the Sequence as Invalid.

The third category contains a single **EOF** delimiter, which is used by both N_Ports and F_Ports to indicate an internal malfunction, such as transmitter failure, which does not allow the entire Frame to be transmitted normally.

EOFa: "EOF Abort" indicates that the Frame content is corrupted or truncated due to a malfunction in a link facility during transmission.The Frame with an **EOFa** delimiter must end on a word boundary and must be discarded by the receiver without transmitting a reply. The transmitter, if it is an N_Port, may retransmit the aborted Frame with the same Sequence count (SEQ_CNT), when it doesn't receiver an ACK before a timeout period expires.

Class 6 Service re-uses the delimiters used for Class 1, and the other varieties of Frame transmission (multicast, dedicated simplex, prioritized routing, data compression, etc.) use signalling methods other than Frame delimiters, so these delimiters listed are the only ones required.

Primitive Signals — Idle and R_RDY

Two Primitive Signals are defined for use over a single link for indicating events at the transmitting Port: Idle and R_RDY. The formats of these Primitive Signals and the Primitive Sequences described below are shown in Figure 6.2. The Idle is used to keep the link active and initialized during the periods when there is no data to send. The R_RDY Primitive Sequence is used for buffer-to-buffer flow control to indicate readiness for Frame reception at the link level. A Primitive Signal is recognized when a single Ordered Set is received. At an N_Port transmitter, there must be a minimum of six Primitive Signals (Idle or R_RDY Ordered Sets) between successive Frames. This space between Frames allows the Fabric to compensate for differing transmitter and receiver clock frequencies by removing or inserting Idles between Frames. Any intermediate Ports must ensure that at least two Primitive Signals follow each Frame on delivery to a destination N_Port.

Idle. An Idle indicates that the Port is ready for Frame transmission and reception. Idles are transmitted during the times when Frames, Primitive Sequences, or R_RDY Ordered Sets aren't being transmitted. An Idle can be removed or inserted between Frames at intermediate Ports to compensate for differences in clock frequencies.

R_RDY. The R_RDY Primitive Signal indicates that the transmitting Frame is ready for a Frame to be transmitted over the link. This applies to all Class 2, Class 3, and Class 1 connect-request Data and Link Control Frames.

Figure 6.2 Primitive Signal and Primitive Sequence Ordered Sets.

Primitive Signals			
Primitive Signal	**Abbrev.**	**Beginning RD**	**Ordered Set**
Idle	Idle	Negative	K28.5 D21.4 D21.5 D21.5
Receiver_Ready	R_RDY	Negative	K28.5 D21.4 D10.2 D10.2
Virtual Circuit Ready (Class 4)	VC_RDY	Negative	K28.5 D21.7 VC_ID VC_ID
Arbitrate	ARByx	Negative	K28.5 D20.4 y x
Arbitrate (val)	ARB(val)	Negative	K28.5 D20.4 val val
Open full-duplex	OPNyx	Negative	K28.5 D17.4 AL_PD AL_PS
Open half-duplex	OPNyy	Negative	K28.5 D17.4 AL_PD AL_PD
Open broadcast replicated	OPNyr	Negative	K28.5 D17.4 D31.7 D31.7
Open selective replicated	OPNfr	Negative	K28.5 D17.4 AL_PD D31.7
Close	CLS	Negative	K28.5 D5.4 D21.5 D21.5
Mark	MRKtx	Negative	K28.5 D31.2 MK_TP AL_PS
Dynamic Half Duplex	DHD	Negative	K28.5 D10.4 D21.5 D21.5

Primitive Sequences			
Primitive Sequence	**Abbrev.**	**Beginning RD**	**Ordered Set**
Offline	OLS	Negative	K28.5 D21.1 D10.4 D21.1
Meaning: Internal Port failure Transmitter may power-down, perform diagnostics, or perform initialization. Receiver will ignore link errors or link failure Response: LR			
Not_Operational	NOS	Negative	K28.5 D21.2 D31.5 D5.2
Meaning: Link Failure Response: OLS			
Link_Reset	LR	Negative	K28.5 D9.2 D31.5 D9.2
Meaning: Remove Class 1 Connections, Reset F_Port, or OLS Recognized Response: LRR			
Link_Reset_Response	LRR	Negative	K28.5 D21.1 D31.5 D9.2
Meaning: Link Reset Recognized Response: Idle			
Loop Initialization -- F7,F7 no valid AL_PA	LIP(F7,F7)	Negative	K28.5 D21.0 D23.7 D23.7
Loop Initialization -- F8,F7 loop failure, no valid AL_PA	LIP(F8,F7)	Negative	K28.5 D21.0 D24.7 D23.7
Loop Initialization -- F7,x valid AL_PA	LIP(F7,x)	Negative	K28.5 D21.0 D23.7 AL_PS
Loop Initialization -- F8,x loop failure, valid AL_PA	LIP(F8,x)	Negative	K28.5 D21.0 D24.7 AL_PS
Loop Initialization -- reset	LIPyx	Negative	K28.5 D21.0 AL_PD AL_PS
Loop Initialization -- reset all	LIPfx	Negative	K28.5 D21.0 D31.7 AL_PS
Loop Initialization -- reserved	LIPba	Negative	K28.5 D21.0 b a
Loop Port Enable	LPEyx	Negative	K28.5 D5.0 AL_PD AL_PS
Loop Port Enable all	LPEfx	Negative	K28.5 D5.0 D31.7 AL_PS
Loop Port Bypass	LPByx	Negative	K28.5 D9.0 AL_PD AL_PS
Loop Port Bypass all	LPBfx	Negative	K28.5 D9.0 D31.7 AL_PS

The R_RDY mechanism is not used for Class 1 Data and Link Control Frames. R_RDY gives no indication of the validity of any previous Frames — only of resource availability for receiving a new Frame. Each R_RDY must be preceded and followed by two or more Idles.

Primitive Sequences

A Primitive Sequence is an Ordered Set used to control the status of individual links between Ports. Primitive Sequences are used for link initialization, for resetting links in case of link-level errors, for bringing links offline cleanly, or for enabling or bypassing ports in a loop topology. The actual set of Primitive Sequences used will depend on whether the Ports are a pair of N_Ports or F_Ports, in a point-to-point topology or Fabric topology, or are a set of 2 or more NL_Ports or FL_Ports, on a Loop topology.

Primitive Sequences are transmitted continuously to indicate conditions encountered by the receiver logic of a Port while the condition exists. Since Primitive Sequences are transmitted continuously, and because they cause major effects when they are recognized, recognition of a Primitive Sequence requires detection of three consecutive instances of the same Ordered Set without any different transmission words in between (as opposed to Primitive Signals, where each Ordered Set is meaningful).

The Primitive Sequences protocols and state machines used to control the Port states are detailed in the "Port States on Non-Loop Topologies" section, on page 98, and in the "Loop Port State Machine Operation: An Example" section, on page 288.

A Fabric handles reception of a Primitive Sequence differently, depending on whether any pending or existing Dedicated Connections exist. If a Dedicated Connection exists or is pending while the Fabric receives a Primitive Sequence, it will remove the Connection and notify the F_Port attached to the other connected N_Port, which will transmit the Link Reset (LR) primitive Sequence to the other connected N_Port. If no Dedicated Connection exists, the effects of Primitive Sequences are localized to the sending and receiving Ports of the link, and don't propagate through the Fabric.

Following are detailed descriptions of the first four Primitive Sequences listed in Figure 6.2. The other Primitive Sequences in Figure 6.2 are only used in the Arbitrated Loop topology, and will be described in detail in Chapter 16.

Not_Operational (NOS): The Not_Operational Primitive Sequence indicates that the transmitting Port has detected a link failure condition or is offline, and is waiting for the OLS Primitive Sequence to be received.

Offline (OLS): The Offline Primitive Sequence indicates that the transmitting Port is

- initiating the Link Initialization protocol (see the "Link Recovery Protocols" section, on page 104),
- in the NOS Receive (LF1) state, receiving and recognizing the NOS Primitive Sequence, or
- entering the Offline state.

A Port will transmit the OLS Primitive Sequence for a minimum period of 5 ms before further actions are taken, to ensure that the receiving end of the link has had time to receive, recognize, and take action on it. If the maximum transmission distance increases beyond 10 km with transmitter/receiver improvements, this time period will be increased accordingly. A Port enters the Offline state in order to perform internal diagnostics or power off, as described in the "Online to Offline Protocol" section, on page 105.

Link Reset (LR): The Link Reset Primitive Sequence indicates that the transmitting Port is initiating the Link Reset protocol or recovering from a link timeout. Link timeout occurs when Sequence timeouts have been detected for all active Sequences or when transmission is stalled for a zero buffer-to-buffer Credit count and no R_RDY Primitive Sequences have been received for longer than the E_D_TOV timeout period. An N_Port supporting Class 1 service may also transmit the LR Primitive Sequence when it is unable to determine its Connection status, a procedure known as Connection recovery (see the "Dedicated Connection Recovery" section, on page 251).

Link Reset Response (LRR): The Link Reset Response Primitive Sequence is transmitted by a Port to indicate that it is in the LRR Receive (LR3) state, receiving and recognizing the LR Primitive Sequence.

Loop Initialization (LIP): These Primitive Sequences are used to initialize Ports on either a Private or Public Loop topology.

Loop Port Bypass (LPB): These Primitive Sequences are used to bypass ports on a loop topology.

Loop Port Enable (LPE): These Primitive Sequences are used to bring ports on a loop topology out of Bypass.

Port States on Non-Loop Topologies

This section defines the possible states for a Port, the actions taken by the Port while it is in those states, and the conditions that cause transitions between states.

A word of caution. When the attached Ports incorporate the capability to attach to Arbitrated Loop topology, they use a more complex state machine, including many more states than are described in this section, and these states are not normally visited. Since many, if not most, device and switch ports include Arbitrated Loop capability (i.e., they are NL_Ports and FL_Ports, instead of N_Ports or F_Ports), the state machine and protocols described in Figure 16.4 and the "Loop Port State Machine Operation: An Example" section, on page 288, will usually be more important than the material in this section.

Figure 6.3 shows the transition chart for the nine Port states. As an N_Port or F_Port receives and recognizes Primitive Sequences, it will follow the transitions shown in the figure. Current states are shown down the left side, with next states shown across the top. For example, if the Port is in the Active (ac) state, and it receives and recognizes the LR Primitive Sequence, it will transition to the LR Receive (LR2) state. For conditions that are not explicitly shown to cause state changes to occur, the Port remains in the current state. These Ports are used in implementing the Link Recovery and Online to Offline Protocols described in the "Primitive Sequence Protocols" section, on page 104.

Active State

When a Port is in the Active state, it can transmit and receive Frames and Primitive Signals. This is the standard operating state of the Port. A Port enters the Active state when it completes the Link Initialization and Link Reset protocols. The Port may transition out of the Active state on its own initiative as part of either the Link Initialization, Online to Offline, Link Failure, or Link Reset protocols. Receipt of the LRR Primitive Sequence while in the Active state is an error, and it triggers an increment of the Primitive Sequence Error count in the Port's Link Error Status Block.

Link Recovery States (LR1, LR2, and LR3)

A Port enters the Link Recovery states as part of the Link Reset protocol, which is used for resetting a link. They may also be entered when a Port supporting Class 1 is unable to determine its Connection status and wants to terminate all Dedicated Connections. The Link Recovery states have three substates: LR Transmit state (LR1), LR Receive state (LR2), and LRR Receive state (LR3). In all these states, the Loss of Synchronization condition described in the "Transmitter and Receiver States" section, on page 85 need

Figure 6.3
Port state transition chart.

next	Active	Link Recovery			Link Failure		Offline		
current	**AC**	**LR1**	**LR2**	**LR3**	**LF1**	**LF2**	**OL1**	**OL2**	**OL3**
AC - Active sending data	Idle, R_RDY		LR	Err- LRR (4)	Err - NOS (1)	Err- Loss of sync(2) or signal(3)		OLS	
LR1 - LR Transmit sending LR		Idle, R_RDY	LR	LRR	Err- NOS (1)	Err- Loss of sync(2) or signal(3), timeout(1)		OLS	
LR2 - LR Receive sending LRR	Idle		LR	LR	Err- NOS (1)	Err- Loss of sync(2) or signal(3), timeout(1)		OLS	
LR3 - LRR Receive sending Idle			LR		Err- NOS (1)	Err- Loss of sync(2) or signal(3), timeout(1)		OLS	
LF1 - NOS Receive sending OLS			LR		LRR, NOS, Idle	Err- Loss of sync(2) or signal(3), timeout(1)		OLS	
LF2- NOS Transmit sending NOS					NOS	LR, LRR, Idle		OLS	
OL1 - OLS Transmit sending OLS			LR*		Err- NOS* (1)		LRR*, Idle*	OLS*	Loss of sync or signal*, timeout
OL2 - OLS Receive sending LR			LR	LRR	Err- NOS (1)			OLS, Idle	Loss of sync or signal, timeout
OL3 - Wait for OLS sending NOS					NOS	Err- LR, LRR		OLS	Idle, Loss of sync or signal

Note: Transitions marked by "Err" are invalid and trigger updates of a Link Error Status block described in Chapter 12, i.e., increment (1) Link Failure, (2) Loss of Synchronization, (3) Loss of Signal, or (4) Primitive Sequence Protocol error counters in the LESB.

Note: Transitions marked by "*" only occur as part of the Link Initialization Protocol. In the Online to Offline protocol, the transitions are disabled, and the Port may power down.

not be checked, since they will all transition to the Link Failure state following the R_T_TOV period anyway, if no other transition condition occurs.

LR Transmit State (LR1). While in the LR Transmit state, the Port will continuously transmit the LR Primitive Sequence. The timeout condition shown occurs when the Port remains in the LR Transmit state for a period of time greater than the R_T_TOV timeout period, causing the Port to enter the

Link Failure NOS Transmit (LF2) state with a Link Reset protocol timeout condition.

Transmission of the LR Primitive Sequence has different conditions based on Class:

Class 1: All pending or existing Dedicated Connections are removed and all end-to-end Credit values are reset to their Login values. If the attached Port is an F_Port, it will notify the F_Port attached to the remote N_Port of the Dedicated Connection, which in turn transmits the LR Primitive Sequence to the other connected N_Port. This initiates the Link Reset Protocol on both N_Port/F_Port links. When the state is in the LR Transmit state and transmitting LR Primitive Sequences, all active or open Class 1 Sequences are terminated abnormally, and all received Class 1 Frames are discarded.

Class 2 and 3 and Class 1/SOFc1: Buffer-to-buffer Credit within the N_Port or F_Port is reset to its Login value and an F_Port will process or discard any Frames currently held in the receive buffers associated with the outbound fiber of the attached N_Port. Class 2 end-to-end Credit is not affected.

LR Receive State (LR2). While in the LR Receive state (LR2), the Port will continuously transmit the LRR Primitive Sequence. A Port enters the LR Receive state when it receives and recognizes the LR Primitive Sequence while it is not in the Wait for OLS or NOS Transmit state. Normally, the Port will be in the LR2 state for less than an R_T_TOV timeout period for a recovery procedure. If it stays in the state for longer than R_T_TOV, it is a serious error, causing a transition into the Link Failure NOS Transmit (LF2) state with a Link Reset protocol timeout condition.

Reception of the LR Primitive Sequence has different effects based on Class:

Class 1: All pending or existing Dedicated Connections are removed and all end-to-end Credit values are reset to their Login values. All active or open Class 1 Sequences are terminated abnormally, and all received Class 1 Frames are discarded.

Class 2 and 3 and Class 1/SOFc1: A Port which receives and recognizes the LR Primitive Sequence will process or discard any Frames currently held in its receive buffers. Buffer-to-buffer Credit within the N_Port or F_Port is reset to its Login value.

LRR Receive State (LR3). While in the LRR Receive (LR3) state, the Port will continuously transmit the Idle Primitive Signal. A Port enters the LRR receive state when it receives and recognizes the LRR Primitive

Sequence while it is in the Active LR Transmit, LR Receive, or OLS Receive state. The timeout condition shown occurs when the Port remains in the LRR Receive state for a period of time greater than the R_T_TOV timeout period, causing the Port to enter the Link Failure NOS Transmit (LF2) state with a Link Reset protocol timeout condition.

Link Failure States (LF1 and LF2)

A Port enters a Link Failure state when it detects a serious error condition such as Loss of Synchronization for more than the R_T_TOV timeout period, Loss of Signal while not in an Offline state, or an R_T_TOV timeout during the Link Reset protocol (described in the "Link Recovery Protocols" section, on page 104). The Link Failure state has two substates: NOS Receive (LF1) and NOS Transmit (LF2).

NOS Receive State (LF1). While in the NOS Receive (LF1) state, the Port will continuously transmit the OLS Primitive Sequence. A Port enters the NOS Receive state when it receives and recognizes the NOS Primitive Sequence. Entry into the NOS Receive state triggers an update of the Link Failure count in the Port's Link Error Status Block. An R_T_TOV timeout period is started when NOS is no longer recognized and no other events occur that cause a transition out of the NOS Receive state. If the timeout period expires, the Port will enter the NOS Transmit state.

NOS Transmit State (LF2). While in the NOS Transmit state, the Port will continuously transmit the NOS Primitive Sequence. A Port enters the NOS Transmit state when a link failure condition is detected. Upon entry into the NOS Transmit state, the Port will update the appropriate error counter in the Link Error Status Block once per failure. The Port remains in the NOS Transmit state while the condition which caused the link failure exists.

Transmission of NOS by an N_Port to the locally attached F_Port of a Fabric will remove any pending or existing Dedicated Connections. The locally attached F_Port will respond by entering the NOS Receive state and notifying the F_Port attached to the other connected N_Port, which will transmit the Link Reset (LR) primitive to the remote N_Port of the Connection. If a Dedicated Connection does not exist, NOS transmission by an N_Port will be received and recognized by the locally attached F_Port but will not be transmitted through the Fabric. The F_Port will respond by entering the NOS receive state.

Offline States (OL1, OL2, and OL3)

While in the Offline states, a Port will not record receiver errors such as Loss of Synchronization. A Port enters the Offline states under the following conditions:

- after power-up, or internal reset, before the Link Initialization protocol is complete,
- after transmission of the first OLS Ordered Set, or
- after reception and recognition of the OLS Primitive Sequence.

The Offline state has three substates: OLS Transmit state (OL1), OLS Receive state (OL2), and Wait for OLS state (OL3).

OLS Transmit State (OL1). There are two reasons for transitioning through the OLS Transmit state. A Port enters the OLS Transmit state either to:

- perform Link Initialization using the Link Initialization protocol in order to exit the Offline state (see the "Link Recovery Protocols" section, on page 104), or
- transition from Online to Offline using the Online to Offline protocol (see the "Online to Offline Protocol" section, on page 105).

In the first case, the starred transitions in Figure 6.3 are enabled. In the second case, they are disabled, and the Port will stay offline. In either case, the Port must stay in the OLS Transmit state for at least 5 ms before either enabling transitions to other states (in the first case) or going completely offline (in the second case).

Transmission of OLS by a Port causes the attached Port to enter the OLS Receive (OL2) state. If an N_Port transmits OLS to an attached F_Port, then any pending or existing Dedicated Connections are removed. The locally attached F_Port will notify the F_Port attached to the remote N_Port in the Dedicated Connection to enter the OLS Receive state too, and to transmit the LR Primitive Sequence to the remote N_Port, causing removal of all pending or existing Dedicated Connections. If no Dedicated Connections exist, OLS transmission by an N_Port will be received and recognized by the locally attached F_Port, but it will not be transmitted through the Fabric.

OLS Receive State (OL2). While in the OLS Receive state, the Port will continuously transmit the LR Primitive Sequence. While in the OLS Receive state, detection of Loss-of-Signal or Loss-of-Synchronization events will not be counted as Link Failure events in the Link Error Status

Block. An R_T_TOV timeout period is started when OLS is no longer recognized and no other events occur that cause a transition out of the OLS Receive state. If the timeout period expires, the Port will enter the Wait for OLS state.

Wait for OLS State (OL3). While in the Wait for OLS state, the Port will continuously transmit the NOS Primitive Sequence. A Port will enter the Wait for OLS state when it detects the Loss of Signal or Loss of Synchronization for more than the R_T_TOV period.

Primitive Sequence Protocols

Primitive Sequence protocols are used to allow a pair of Ports attached over a link to implement an interlocked mechanism for modifying Port states, as described above. These allow the Ports to cooperate in recovering from various types of serious errors and allow a Port to notify the attached Port of internal state changes, to prevent invalid data transmission.

Again, these protocols are not used when the attached Ports are acting as loop ports, since loop initialization and loop port enable/disable take the place of link recovery and the link offline/online protocol on loop topologies.

Link Recovery Protocols

The state transition chart shown in Figure 6.3 allows three levels of link recovery. These three levels are referred to as the (1) Link Reset, (2) Link Initialization, and (3) Link Failure protocols. They are hierarchical: the Link Failure protocol includes Link Initialization, which includes Link Reset.

The first level of link recovery, Link Reset, is initiated when one Port of the link sees a link timeout (Sequence timeouts on all active Sequences) or becomes unable to determine its Class 1 Connection status. It transitions into the LR Transmit (LR1) state and starts sending LR Primitive Sequences. As shown in Figure 6.2, this removes all Class 1 Connections, resets an attached F_Port, and causes the other Port to transition into the LR Receive (LR2) state, sending LRR Primitive Sequence. Following the transitions shown, both Ports transition through the LRR Receive (LR3) state, sending Idles, then go to the Active state. At this stage, both Ports are active, with all Class 1 Connections removed and all Sequences timed out. Both Ports can then begin resending Active Sequences and following normal data transmission procedures.

The second level of link recovery, Link Initialization, is initiated when a Port encounters a severe internal error, requiring it to go offline or halt transmission temporarily. This level of recovery is the normal level of recovery required in the initial power-on bring-up of a Port. The Port initiates its own transition into the OLS Transmit (OL1) state. It will be continuously sending OLS Primitive Sequences, causing the attached Port to enter the OLS Receive (OL2) state. The local Port must stay in the OLS Transmit state for a minimum of 5 ms. During this time, it will be transmitting OLS Primitive Sequences and receiving LR Primitive Sequences, but it will disable the transition to the LR Receive (LR2) state shown in Figure 6.3. It may be doing its own internal diagnostics during this time, or performing initialization. When it is ready, it enables the OLS Transmit (OL1) to LR Receive (LR2) transition, which according to normal flow follows the Link Reset protocol described above to get to the Active state on both Ports.

The third level of link recovery, Link Failure, occurs when a very serious error occurs, such as Loss of Synchronization for more than the R_T_TOV timeout period while not in an Offline state, Loss of Signal while not in the Offline state, or R_T_TOV timeout during the Link Reset protocol. In these cases, the Port transitions to the NOS Transmit (LF2) state, sending the NOS (Not_Operational) Primitive Sequence. This transition is serious enough to cause an error to be recorded in the Port's Link Error status block. The other Port will then enter the NOS Receive (LF1) state, sending the OLS Primitive Sequence. The local Port, receiving the OLS Primitive Sequence, transitions to the OLS Receive (OL2) state, and state transitions from here follow the Link Initialization and Link Reset procedures.

In all three levels of the Link Recovery hierarchy, both Ports stay in their respective states until receipt of three identical Primitive Sequences from the other side causes a transition to a more functional state. The protocols are fully interlocked and acknowledged at all stages, and an R_T_TOV timeout at any stage causes a transition to the lowest level of the hierarchy, for recovery to be attempted from there.

Online to Offline Protocol

The other situation where Primitive Sequences are used is when a Port wants to go Offline. This may occur prior to power-down or when a Port wants to perform internal diagnostics or significant internal initialization. The Port cannot just stop sending data and go do its own internal work, because the attached Port will continue sending data and will become confused when it never gets any ACKs back. The Online to Offline protocol allows the Port to warn the attached Port that it is going to go offline, so that the attached Port can be prepared.

In the Online to Offline protocol, the Port will enter the OLS Transmit (OL1) state, transmitting the OLS Primitive Sequence, for a minimum time of 5 ms. This causes the attached Port to enter the OLS Receive (OL2) state and stop sending data. The Port then goes offline and can perform diagnostic procedures, disable its transmitter, power-down, or transmit any bits that will not cause the attached Port to leave the OLS Receive or Wait for OLS states. The attached Port will not detect any errors during this period.

To exit the Offline state, the Port will perform the Link Initialization protocol described in the "Link Recovery Protocols" section, on page 104, transitioning to the OLS Transmit state, and allowing the starred transitions in Figure 6.3. Normal responses from an attached Port which is also allowing these transitions will bring the Port through the LR Receive, LRR Receive, and Active states.

Chapter 7

FC-2: Frames

Introduction

This chapter covers the various types of Frames used for data transmission and protocol control under the Fibre Channel Standard protocol. All Frames follow the format shown in Figure 7.1.

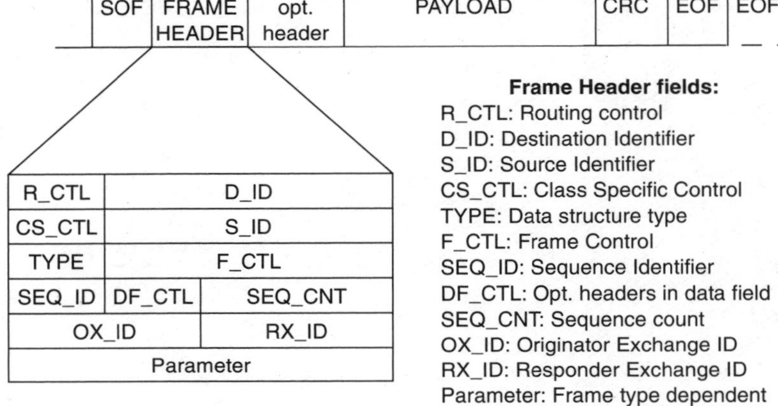

Figure 7.1
Frame and Frame Header formats.

There are several different types of Frames, distinguished by the values in the R_CTL, TYPE, and Parameter Header fields. They are broadly classed as either Data or Link Control Frames, with a number of sub-categories for each type:

- FC-4 Device Data Frame: Used for transmitting data between N_Ports
- Link Control Frames, including the types
  - Acknowledgment (ACK): Used to acknowledge the successful reception of either one Frame (ACK_1), a set number of Frames N (ACK_N), or all the Frames in a Sequence (ACK_0)
  - Link Response: Used for indicating Frame-level non-acknowledgment if available, including Fabric and Port Busy (F_BSY and P_BSY) and Reject indications (F_RJT and P_RJT)
  - Link Command: Only one defined, which is used for resetting the link credit values (LCR)
- Basic Link Data: Used for Basic Link Service commands, such as aborting Sequences and abnormally terminating Class 1 Dedicated connections.
- Extended Link Data: Used for Extended Link Service commands, such as Port and Fabric Logins and Logouts, aborting Exchanges, requesting transmission status, etc.

- Video Data: Used for transmitting data marked as video data, which may be treated differently from other data.

A complete description of the different types of Frames is given in the "Frame Header Fields" section, on page 112.

Frame Format

This section describes the fields shown in Figure 7.1.

Idle and R_RDY Primitive Signals

Idle and R_RDY are Primitive Signals that are used between Frames to maintain bit, byte, and word synchronization, to keep links active between Frames, and to do link level flow control. They are also used to allow the receiving circuitry time to finish processing of a Frame before the next Frame arrives. The R_RDY signal is sent from a receiver to a transmitter to notify of the availability of a Frame buffer and any other required resources for buffer-to-buffer Credit for receiving a Frame, as described in the "R_RDY" section, on page 95. The Idle is a true idle, which is transmitted to maintain activity on the link while data or control information isn't being sent, and to provide processing time between Frames. Idles are discarded at the receiver as soon as they are recognized.

To ensure that the receiver has sufficient processing time between Frames, each transmitter must transmit a minimum of six Primitive Signals between each Frame, of which the first two and last two must be Idles (allowing processing of any R_RDY signals). The Fabric may insert or remove Idles between Frames as long as the destination receives at least two Idles preceding each Frame. This is done for speed-matching Ports with different clock frequencies, and for Intermix, when Frames are inserted or removed by the Fabric.

Start of Frame Delimiter

The Start of Frame (SOF) delimiter is an Ordered Set, transmitted on a word boundary, that immediately precedes the Frame content. Different SOF delimiters are defined for the different types of Frames, as described in the "Start of Frame Delimiters" section, on page 91.

The maximum allowable size for Frames may depend on the type of Frame transmitted, and the Class used for the Frame. For example, Frames begun with the **SOFc1** delimiter are limited to a maximum size specified by the destination N_Port or the Fabric, whichever size is smaller. This may be different for different service Classes. The maximum Data Field size (Payload plus any included optional headers) is determined during the Fabric and N_Port Login procedures, as described in Chapter 9.

Frame Header Fields

The Frame Header is the first field of the Frame content and immediately follows the SOF for all Frames, on a word boundary. The Frame Header is used by the N_Port to control link operations, to route Frames through the network, to do protocol processing on received Frames, and to detect missing or out-of-order Frames.

The various fields of the Frame Header are described in detail in the "Frame Header Fields" section, on page 112.

Optional Headers

Fibre Channel provides for the insertion of up to three different types of optional headers between the Frame Header and the Payload. These include (1) a Network Header, which can contain header information for some specific network-related ULPs, (2) an Association Header, which is used for associating Frames with different Exchange IDs together, and (3) a Device Header, which can contain other ULP information. (A previously defined Expiration Security Header, specifying a "kill" time for the Frame, was removed, as the function is better implemented in a different way.) The presence or absence of each optional header is indicated in the DF_CTL Frame Header fields. The lengths of each of these optional headers are fixed.

Payload

The Payload follows the Frame Header. The Payload must be aligned on word boundaries and must be a multiple of 4 bytes in length. Link Control Frames have a Payload of length 0, and other types of Frames may contain between 0 and 2,112 bytes in the combination of Payload plus any included optional headers. If the ULP supplies a Payload size which is not divisible by 4, between 1 and 3 fill bytes are appended to round the Payload to a word boundary. The number of fill bytes is indicated in the F_CTL Frame Header field, described below. Fill bytes can only be included on the last Data Frame of a series of consecutive Data Frame of a single Information Category within a single Sequence.

CRC Field

The Cyclic Redundancy Check (CRC) is a word-aligned 4-byte field that immediately follows the Payload and is used to verify the data integrity of the Frame Header and Payload. The CRC field is calculated on the Frame Header and Payload before encoding for transmission and after decoding upon reception. SOF and EOF delimiters are not included in the CRC value calculation and verification.

For the purpose of CRC computation, the bit of the word-aligned 4-byte field that corresponds to the first bit transmitted is the highest-order bit. The least-significant bit of the first character of the Frame Header is the first bit of the transmission word transmitted. The CRC used in Fibre Channel is specified in the document "ANSI X3.139 Fiber Distributed Data Interface — Media Access Control" — the same CRC is used in the FDDI protocol. The CRC uses the following 32-bit polynomial:

$$X^{32} + X^{26} + X^{23} + X^{22} + X^{16} + X^{12} + X^{11} + X^{10} + X^{8} + X^{7} + X^{5} + X^{4} + X^{2} + X + 1$$

This CRC polynomial allows detection of all single- and double-bit errors and most errors over a higher number of bits, including all errors over an odd number of bits.

End-of-Frame Delimiter

The EOF delimiter is an Ordered Set that immediately follows the CRC. The **EOF** delimiter designates the end of the Frame content and must be immediately followed by Idles. There are different categories of EOF delimiters, describing the contents and handling of the Frame, as described in the "End-of-Frame Delimiter" section, on page 93.

Frame Header Fields

Each Frame contains a 24-byte Frame Header, which contains fields describing the contents and handling of the Frame. The Frame Header format is shown in Figure 7.1 and includes the following fields:

R_CTL: Routing Control — used for categorizing the Frame function

D_ID: Destination Identifier — address identifier of the Frame's destination Port

S_ID: Source Identifier — address identifier of the Frame's source Port

CS_CTL: Class Specific Control — control information on Frame handling that depends on the Class of transmission.

TYPE: data structure type — categorization of the Frame's data

F_CTL: Frame Control — control information on Frame handling

SEQ_ID: Sequence identifier — unique identifier for the Frame's Sequence

DF_CTL: Data Field control — indication of optional header inclusion

SEQ_CNT: Sequence count — number of the Frame within its Sequence or Exchange

OX_ID: Originator Exchange ID — identification of Frame's Exchange at Originator — also may be used for frame prioritization and preemption of dedicated connections.

RX_ID: Responder Exchange ID — identification of Frame's Exchange at Responder

Parameter: Relative Offset in Data Frame, Frame information in Link Control Frame

A major function of the Frame Header fields is to uniquely identify Frames. No two Frames active at the same time can have the same identifier fields, to prevent the possibility of data corruption through Frame mis-identification. Each Frame is uniquely identified by the values in the S_ID, D_ID, OX_ID, RX_ID, SEQ_ID, and SEQ_CNT field values. The set of Frame ID values (S_ID, D_ID, OX_ID, RX_ID, and SEQ_ID) is termed the "Sequence Qualifier." The N_Ports use these values to uniquely identify Active and Open Sequences. The SEQ_CNT field identifies Frames within a Sequence.

Figures 7.2 and 7.3 show how the various kinds of Fibre Channel Frames are distinguished, based on the values found in the R_CTL, TYPE, and Parameter Header fields. Further usage of the various header fields is described in the following sections.

Figure 7.2
Link Control Frame definitions.

R_CTL (Word 0, bits 31-24) 31-28	27-24		Definition	TYPE (Word 2, bits 31-24) and/or Parameter (Word 5)	
x'C' Basic Link Service Data	x'0'	Ack-nowledge (ACK)	ACK_1	TYPE: rsrvd	Parameter: Bits15-0=1
	x'1'		ACK_0 ACK_N	TYPE: rsrvd	Parameter: Bits15-0=0 Parameter: Bits15-0=N
	x'2'	Link Response	P_RJT: N_Port Reject	Frame reject reason codes - see Figure 7.10.	
	x'3'		F_RJT: Fabric Reject		
	x'4'		P_BSY: N_Port Busy	TYPE: rsrvd	31-24: Action code 23-16: Reason code
	x'5'		F_BSY: Fabric busy to Data Frame	reason codes: x'1x': Fabric busy x'3x': N_Port busy	Word 2[27-24] holds the Link Control code (R_CTL Wd0[27-24]) of the busied Frame Word 5: reserved
	x'6'		F_BSY: Fabric busy to Link Control Frame		
	x'7'	Link Command	LCR: Link Credit Reset	TYPE: rsrvd	Parameter: reserved
	x'8'		NTY: Notify	TYPE: rsrvd	Parameter: reserved
	x'9'		END: End	TYPE: rsrvd	Parameter: reserved

Routing Control Field: R_CTL

The "Routing Control" field (word 0, bits 31 to 24) is a 1-byte field that contains routing bits and information bits to categorize the Frame function. This field provides the first-level (highest-level) distinction between different types of Frames. The currently defined valid Routing Control bits are as shown in Figures 7.2 and 7.3. Of the 8 R_CTL bits, the first 4 bits denote the routing of the Frames and their general usage and are termed "Routing Bits":

0000 — FC-4 Device Data: These are the most common Data Frames containing Payload information related to a specific ULP. These Frames are routed to the FC-4 interface to the ULP.

0010 — Extended Link Data: These Frames are directed to the N_Port to provide N_Port Extended Link Services, such as Login, which are common to multiple FC-4s. These Frames are routed to the N_Port as either unsolicited ("Request") or solicited ("Reply") control Frames and are used for Port configuration and status determination.

Figure 7.3
Data Frame definitions.

R_CTL (Word 0, bits 31-24)		TYPE (Word 2, bits 31-24) Unlisted values are reserved
31-28: Routing	27-24: Information Category	
x'0' FC-4 Device Data Frames	x'0': Uncategorized x'1': Solicited Data x'2': Unsolicited Control x'3': Solicited Control x'4': Unsolicited Data x'5': Data Descriptor x'6': Unsolicited Command x'7': Command Status others unspecified	x'04': ISO/IEC 8802 - 2 LLC/SNAP (In order) x'05': ISO/IEC 8802-2 LLC/SNAP (Mis-ordered) -IP/ARP x'08': SCSI - FCP x'09': SCSI - GPP x'0A-x'0F': reserved for SCSI x'10': reserved for IPI-3 x'11': IPI-3 Master x'12': IPI-3 Slave x'13': IPI-3 Peer x'14': reserved for IPI-3 x'15': CP IPI-3 Master x'16': CP IPI-3 Slave x'17': CP IPI-3 Peer x'18': reserved for SBCCS x'19': SBCCS - Channel x'1A': SBCCS - Control Unit x'1B'–x'1F': Rsrvd for SBCCS/ESCON x'20': Fibre Channel Services x'21': FC-FG x'22': FC-SW - Internal Link Service x'23': FC-AL x'24': SNMP x'25'–x'27': reserved, Fabric Services x'28'–x'2F': Futurebus x'30'–x'33': SCI x'34'–x'37': MessageWay x'40': HIPPI - FP x'41'–x'47': reserved - HIPPI x'48'-x'4F': reserved for FC-AE x'5D': Fabric Controller x'E0'–x'FF': Vendor Unique
x'2' Extended Link Service Data	x'2': Unsolicited Cntrl-"Rqst" x'3': Solicited Cntrl- "Reply"	x'00': invalid (used for basic) x'01': Extended Link Service x'D0'–x'FF': Vendor Unique
x'3':FC-4 Link Service Data	x'2': Unsolicited Cntrl-"Rqst" x'3': Solicited Cntrl- "Reply"	same as FC-4 Device Data Frames
x'4' Video Data	same as FC-4 Device Data Frames	x'D0'–x'FF': Vendor Unique
x'8' Basic Link Service Data	x'0': No Operation NOP x'1': Abort Seq. ABTS x'2': Remove Conn. RMC x'4': Basic Accept BA_ACC x'5': Basic_Reject BA_RJT	x'00': Basic Link Service x'01': invalid (used for extended) x'D0'–x'FF': Vendor Unique

0011 — FC-4 Link Data: These Frames are used for providing FC-4 services to help in the processing of FC-4 Device_Data Frames. These Frames are routed to the FC-4 interface to the ULP as either unsolicited or solicited control Frames.

0100 — Video Data: This routing is provided to allow special processing of video data, such as direct routing into a display buffer.

1000 — Basic Link Data: A 12-byte maximum Payload provides low-level Basic link services, such as aborting Sequences, removing Connections, and transmitting No-ops. These Frames are directed to the FC-2 N_Port.

1100 — Link Control: These Frames are used for link control, such as acknowledgments, Frame rejects or busies, and for resetting link credits. These Frames are generated either by the Fabric or by an N_Port in response to a transmitted Frame and are directed to the N_Port which transmitted it.

The next 4 bits (bits 27 to 24) are termed "Information field" bits, and their interpretation is dependent on the Routing bits (bits 31 to 28) field value. For Routing = b'1000' and b'1100,' the Information field contains a command identifier. For all other R_CTL values, the Information field of the R_CTL field specifies one of the Information Categories specified in Figures 7.2 and 7.3. An "Information Category" serves two purposes. It describes the usage of the data, indicating for example whether it is control or command data and whether it was solicited or not. The Information Category can also be used to divide the Frames within a Sequence into independent data streams, to be stored to different buffers at the destination Node. The default is for only one Information Category per Sequence, but multiple Information Categories per Sequence can be allowed, if agreed upon in the N_Port Login Class Service Parameters described in the "N_Port and F_Port Class Service Parameters" section, on page 171.

Several Information Categories have specified Payload formats. These are shown in Figure 7.4. The format of the Payload for the other Information Categories is unspecified and is FC-4 dependent.

Address Identifiers: S_ID and D_ID

Each N_Port has a 3-byte N_Port Identifier which is unique within the address domain of a Fabric. N_Ports either negotiate their own N_Port identifiers or are assigned them by the Fabric during Fabric Login. The N_Port Identifier x'00 0000' is reserved to indicate that an N_Port is unidentified, and does not have an N_Port identifier. When an N_Port is unidentified, it will accept only Basic Link Service or Extended Link Service Frames routed

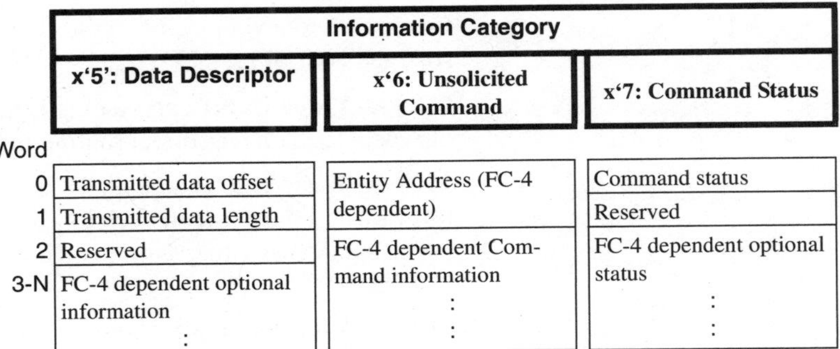

Word	x'5': Data Descriptor	x'6: Unsolicited Command	x'7: Command Status
0	Transmitted data offset	Entity Address (FC-4 dependent)	Command status
1	Transmitted data length		Reserved
2	Reserved	FC-4 dependent Command information	FC-4 dependent optional status
3-N	FC-4 dependent optional information		

Information Category

Figure 7.4
Frame Payload formats for Data Descriptor, Unsolicited Command, and Command Status Information Categories.

to it, with any D_ID value. Other Frames will be rejected using the "Login Required" P_RJT code. F_Ports as well as N_Ports have unique native Address Identifiers, which may be used in directing Frames to Ports on the Fabric.

Although the Fibre Channel standards documents don't actually require a particular partitioning of the 3-byte address space, Fabric vendors have followed the partitioning shown in Figure 7.5 to aid in interoperability between different switch vendors. Address Identifiers actually used for N_Ports or F_Ports are in the range x'01 0000' to x'EF EFFF.'

The 3-byte address range is partitioned into domains, areas, and Port_IDs, to simplify and hierarchically organize Address Identifier assignment. Domains are the highest hierarchical level in a three-level addressing scheme, and there are 256 possible domains within the address space, distinguished by the 8 most significant address bits. Areas are the next level of hierarchy, divided into a maximum of 240 areas per domain. Within each area can be up to 256 possible Ports, distinguished by the least-significant 8 bits of the Port address. This hierarchy allows different parameters of functionality within different parts of the network and simplifies address translation and routing.

The address identifiers in the range of x'FF FFF0' to x'FF FFFF' are well-known and reserved for the following uses. These are described in more detail in the "Well-Known Generic Services" section, on page 309.

x'FF FFF0' to **x'FF FFF4':** Reserved for future usage as well-known addresses.

x'FF FFF5': Multicast Server. Service used in Class 6 Uni-Directional Dedicated Connection which acts as the Destination Port for responses and aggregates them properly to provide a single response to the multicast originator to provide a reliable, acknowledged unidirectional multicast.

Figure 7.5
S_ID/D_ID address space partitioning.

Address Identifier			Description	
23....16	15.....8	7......0	Usage	(# of addresses)
00000000	00000000	00000000	Undefined, or FL (Loop)	(1)
00000000	00000000	00000001	Rsvd (Private Loop)	(1)
00000000	00000000	0000001x	Rsvd (Private Loop)	(2)
00000000	00000000	000001xx	Rsvd (Private Loop)	(4)
00000000	00000000	00001xxx	Rsvd (Private Loop)	(8)
00000000	00000000	0001xxxx	Rsvd (Private Loop)	(16)
00000000	00000000	001xxxxx	Rsvd (Private Loop)	(32)
00000000	00000000	01xxxxxx	Rsvd (Private Loop)	(64)
00000000	00000000	1xxxxxxx	Rsvd (Private Loop)	(128)
00000000	00000001	xxxxxxxx	Rsvd	(256)
00000000	0000001x	xxxxxxxx	Rsvd	(512)
00000000	000001xx	xxxxxxxx	Rsvd -----------	(1,024)
00000000	00001xxx	xxxxxxxx	Rsvd	(2,048)
00000000	0001xxxx	xxxxxxxx	Rsvd	(4,096)
00000000	001xxxxx	xxxxxxxx	Rsvd	(8,192)
00000000	01xxxxxx	xxxxxxxx	Rsvd ----------	(16,384)
00000000	1xxxxxxx	xxxxxxxx	Rsvd	(32,768)
[Domain]	[Area]	[Port_ID]	Port Identifiers	(14,684,160)
[Domain]	1111	[Special_ID]	FABRIC assists	(987,944)
11110xxx	xxxxxxxx	xxxxxxxx	Rsvd ---------	(524,288)
111110xx	xxxxxxxx	xxxxxxxx	Rsvd	(262,144)
1111110x	xxxxxxxx	xxxxxxxx	Rsvd	(131,072)
11111110	xxxxxxxx	xxxxxxxx	Rsvd	(65,536)
11111111	0xxxxxxx	xxxxxxxx	Rsvd ----------	(32,768)
11111111	10xxxxxx	xxxxxxxx	Rsvd	(16,384)
11111111	110xxxxx	xxxxxxxx	Rsvd	(8,192)
11111111	1110xxxx	xxxxxxxx	Rsvd	(4,096)
11111111	11110xxx	xxxxxxxx	Rsvd -----------	(2,048)
11111111	1111100x	xxxxxxxx	Rsvd	(512)
11111111	11111010	xxxxxxxx	Rsvd	(256)
11111111	11111011	xxxxxxxx	Vendor Unique	(256)
11111111	11111100	00000000	Rsvd --------------	(1)
11111111	11111100	[Domain_ID]	Domain Controllers	(239)
11111111	11111100	1111xxxx	Vendor Unique	(16)
11111111	11111101	[Area_ID]	Area Controllers	(240)
11111111	11111101	1111xxxx	Vendor Unique ---------	(16)
11111111	11111110	[Port_ID]	Port Controllers	(256)
11111111	11111111	0xxxxxxx	Vendor Unique	(128)
11111111	11111111	10xxxxxx	Rsvd	(64)
11111111	11111111	110xxxxx	Rsvd --------------	(32)
11111111	11111111	1110xxxx	Rsvd	(16)
11111111	11111111	111100xx	Rsvd	(4)
11111111	11111111	11110100	Rsvd	(1)
11111111	11111111	11110101	Multicast Server	(1)
11111111	11111111	11110110	Clock Sync. Server	(1)
11111111	11111111	11110111	Sec. Key Distrib. Server	(1)
11111111	11111111	11111000	Alias Server	(1)
11111111	11111111	11111001	Quality of Service Fac.	(1)
11111111	11111111	11111010	Management Server ------	(1)
11111111	11111111	11111011	Time Server	(1)
11111111	11111111	11111100	Name Server	(1)
11111111	11111111	11111101	Fabric Controller	(1)
11111111	11111111	11111110	F_Port ----------------	(1)
11111111	11111111	11111111	Broadcast Alias_ID	(1)

x'FF FFF6': Clock Synchronization Server. Used for requesting a Fabric-wide real time to synchronize clocks to.

x'FF FFF7': Security Key Distribution Server. Used for securely distributing authenticated private keys for Port pairs so that they may securely exchange encrypted data.

x'FF FFF8': Alias Server. When Alias addresses are implemented, the entity addressed at this address will maintain Alias identifier mappings. This facility may be located in the Fabric or in a Node addressed to the Fabric. Extended Link Service requests dealing with multicast (GAID, FACT, FDACT, NDACT) are sent to this destination address.

x'FF FFF9': Quality-of-Service Facilitator for Class 4 Service. When implemented, Extended Link Service commands sent to this address is used to request a set of Quality-of-Service parameters for a Class 4 Virtual Circuit. A Class 4 Virtual Circuit is an extension of Class 1 service which allows reservation of some fraction of the available link bandwidth and/or allows guarantee of maximum delivery latency, with the Quality-of-Service for these parameters negotiated between the requestor and the Quality of Service Facilitator.

x'FF FFFA': Management Server; an optional entity which collects and reports information on link usage and quality, errors, etc.

x'FF FFFB': Time Server; an optional entity used to distribute synchronized time values.

x'FF FFFC': Directory Server or Name Server; an optional entity contained either within the Fabric or at an N_Port that maintains tables correlating N_Port Address Identifiers with N_Port Name Identifiers and possibly many other Port characteristics.

x'FF FFFD': Fabric Controller; a required entity within the Fabric (possibly distributed across multiple switch elements) that controls the general operation of the Fabric, including Fabric initialization, Frame routing, generation of F_BSY and F_RJT link responses, and setup and tear down of Dedicated Connections.

x'FF FFFE': Fabric F_Port; a required entity within the Fabric that provides access to the Fabric for Fabric Login (FLOGI). This entity assigns, confirms, or reassigns N_Port address identifiers and notifies N_Ports of the operating characteristics of the Fabric, if present.

x'FF FFFF': Broadcast address; If this optional function is supported, the Fabric will route a Class 3 Frame with this destination ID to every connected N_Port.

Figure 7.5 is a bit misleading in showing 255 addresses as being used for "Private Loop." There are actually only 126 addresses in the range between x'01' and x'FF' which can be used as Physical Addresses for NL_Ports on

Arbitrated Loop topologies. The addresses used in this range are the single-character values with neutral disparity 8B/10B code points, as shown in Figure 16.3. These issues are discussed in detail in Chapter 16; it is enough to say here that some of the addresses between x'01' and x'FF' are used in Arbitrated Loop networks.

On a Public Loop, with a switch's FL_Port attached, the switch will assign Domain_ID and Area_ID 8-bit values for the loop, and the NL_Ports will take the Address Identifiers x'Domain_ID Area_ID AL_PA' after the Arbitrated Loop Physical Address values have been negotiated.

An N_Port may also optionally have one or more alias Address Identifiers which may be shared across multiple N_Ports, through registration with an Alias Server in the fabric. This way, either multiple N_Ports may respond to the same Address Identifier, or a single Port may perform independent functions for Frames directed to different Port identifiers of the same physical Port.

Class Specific Control: CS_CTL

This field is used for Class-dependent Frame handling, as shown in Figure 7.6, on page 119. The field is meaningful in Class and Class 4 Frames. In Class 2 and Class 3 Frames, this field must be set to x'00'.

Figure 7.6
CS_CTL Class Specific Control Field

CS_CTL bit(s)	Usage	
Class 1 Usage		
31: Simplex - Dedicated simplex connection request	0: Normal Class 1 (Duplex) 1: Class 1 Dedicated Simplex	Only on connect-request frames
30: SCR - Stacked Connect Request	0: SCR Not requested 1: SCR Requested	
29: COR - Camp-On Requested	0: COR Not requested 1: COR requested	
28: BCR - Buffered Class 1 Request	0: Normal Class 1 (unbuffered) 1: Buffered Class	all frames
27:24: reserved		
Class 4 Usage		
31:24: VC_ID	Virtual circuit Identifier	

In Class 1, the field is used for requesting specific options in either the connect-request (as described in the "Class 1 Dedicated Connection: Detailed Operation" section, on page 254), or in the packet buffering on multi-speed Fabrics (as described in the "Variations on Class 1 Service" section, on page 187).

In Class 4, which allows quality of service across a Fabric at less than a full link bandwidth using Virtual Circuits (VCs) with pre-allocated QoS parameters, this field holds the identifier for the Frame's VC.

Data Structure Type: TYPE

The data structure type (TYPE) is a 1-byte field (Word 2, bits 31–24) that further identifies the kind of Frame. The most common usage is in Data Frames for distinguishing between FC-4 ULP interfaces. The codes specified for the various FC-4s already defined are shown in the FC-4 Device Data rows. Specific vendors can use particular values of this field to send private control information through the equivalent of a distinct FC-4.

In Link Control Frames, the TYPE field is reserved, except in F_BSY Frames, where it is used (a) to indicate whether the Fabric or the N_Port was the busy entity, and (b) to return the Information Category for a busied Link Control Frame, to simplify retransmission of a busied Link Control Frame.

Frame Control: F_CTL

The Frame Control (F_CTL) field (word 2, bits 23 to 0) is a 3-byte field that contains control information relating to the Frame content. Most of the other Frame Header fields are basically used for Frame identification. F_CTL is the most important field for controlling Frame processing. The usage of the bits is shown in Figure 7.7 and is described as follows:

Bit 23 — Exchange_Context: This bit indicates whether the Frame was sent by the Originator or Responder N_Port of the Frame's Exchange.

Bit 22 — Sequence_Context: This bit indicates whether the Frame was sent by the Initiator or Recipient of the Frame's Sequence. Besides simplifying differentiation of data from the Link Control Frames, this bit simplifies distinguishing between a BSY Frame received in response to a Data Frame from the Sequence Initiator or to an ACK Frame from the Sequence Recipient.

Bit 21 — First_Sequence: This bit indicates whether the Frame is in the first Sequence of the Frame's Exchange.

Bit 20 — Last_Sequence: This bit indicates whether the Frame is in the last Sequence of the Frame's Exchange. This bit must be set to 1 on the last Frame of the last Sequence of the Exchange. It can also be set to 1 on previous Frames within the Sequence, to provide an advance indication

Figure 7.7
Bit definitions for F_CTL Frame header field.

F_CTL bit(s)	Usage	
23: Exchange_Context	0: Exchange Originator	1: Exchange Responder
22: Sequence_Context	0: Sequence Initiator	1: Sequence Recipient
21: First_Sequence	0: not first Seq. of Exchange	1: first Seq. of Exchange
20: Last_Sequence*	0: not last Seq. of Exchange	1: "last" Seq. of Exchange*
19: End_Sequence	0: not last Frame of Sequence	1: last Frame of Sequence
18: End_Connection or Deactivate Class 4 circuit	0: Connection active (Class 1) or Circuit Active (Class 4)	1: End of Connection pending or End of live Class 4 circuit.
17: Reclaimed and Reserved (was Chained Sequences)		
16: Sequence_Initiative	0: hold Sequence Initiative	1: transfer Seq. Initiative
15: X_ID reassigned	0: X_ID assignment retained	1: X_ID reassigned
14: Invalidate X_ID	0: X_ID assignment retained	1: Invalid X_ID
13-12: ACK_Form	00: No assistance provided 10: ACK_N required	01: ACK_1 required 11: ACK_0 required
11: Data Compression	0: Uncompressed Payload	1: Compressed Payload
10: Data Encryption	0: Unencrypted Payload	1: Encrypted Payload
9: Retransmission	0: Original Seq. (Class 1)	1: Retransmitted Sequence
8: Unidirectional Transmit (Class 1) or Remove circuit (Cl. 4)	0: Bidirectional Connection (Class 1), or Retain or deactivate circuit (Class 4)	1: Unidirectional Connection (Class 1) or Remove circuit (Class 4)
7-6: Continue Sequence Condition	00: No info on next Seq. 10: Seq. to follow - soon	01: Seq. to follow - immed. 11: Seq. to follow - delayed
5-4: Abort Sequence Condition Usage is different for ACK Frames than for Data Frames	ACK Frame - Sequence Recipient	
	00: Continue Sequence 10: Stop Sequence	01: Abort Seq., do ABTS 11: Immediate Sequence retransmission requested
	First Data Frame of Exchange - Exchange Originator	
	00: Abort, discard multiple Sequences 10: Process with infinite buffering	01: Abort, discard a single Sequence 11: Discard multiple Sequences with immediate retransmission
3: Relative Offset present	0: Param. field not meaningful	1: Param. field = Rel. Offset
2: Reserved for Exchange Reassembly		
1-0: Fill Bytes at end of payload	00: 0 bytes of fill 10: 2 byte of fill	01: 1 byte of fill 11: 3 bytes of fill

* Bit 20 may be set to 1 in multiple final Frames of Seq., to give advance warning of Exchange termination. Once set, it cannot be unset within the Exchange.

that the Exchange is nearing completion. It is set to the same value in Link Control Frames as in Data Frames, and once it is asserted to 1 in one Frame, it must be asserted in all following Frames until the end of the Exchange.

Bit 19 — End_Sequence: In Data Frames, this bit indicates whether the Frame is the last Data Frame of the Sequence. In Link Control Frames, the usage depends on Class of service, since the Classes have different ordering properties. In Class 1, the Sequence Recipient sets this bit to 1 in the ACK for the last Data Frame, confirming reception of the Sequence's last Data Frame. In Class 2, without in-order Frame delivery, the last ACK transmitted for the Sequence may or may not correspond to the last Data Frame transmitted. This bit is set to 1 in the last ACK for the Sequence, which may correspond either to the last Data Frame transmitted or the last Data Frame received. Handling of the End_Sequence bit for Sequence termination in the case of out-of-order Frame reception is covered in detail in the "Normal Sequence Completion" section, on page 147.

Bit 18 — End_Connection (E_C): In a Data Frame, this bit indicates whether the Initiator is beginning the disconnect procedure to remove a Class 1 Dedicated Connection and requesting that Fabric, if present, and the receiving N_Port do the same. The receiving N_Port will normally respond by assuring that all active Sequences have been completed and by transmitting an ACK Frame delimited with EOFdt to disconnect the Connection, as described in the "Removing a Connection" section, on page 258. The E_C bit is only applicable on the last Data Frame of a Class 1 Sequence and cannot be set to 1 on a connect-request Frame, to avoid ambiguous error scenarios where the ACK to the connect-request cannot be returned to the Connection Initiator. For Class 4 circuits, described in the "Class 4 — Fractional" section, on page 192, this bit is used to request that the Class 4 circuit be deactivated, such that no new Sequences are started.

Bit 17 — Reserved: This bit was reclaimed from the "Chained Sequence" usage, which required a reply Sequence before the Dedicated Connection could be removed. It is now reserved.

Bit 16 — Sequence_Initiative: This bit indicates whether the Sequence Initiator of the Frame's Sequence is passing a Sequence Initiative for the next Sequence of the Exchange to the Recipient of the current Sequence. This bit is only meaningful with End_Sequence = 1. In Class 1 and 2, the Sequence Initiator does not consider Sequence Initiative successfully transferred until the Data Frame's ACK returns with Sequence_Initiative = 1.

Bit 15 — X_ID reassigned: This bit indicates whether an X_ID value has been reassigned. This bit is only meaningful if an N_Port requires or supports X_ID reassignment, as described in the "Association_Header" section, on page 244.

Bit 14 — Invalidate X_ID: This bit indicates whether an X_ID value has been invalidated. This bit is only meaningful if an N_Port requires or supports X_ID reassignment, as described in the "Association_Header" section, on page 244.

Bit 13 to 12 — ACK_Form: The ACK_Form bits can optionally be implemented to let the Sequence Initiator help the Sequence Recipient by notifying it whether to use ACK_1, ACK_N, or ACK_0 for the Sequence. This information is available at both N_Ports from the N_Port Login Class Service Login Parameters, but it can be included with the Frames to possibly simplify Frame reception by saving a table look-up. Doing the look-up at the Initiator instead helps distribute the work load, since normal protocol processing requires more work per Frame by the Sequence Recipient than by the Initiator. This is a proposed FC-PH-2 construct.

Bit 11 — Data Compression: This bit indicates whether the frame is transmitted in a compressed format. The effective data transmission bandwidth can be increased by compressing the data before transmitting it. The format for Fibre Channel data compression is the ALDC LZ–1 (Adaptive Lossless Data Compression Lempel-Ziv–1), as described in Chapter 13.

Bit 10 — Data Encryption: This bit indicates whether the frame is transmitted in an encrypted format. The actual format of data encryption is not specified, since different applications have different requirements.

Bit 9 — Retransmission: This bit indicates that the Sequence is a retransmission of a Sequence transmitted earlier. It is only used in the "Discard Multiple Sequences with immediate retransmission" Exchange Error Policy and is only set to 1 in Sequences retransmitted following receipt of an ACK with Abort Sequence Condition bits set to "Immediate Sequence retransmission requested" (b'11'). If multiple Sequences have to be retransmitted, this bit is only set to 1 on the first retransmitted Sequence. If the Sequence Initiator cannot determine which Sequences need to be retransmitted, it can use the Read Exchange Status Block (RES) Extended Link Service Request (see the "Read Exchange Status Block (RES)" section, on page 207) or another such method to find out.

Bit 8 — Unidirectional Transmit: This bit indicates whether the Connection Initiator is requesting a unidirectional Class 1 Dedicated Connection. It can be set to 1 in a Class 1 connect-request Frame to request a unidirectional Dedicated Connection, where only the N_Port that transmitted the connect-request can transmit Data Frames for the Connection.

Usage of this bit and procedures for optionally converting a unidirectional Class 1 Dedicated Connection to bidirectional are described in the "Unidirectional versus Bidirectional Connections" section, on page 255.

Bits 7 to 6 — Continue Sequence Condition: These bits can be optionally used to indicate how soon the next Sequence of the Exchange will be transmitted. These bits are only valid on the last Frame of the Sequence sent by the N_Port which will initiate the next Sequence, i.e., (1) on the last Data Frame of a Sequence which doesn't transfer Sequence Initiative or (2) on the ACK to the last Data Frame of a Sequence which does. The distinction between "soon" and "delayed" is referenced relative to the time required to remove and reestablish a Class 1 Dedicated Connection, regardless of the Class of service used for the Sequence.

Bits 5 to 4 — Abort Sequence Condition: These bits are used differently on Data Frames than on ACK Frames. In Data Frames, these bits are only valid on the first Data Frame of an Exchange and are used by the Exchange Originator to specify one of the four Exchange Error Policies for the Exchange: (a) Abort, discard multiple Sequence, (b) Abort, discard a single Sequence, (c) Process with infinite buffering, or (d) Discard multiple Sequences with immediate retransmission

Usage of these policies is as follows. If the delivery order of Sequences to an FC-4 or upper level at the destination must match the transmission order of Sequences within an Exchange, then one of the two "Discard multiple Sequences" Error Policies is required. In the "Abort, discard a Single Sequence" Error Policy, out of order Sequence delivery can occur on occasion and is handled by the FC-4 or ULP. In the "Process with infinite buffering" Error Policy, only Class 1 transmission is allowed, Frames must be delivered in the order transmitted, the ACK_0 acknowledgment form is used, and a Sequence with valid first and last Frames may be delivered to the FC-4 or upper level even if intermediate Frames are invalid or missing.

All three "Discard" policies must be supported by all N_Ports, while support for the "Process with infinite buffering" Error Policy is optional. Both N_Ports learn which Exchange Error Policies the other can support during N_Port, and the Exchange Originator cannot indicate a policy which the Responder does not support. Characteristics and usage of the various Exchange Error Policies are described in the "Exchange Error Policies for Class 1 and Class 2" section, on page 229.

In ACK Frames, these bits are used to indicate whether an error or abnormal condition has been detected by the Recipient. The codes for these bits are shown in Figure 7.7. The non-zero settings indicate that the Recipient cannot continue receiving Frames for the Sequence.

The "Abort Sequence, perform ABTS" indication requests that the Initiator begin the Abort Sequence protocol, described in the "Abort Sequence (ABTS) Command" section, on page 199. "Stop Sequence," described in the "Stop Sequence Protocol Overview" section, on page 239, requests the Initiator to terminate the Sequence normally without invoking any recovery procedures, usually due to a request from the FC-4 or ULP at the Sequence Recipient. "Immediate Sequence retransmission requested" can only be used in the "Discard multiple with immediate retransmission" Exchange Error Policy and requests that the Initiator retransmit a Sequence with the Retransmission (F_CTL bit 9) flag set.

Bit 3 — Relative Offset present: This bit indicates whether the Parameter field (Frame Header word 5) contains the Relative Offset value for the Frame Payload. This bit is only meaningful on Data Frames.

Bit 2 — reserved for Exchange Reassambly: This bit is reserved for future use to indicate that the Frame is in an Exchange which is being managed over multiple N_Ports at a single Node, using as-yet-undefined FC-3 operations.

Bits 1 to 0 — Fill bytes: These bits indicate how many fill bytes are included in the last Frame of the Sequence to round out the Payload length to Transmission Word boundaries. These bits are only meaningful on the last Data Frame of a single Information Category within a single Sequence. Fill bytes are inserted to round the Frame length to a 4-byte multiple and allow CRC calculation over full word boundaries. The fill bytes may be any valid data characters, and they are ignored at the receiver, except in CRC calculation.

Summary of F_CTL Bit Usage. The usage of and interaction between the F_CTL bits is fairly complicated, since specific bits may only be meaningful or valid under specific conditions. For example, Abort Sequence Condition (F_CTL bits 5 and 4), which sets the Exchange Error Policy, is only meaningful on the first Data Frame of the first Sequence of the Exchange, where First_Sequence (F_CTL bit 21) is 1. Similarly, Sequence Initiative (F_CTL bit 16) is only meaningful if End Sequence (F_CTL bit 19) is set to 1. It is possible to infer the validity of the various F_CTL bits from the descriptions above, but it is worth summarizing the validity of the bits in Data, ACK, BSY, and RJT Frames, as is shown in Figure 7.8.

The bits called End_Sequence, End_Connection, Chained_Sequence, and Sequence_Initiative in a Data Frame are retransmitted by the Sequence Recipient in the corresponding ACK Frame to reflect confirmation or denial of those indications. For example, consider the situation when a Recipient receives a Data Frame with End_Sequence = Sequence Initiative = 1, indicating that the Initiator is requesting that the current Sequence's Recipient

Figure 7.8 Summary of F_CTL bit usage and validity in Data and Link Control Frames.

F_CTL bit Frame order within Seq.	23	22	21 =1	20 =1	19 =1	18 =1	17 =1	16 =1	15	14	9 =1	8 =1	7-6	5-4	3 =1	1-0
F_CTL bit validity on Data Frames																
First Frame	M	M	M						M+		M	M		MF	M	M
Last Frame	M	M	M		M				M+		M	M			M	M
any frame	M	M	M						M+		M				M	M
connect-request Frame	M	M	M	ML	ML		ML		M+ ML	ML		M	ML		M	M
The bits above are meaningful on Data Frames when the bit on the left is:																
21:First_Seq=1														MF		
19:End_Seq =1				ML		ML	ML	ML		ML			ML			
18:End_Cnctn=0							ML									
16:Seq.Init=0							ML									
F_CTL bit validity on ACK, BSY or RJT Frames																
ACK to First	V	V	E						M+		M	M		Ma		
ACK to Last	V	V	E						M+		M	M	ML	Ma		
ACK to any	V	V	E						M+		M	M		Ma		
ACK to connect-rqst	V	V	E	E	ML		ML	ML	M+	ML		M	ML	Ma		
The bits above are meaningful on Link Control Frames when the bit on the left is:																
23:Exch. Cntxt	V															
22: Seq. Cntxt		V														
21: First_Seq.			E													
20: Last_Seq.				E												
19: End_Seq.=0 =1					E ML	ML	ML	ML		ML			ML			
18:End_Cnctn=0 =1						E ML										
16:Seq. Init=0 =1							ML	E ML								

Notes on Data Frame usage:
M = Meaningful
MF = Meaningful on first Frame only
ML = Meaningful on last Frame only
M+ = Meaningful on first and all following Data Frames of a Sequence until an ACK is received to reassign the X_ID

Notes on Link Control Frame usage:
M = Meaningful
Ma = Meaningful on ACK Frames
ML = Meaningful on ACK, BSY, or RJT to last Frame of Sequence
E = Echo corresponding bit in Frame
V = Invert corresponding bit in Frame
M+ = Meaningful on first and all following ACKs until Data Frame has RX_ID

initiate the next Sequence of the Exchange. If the Recipient returns an ACK with “Abort Sequence Condition” set to “Abort Sequence, perform ABTS” (b‘01’) to request that the Initiator abort the Sequence, it will set the End_Sequence and Sequence Initiative bits to 0 in the ACK Frame to indicate that the Data Frame was not processed as the last Data Frame and that Sequence Initiative was not accepted by the Recipient of the Data Frame.

Since a connect-request Frame is the first Frame of a Class 1 Sequence, but follows Class 2 and 3 Frame buffer size and flow control rules, it acts in some respects as a single-Frame Sequence. The ACK to a connect-request may therefore reflect settings for either the first Data Frame of a Sequence, the last Data Frame of a Sequence, or both first and last, with no ambiguity, depending on implementation.

Sequence ID: SEQ_ID

The SEQ_ID field is used to uniquely identify Sequences within an Exchange. The value is assigned by the Sequence Initiator and is part of the "Sequence Qualifier" combination of OX_ID, RX_ID, S_ID, D_ID, and SEQ_ID which always uniquely identifies any active Sequence. Within a single Exchange, a consecutive Sequence from the same Initiator cannot have the same SEQ_ID value, but every other Sequence can, as long as two Sequences with identical SEQ_ID values are not active simultaneously. Also, consecutive Sequences within an Exchange can have the same SEQ_ID value if they have different Initiators (i.e., they are separated by a transfer of Sequence Initiative). The only requirement is that the SEQ_ID value, in combination with other values in the Sequence Qualifier, be unique while the Sequence is open.

If the Sequence Initiator initiates a new Sequence for the same Exchange before receiving the final ACK (**EOFt** or **EOFdt**) for the previous Sequence in Class 1 and 2, or before R_A_TOV has expired for all Frames of a Class 3 Sequence, it is termed a "streamed Sequence." If streamed Sequences are used, and an Initiator has established at N_Port Login that it can have *X* open Sequences per Exchange, then it must use at least *X* + *1* different SEQ_ID values to ensure uniqueness.

Data_Field Control: DF_CTL

The Data_Field Control field specifies the presence or absence of optional headers between the Frame Header and Frame Payload. These optional headers are included in the architecture for (1) upper-level network addressing, (2) inter-exchange Frame association, and (3) generic ULP usage — actual usage will depend on protocol implementation. The definitions of the DF_CTL bits are shown in Figure 7.9. The definitions and usages of the optional headers, including their validity in the various types of Data and Link Control Frames, are described in detail in Chapter 13.

Figure 7.9
Bit definitions for DF_CTL (Data Field Control) Frame Header field.

Word 3 Bits(s)	DF_CTL bit definition	
23	reserved	
22	reserved	
21	0: No Network_Header	1: Network_Header
20	0: No Association_Header	1: Association_Header
19–18	reserved	
17–16	00: No Device_Header	01:16 Byte Device_Header
	10: 32 Byte Devlce_Header	11: 64 Byte Device_Header

Sequence Count: SEQ_CNT

SEQ_CNT is used to uniquely identify Frames within a Sequence, to assure sequentiality of Frame reception, and to allow unique correlation of Link Control Frames with their related Data Frames. The SEQ_CNT value included in every Frame Header is incremented by one from the value of the previous Frame in the Sequence, wrapping to x'0000' from x'FFFF.' Each ACK or Link Response Frame is identified with the SEQ_CNT of the corresponding Data Frame (which in the case of ACK_N or ACK_0 is the highest SEQ_CNT being acknowledged).

The assignment of the SEQ_CNT value in a Sequence's first Frame is slightly complicated. The first Data Frame of an Exchange (which will always be the first Frame of a Sequence) is always assigned the value x'0000.' If a Sequence is streamed (i.e., initiated before the previous Sequence is completed), then the first Data Frame's SEQ_CNT must be incremented by 1 from the SEQ_CNT of the last Data Frame of the previous Sequence, so that SEQ_CNT values continuously increase through the Exchange. If the Sequences are not streamed, the first Data Frame SEQ_CNT value may be either continuously increasing or may be set to x'0000.' Using continuously increasing SEQ_CNT values within an Exchange is somewhat preferable since it allows streamed Sequences, and because it can make error recovery simpler, since Frame uniqueness can be assured without examination of the SEQ_ID field. Further SEQ_CNT usage is described in the "Detection of Missing Frames" section, on page 226.

Originator Exchange_ID: OX_ID

Each Frame transmitter must assure that all Frames can be uniquely identified. It is possible to do this using unique S_ID, D_ID, SEQ_ID, and

SEQ_CNT values, but doing this would not allow association of Frames within a Sequence to an Exchange without some alternate tracking mechanism. The OX_ID value provides association of Frames with specific Exchanges originating at a particular N_Port.

If the alternate tracking mechanism is used, the Originator may set the OX_ID field to x'FFFF,' meaning "unassigned," and must ensure that only one Exchange is active at once for the S_ID, D_ID pair. Otherwise, each open Exchange is assigned an identifier unique to the Originator or Originator-Responder pair, while the Exchange is open. The OX_ID value is indicated to the Exchange Responder in the first Data Frame of the first Sequence of the Exchange. The OX_ID value has no meaning for the Exchange Responder, which merely echoes the OX_ID value back to the Originator for all Frames within the Exchange.

Priority and Preemption. FC-PH-3 added the capability for prioritizing Frames in the network, and for preempting existing Dedicated connections to establish new dedicated connections. These are both most useful in interconnection networks supporting real-time systems, such as systems distributing real-time audio and video. If priority is supported by both the N_Port and the Fabric, and is enabled during Fabric Login, then the 16-bit OX_ID field is replaced by (1) a 1-bit flag, meaningful only on **SOFc1** frames, indicating that the frame is a preemption request, (2) a 7-bit priority, increasing from x'01' to x'3F', which is generally constant during the life of a Class 1 or 6 dedicated connection, a Class 2 Exchange, or a Class 4 virtual circuit, and is not used in Class 3, and (3) an 8-bit OX_ID, which may have 254 valid values.

Responder Exchange_ID: RX_ID

The RX_ID field provides the same function for the Exchange Responder that the OX_ID provides for the Exchange Originator. The OX_ID field is meaningful on the Exchange Originator, and the RX_ID field is meaningful on the Responder. The Responder can indicate to the Originator a unique RX_ID value, other than x'FFFF,' in the first Frame it transmits for the Sequence. In Class 1 and Class 2, this is the first ACK transmitted for a Data Frame in the first Sequence, and in Class 3 this is in the first data Frame transmitted by the Exchange Responder, following a transfer of Sequence Initiative.

An Exchange Responder can use an alternate to the RX_ID tracking mechanism to ensure uniqueness, in which case the RX_ID is set to x'FFFF' for the whole Exchange. If the Responder is going to use the RX_ID mechanism for tracking Exchanges, it must assign the RX_ID value before the end of the first Sequence.

In most implementations that use the RX_ID mechanism for tracking Exchanges, it will be preferable that the RX_ID value be assigned immediately after the first Frame of a new Exchange. This ensures that the Responder can build any Exchange tracking mechanisms required to maintain the Exchange before any further Frames for the Exchange are received. The behaviour is actually the default, and is called "X_ID Interlock." Under X_ID Interlock, the Originator of a new Exchange can only send a single Frame. The Responder then must return a Link Control Frame with the RX_ID field value assigned before the Originator can send any more Frames, and all following Frames will include the RX_ID value assigned. Naturally, this mechanism is only applicable to Class 1 and Class 2 Sequences, since Class 3 Sequences do not use Link Control Frames.

Parameter

The Parameter field has two meanings, depending on Frame type. For Link Control Frames, the Parameter field indicates the specific type of Link Control Frame, as shown in Figure 7.2. In ACK Frames, the Parameter field is used to indicate how many Frames are being acknowledged, and whether all Frames with lower SEQ_CNT values have been acknowledged.

For Data Frames, the Parameter field contains the Relative Offset value. This specifies an offset from a ULP buffer from a ULP base address. This field is optional, but it is very useful if implemented, because of the variable Frame length of Fibre Channel Frames.

To illustrate the utility of the Parameter field, assume that a Sequence is being sent in Class 2 from a 1-MByte buffer at N_Port A to a 1-MByte buffer at N_Port B. Since Class 2 Frames are of variable size, if no Relative Offset were implemented, the correct location to store a Frame's Payload would not be knowable until all previous Frames for the Sequence had been received and stored. If, for example, the Frame with SEQ_CNT 5 has been received but SEQ_CNT 4 has not, then the receiver does not know where to store the data in the receive buffer and must temporarily buffer the received data until SEQ_CNT 4 and all other previous Frames have been received, so that their lengths are known.

If Relative Offset were implemented in the above situation, the Frame with SEQ_CNT 5 would contain an Offset field, specifying how many bytes were transmitted in previous Data Frames of the Sequence, and the Recipient would know where to store the data. It would then only have to keep track of out-of-order Frame Headers, without buffering all the out-of-order data received as well.

The Offset value is relative to an Information Category within a Sequence for the Exchange and can have a maximum value of $(2^{32}-1)$. Its presence is indicated in the "Relative Offset Present" flag at F_CTL bit 3.

Data Frames and Link Control Frames

As shown in Figures 7.2 and 7.3, Frames can be classified into "Data Frames," including Device Data, Basic and Extended Link Data, Video Data, and FC-4 Link Data, and into "Link Control Frames," including ACKs, the BSY and RJT Link Responses, and the Link Credit Reset (LCR) Link Command. The rules of transmission of Data Frames and corresponding response Frames (Link Control Frames) are equally applicable to each type of Data Frame.

Data Frames contain Payloads of between 0 and 2,112 bytes, with data, control, header, or status information from an ULP or FC-4 level. Each Frame is transmitted as a unit from a source N_Port, through a Fabric or Arbitrated Loop, to the destination N_Port indicated in the Frame Header.

Link Control Frames are returned to the source N_Port for Frames transmitted in Class 1 and Class 2 Sequences. Link Control Frames are divided into three types. Acknowledgments, or ACKs, have normal Frame Headers, with no Payload. ACKs acknowledge successful receipt and storage of Data Frames at the destination N_Port. There are three types of ACKs. An ACK_1 acknowledges one Data Frame, an ACK_N acknowledges N Data Frames, and an ACK_0 can be used to acknowledge all the Data Frames of a Sequence. The three ACK forms provide increasingly efficient acknowledgment traffic, with correspondingly less control over Data Frame flow control. Each ACK Frame contains Sequence Qualifier Header field values that are identical to those in the Data Frames being acknowledged, and the SEQ_CNT field in an ACK is set to the value of the highest SEQ_CNT Data Frame being acknowledged.

Several types of Link Control Frames, termed "Link Response Frames," are provided for use when a Data Frame cannot be delivered successfully. There are a number of reasons why Frames may not be deliverable, including data errors occurring during transmission, collision or congestion with other Frames for longer than a timeout period, incompatible or unrecognized Frame Header fields, or invalid types of service requests. Also, there may be problems with delivery at the destination N_Port, or there may be problems for a Fabric in delivering the Frame to the destination N_Port.

For these reasons, there are four different types of Link Response Frames, each of which contains a reason code explaining the problem with the transmission or Frame delivery. An "F_BSY" Link Response Frame indicates that the Fabric is too busy handling other traffic or other requests for service to deliver the relevant Frame. Similarly, a "P_BSY" indicates that the destination N_Port is too busy to receive the Frame. Two Link Response Frames are defined for rejecting Frame delivery: "F_RJT" and "P_RJT." In each RJT Frame, a reason code describes the reason for the rejection, as described in the "F_RJT and P_RJT Frames" section, on page 140. These include problems such as "Invalid D_ID," for when the D_ID field does not specify an N_Port on the network, or "Invalid F_CTL," for when the bits of the F_CTL field are not internally consistent with each other.

Link Control and Link Response Frames are generated for Data Frames sent in the Class 1 and Class 2 Classes of service. In Class 3, the operations handled by the Link Control and Link Response Frames are expected to be handled by an ULP, so no Link Control Frames are used. If a Data Frame can be delivered, it is delivered, and if it can't, it is discarded with no notification given to the source N_Port.

The last type of Link Control Frame is used to reset the end-to-end flow control mechanism when, for one of a variety of reasons, the end-to-end Credit is lost. This normally happens when Data Frames or other Link Control Frames are corrupted during transmission. If a Sequence Initiator sends out a Data Frame, and either the Data Frame or its corresponding Link Control Frame is corrupted, then no response is returned, and the available end-to-end Credit is (mistakenly) decreased by 1. If this happens, it can degrade transmission efficiency greatly or even stop transmission altogether. The "Link Credit Response (LCR)" Link Command Frame allows the source N_Port to request that any outstanding Frames be discarded so that the source can reset its end-to-end Credit and start flow control over again.

When an N_Port transmits a Data Frame, it must provide resources for handling any Link Control Frame that may be returned for it. This means that while an N_Port can return a P_RJT for a Link Control Frame with some types of protocol errors, it cannot transmit a P_BSY Frame in response to any of the Link Control Frames.

Data Frame Types and Handling

As shown in Figure 7.3, there are five different types of Data Frames. The most commonly understood type is an FC-4 Device_Data. This type is used for moving data from a source N_Port to a destination N_Port for an ULP. Two different types of Data Frames are used for performing link-level service functions and for doing protocol-related jobs like determining the network configuration, determining the status of operations at other Nodes, and performing error recovery procedures. Basic Link Service Frames, which all have R_CTL[31:28] = b'1000,' are sent inside of already-existing Sequences and Exchanges, to provide basic link service such as aborting existing Sequences and removing Class 1 Dedicated Connections. Extended Link

Service Frames, identified with R_CTL[31:28] = b'0010,' are sent in their own Sequences (and generally their own Exchanges) and are used for more complex functions such as determining network configuration, notifying the network of the local configuration, and determining the status of operations at remote Nodes.

FC-4 Link Service Frames, identified with R_CTL[31:28] = b'0011,' are used at the discretion of an FC-4 or upper level for performing link service functions that are effective at the FC-4 level. These Frames are defined as a method of providing help to the FC-4 level, and the Frames are passed through the Fibre Channel levels equivalently to Device_Data Frames. Finally, Video_Data Frames, with R_CTL[31:28] = b'0100,' are distinguished from other types of Data Frames since they may be handled quite differently. Video Data must be handled isochronously (the network must provide a fixed and guaranteed bandwidth), and the error recovery requirements are much less stringent than for standard data communications, which don't require isochronous delivery and must guarantee intact data delivery. In addition, video data may be stored differently at the destination N_Port, such as to a video frame buffer, than Device_Data Frames, which are typically stored to memory.

Flexibility in transmission between these five types of Data Frames, along with the flexibility in choice of Class of service, and Exchange Error Policy allow a wide range of possible data transmission characteristics for handling multiple types of traffic over the same hardware and data delivery protocols.

There are some rules which specify how the Data Frames within Sequence can be transmitted and how the Data Frames relate to other Frames within the Sequence and in other Sequences.

Data Frames may be streamed, i.e., multiple Frames may be transmitted before any response Frames are returned. The number of these outstanding Frames is determined by the two flow control mechanisms. End-to-end flow control governs the number of Frames that may be outstanding to a particular destination N_Port within a particular Class of service at once, and buffer-to-buffer flow control governs the number of Frames that may be outstanding over a particular link at once. Negotiation for these values is carried out during N_Port Login and Fabric Login, respectively.

Data Frames do not have to be transmitted directly following each other — there may be a delay following one Data Frame before the next Frame of the same Sequence or before another Sequence is transmitted. However, within a Sequence, there cannot be more than an E_D_TOV timeout period between sequential Frame transmission, or the Sequence Recipient will assume that the following Frame is lost and will never be delivered.

There are a number of mechanisms for handling errors in Frame delivery. Class 1 connect-request Frames, Class 2, and Class 3 Frames can be retransmitted in response to an F_BSY or P_BSY Link Response Frame. Class 1

Data Frames can be retransmitted as a sub-set of retransmission of the entire Sequence if the "Discard multiple Sequences with immediate retransmission" Exchange Error Policy is used for the Frame's Exchange. Other than this, the data within a corrupted Sequence can be retransmitted under control of the FC-4 level as an independent Sequence.

Link Credit Reset (LCR) Frame

The LCR Frame is a Link Control Frame termed a "Link Control Command." Link Control Commands initiate a low-level action at the destination N_Port specified by the D_ID, such as reset. LCR is currently the only defined Link Command Frame, although others may be defined in the future.

The LCR Frame is used to recover from hang or deadlock situations that arise due to a loss of credit, where a source N_Port mistakenly stops sending Data Frames to a destination N_Port because it perceives that there is no more credit. This can occur as a secondary effect due to the loss of either Data Frames or Link Control Frames on a link, over a period of time.

The LCR Frame commands the destination N_Port to reset and clear any buffers containing Data Frames from the source N_Port. This allows the source N_Port to reset its end-to-end Credit to the value determined at Login.

Since it is a low-level reset function for the end-to-end Credit, the LCR Frame is sent outside of any existing Exchanges or Sequences and isn't itself controlled by any end-to-end flow control mechanisms. All fields other then R_CTL, D_ID, and S_ID are reserved and ignored by the Recipient except for CRC calculation, and the **SOFn2** and **EOFn** delimiters are used. Dedicated Connections are not affected by the LCR Frame and should be removed prior to LCR transmission, if possible.

Following transmission of the LCR Frame, all active Sequences with the same Initiator and Recipient are abnormally terminated for all Classes by both N_Ports. Exchange and Sequence recovery may or may not be performed, depending on the FC-4 ULP at the source N_Port. No more Sequences can be initiated until after an R_A_TOV timeout period, to prevent collision with previously existing Frames.

The LCR Frame may only be responded to with the R_RDY Primitive Signal (normal case as **SOFi2** or **SOFn2**), or the F_RJT, P_RJT, F_BSY Link Control Frames. F_RJT can be returned for any normal reasons, but P_RJT can only be returned for the "Invalid D_ID" or "Class not supported" reasons.

Detailed Responses to Data and Link Control Frames

When an N_Port or F_Port transmits a Data Frame or Link Control Frame, a number of responses can be returned. These include the R_RDY response, for link level buffer-to-buffer flow control, and the Link control Frames, which can be returned for Data Frames and sometimes for other Link Control Frames as well. This section describes in more detail the responses to Frame transmission and the detailed handling of the various Link Control Frames.

R_RDY Response

As described in the "R_RDY" section, on page 95, the R_RDY Primitive Signal is transmitted in response to a Frame when enough processing has been done at the receiver that another Frame can be accepted. R_RDY is used for buffer-to-buffer flow control between two attached Ports for Class 2, Class 3, and Class 1 connect-request Data Frames and Link Control Frames. For the Class 1 connect-request, R_RDY is only used for flow control and does not indicate whether the Connection has been made. When an R_RDY is received, it indicates that an interface buffer and associated resources are available for receiving one more Frame. There may be multiple interface buffers supported, and the number of buffers is specified during Fabric Login.

ACK Frames

The ACK Frame is used for end-to-end flow control in Class 1 and Class 2 Sequences, as described in Chapter 15. Return of an ACK Frame indicates that one or more Data Frames were received by the destination N_Port and transferred out of the N_Port to the FC-4 or upper level. It also indicates that receive buffers and associated resources are available for receiving more Frames for that Class at the destination N_Port. ACK Frames are transmitted in the same Class (identified by Frame delimiters) as the Data Frames they acknowledge, and have the same Sequence Qualifier values. Each Data Frame is acknowledged only once. Generally, ACK Frames will be transmitted in the same order that the corresponding Data Frames were transmitted, but for Class 2 operation, where there is possible reordering of both Data and Link Control Frames in the Fabric, ACK Frames may not be received at the source N_Port in the same order in which the corresponding Data Frames were transmitted.

In addition to end-to-end flow control, ACK Frames are used for a number of other protocol level functions:

- X_ID assignment, as described in the "Exchange Origination and X_ID Assignment" section, on page 150
- X_ID reassignment, as described in the "Association_Header" section, on page 244
- X_ID interlock, as described in the "Exchange Management" section, on page 149
- Terminating a Sequence, as described in the "Normal Sequence Completion" section, on page 147
- Establishing a Dedicated Connection, as described in the "Class 1 Service" section, on page 184 and the "Establishing a Connection" section, on page 249
- Removing a Dedicated Connection, as described in the "Class 1 Service" section, on page 184 and the "Removing a Connection" section, on page 250
- Notification of Abort Sequence conditions, as described in the "Rules Common to All Discard Policies" section, on page 229
- Setting a Stop Sequence condition, as described in the "Stop Sequence Protocol Overview" section, on page 239
- Abnormally terminating sequences, as described in the "Sequence Recovery" section, on page 233
- Requesting of sequence retransmission, as described in the "Discard Multiple with Immediate Retransmission" section, on page 231.

ACK Frame Headers basically correspond to the headers of the corresponding Data Frames. The D_ID and S_ID fields are interchanged, so that the D_ID field indicates the source of the Data Frame being acknowledged. In the F_CTL field both the Sequence and Exchange context bits are inverted. Other bits are as shown in Figure 7.8. The SEQ_ID and SEQ_CNT fields match those of the highest number Frame being acknowledged. In the Parameter field, the History bit (bit 16) being set to 1 indicates that at least one previous ACK in the Sequence may not have been transmitted, as described in the "Use of the History Bit in ACK Frames" section, on page 138. Bits 15 to 0 are either set to the number of Frames being acknowledged or are set to 0 to indicate acknowledgment of all Frames in the Sequence.

ACK Frames can be responded to with F_RJT, P_RJT, or F_BSY Frames, but the destination N_Port has to guarantee that it is not too busy to receive an ACK and cannot respond with P_BSY. Each ACK must be transmitted within an E_D_TOV timeout period of the event which prompted the ACK transmission.

The ACK_1 form of ACK (indicated using R_CTL bits 27–24 = b'0000') is supported by all N_Ports as the default. This will be the most commonly used mechanism and is the simplest to implement. The ACK_N form of ACK is useful to reduce link traffic when an N_Port has more than one sequential Data Frame to acknowledge. Usage of the ACK_N facility is optional, as specified in the Class Service parameters during N_Port Login, and the value of N for any particular ACK can be 1. Decision of how often to transmit ACK_N Frames is somewhat implementation-dependent. Transmitting an ACK_N with N = 1 doesn't provide any better performance than using ACK_1. However, saving up ACKs for grouped transmission exposes the Initiator to loss of end-to-end Credit while waiting for acknowledgments to return. The Recipient cannot delay for too long, since an Initiator which does not receive an ACK within the E_D_TOV timeout period will not expect the ACK ever to be returned. Usage of ACK_N requires a little care.

ACK_0 is the designation used when R_CTL[27 to 24] = b'0001' and bits 15 to 0 of the Parameter field contain a value of 0, to indicate that an entire Sequence is being acknowledged. The ACK_0 mechanism provides no Frame level flow control and only indicates that a full Sequence has been delivered. Use of ACK_0 essentially assumes that the FC-4 level is providing its own flow control and that the receiving N_Port can receive the transmitted data at whatever rate it arrives.

There are two cases where two ACK_0 Frames are needed per Sequence. First, if X_ID interlock is required, where the Exchange Responder requires that all Frames after the first Frame of the Exchange contain the Responder-assigned RX_ID value, then an ACK_0 must be returned following the first Frame of the first Sequence of a new Exchange, containing the RX_ID value. Second, if the first Sequence of the Exchange is transmitted in Class 1, then an ACK must be returned for the connect-request, to indicate establishment of the Dedicated Connection. Since the standard specifies that ACK forms cannot be mixed within a single Sequence, if ACK_0 is being used for the whole Sequence, it must be used for these first Frames as well.

Sequences transmitted with an ACK_0 expected don't participate in end-to-end Credit management, as described in the "End-to-End Flow Control" section, on page 265, so the N_Port transmitting a Sequence using ACK_0 doesn't need to track end-to-end Credit. The only flow control performed is at the buffer-to-buffer level. ACK_0 can be used for all four Exchange Error Policies described in Chapter 12, and support for ACK_0 processing must imply support for the "Process with infinite buffering" Exchange Error Policy.

Use of the History Bit in ACK Frames. The History bit is used as a method of simplifying ACK processing by indicating whether or not all previous Frames of a Sequence have been receiving intact or not.

For all forms of ACK, when the History bit of the Parameter field (word 5, bit 16), is set to 0, it indicates that the Sequence Recipient N_Port has transmitted all previous ACKs, if any, for this Sequence. When the History bit is set to 1, it indicates that at least one previous ACK for this Sequence has not been transmitted by the Sequence Recipient N_Port. This may be due to a withheld ACK, a Data Frame not processed, or a Data Frame not received. Using this historical information allows an N_Port to reclaim end-to-end Credit for a missing ACK Frame, if the History bit is set to 0, or to simplify processing to be sure of when all ACK Frames have been processed.

ACK Form Usage. When multiple ACK forms are supported by both Sequence Initiator and Recipient N_Ports, as described in the N_Port Login parameters, ACK_0 usage is expected to take precedence over ACK_N, and ACK_N usage will take precedence over ACK_1. ACK_1 is the default, if no other common ACK form is supported by both ends. However, ACK_0 usage provides very little flow control and has some ambiguities relative to Class 1 service, and it can only be used with the "Process with Infinite buffering" Exchange Error Policy, so a significant number of N_Ports may not implement ACK_0 support.

The usage of the ACK Forms is not required to be symmetrical (an N_Port may be able to receive ACK_N Frames, but only transmit ACK_1 Frames) or equal (an N_Port acting as Recipient may reply with ACK_1 Frames to one Sequence Initiator and with ACK_N to a different Initiator).

Mixing of ACK forms within a Sequence is not allowed. ACK usage is somewhat tricky during X_ID interlock and in response to a connect-request. It is possible for the Sequence Recipient to transmit an ACK_1, and ACK_N with an ACK_CNT of 1, or an ACK_0, as shown in Figure 7.2 in these two cases, although using the ACK_N with ACK_CNT = 1 form generally makes the most sense.

Handling of ACK Frames in the presence of data integrity problems is described in detail in Chapter 12.

F_BSY Frames

The F_BSY Link Response indicates that the Frame can't be delivered, because either the Fabric or the destination N_Port is temporarily busy. On receipt of an F_BSY in response to a Frame transmitted, the source N_Port is expected to attempt Frame retransmission, up to some implementation-dependent number of retries. Recovery after retry is exhausted is dependent on the FC-4 ULP and the Exchange Error Policy.

An F_BSY can only be sent in response to a Class 1 connect-request (where it uses the **EOFdt** delimiter to indicate that no Connection was made) or to a Class 2 Data or ACK Frame. A Link Control Frame which runs into a busy Fabric or destination N_Port link is discarded to prevent two N_Ports from continuously transmitting BSY Frames to each other.

There are two different Link Response codes defined for F_BSY, as shown in Figure 7.2, for responses to Data Frames and Link Control Frames. This is to simplify retransmission of Link Control Frames, where the F_BSY to a Link Control Frame contains enough information internally to allow retransmission. Retransmission of a Data Frame requires information from the retransmitting N_Port, which the N_Port must track until the Frame is acknowledged or retry is exhausted.

In Class 2, if an N_Port receives an F_BSY in response to an ACK Frame, it will discard the F_BSY Frame. If it were retransmitted, a data integrity problem might occur. This is because the ACK may finally be received by the Sequence Initiator after R_A_TOV has expired for that Sequence. Note that, after R_A_TOV has expired, the Sequence Initiator is free to reuse the Sequence Qualifier for a different Sequence. Therefore, the reception of the retried ACK may be mistaken for an ACK for the current Sequence, causing a potential data integrity problem.

When forming an F_BSY, the Fabric will interchange the S_ID and D_ID fields, invert both the Exchange and Sequence context bits, and set the R_CTL Information Category (bits 27 to 24) to b'0101' for "F_BSY to a Data Frame" or b'0110' for "F_BSY to a Link Control Frame." The SEQ_ID and SEQ_CNT fields are the same as for the Frame being busied. The Parameter field is returned unchanged.

In the TYPE field, bits 31 to 28 indicate either "Fabric busy" (b'0001') or "N_Port busy" (b'0011'). Bits 27 to 24 are reserved for F_BSY to a Data Frame. For F_BSY to a Link Control Frame, R_CTL[27:24] of the busied Frame are copied to TYPE[27:24] of the F_BSY, to simplify retransmission.

P_BSY Frames

The P_BSY Link Response indicates that the Frame can't be delivered because the destination N_Port is temporarily busy. P_BSY cannot be sent in response to any Link Control Frame, since N_Ports are required to be able to handle Link Control Frames for any outstanding Data Frames.

The Parameter field of the P_BSY Frame contains an "action code" in bits 31 to 24, a "reason code" in bits 23 to 16, and vendor-dependent reason codes in bits 7 to 0. The possible action codes and reason codes defined are:

Action Code x'01' — Sequence terminated — retry later: This action code indicates that the Sequence Recipient has terminated the Sequence using the **EOFt** or **EOFdt** delimiter. This will only occur in response to a connect-request or to an interlocked Data Frame associated with X_ID assignment or reassignment. The Frame and Sequence can be retried at a later time.

Action Code x'02' — Sequence active — retry later: This action code indicates that the Sequence Recipient has busied a Class 2 Frame and that the Sequence has not been terminated (**EOFn**). The Frame can be retried at a later time.

Reason Code x'01' — Physical N_Port busy: This reason code indicates that the destination N_Port link facilities are currently busy and the N_Port is unable to accept the Payload of the Frame.

Reason Code x'03' — N_Port Resource busy: This reason code indicates that the destination N_Port is unable to process the Data Frame at the present time.

Reason Code x'FF' — Vendor unique busy: This code indicates that a special vendor code is present in bits 7 to 0.

In other respects, operation of P_BSY Frames is identical to F_BSY Frames at both the Frame destination and source N_Ports.

F_RJT and P_RJT Frames

The two different Reject Link Responses to a Frame indicate that delivery of that Frame is being "rejected." Rejection indicates that the Frame contents are intact (i.e., no transmission errors) but that the Frame can not be received for some protocol-related reason, such as non-support of a service or inconsistent Frame Header fields. A 4-byte reject action and reason codes are contained in the Parameter field, as shown in Figure 7.10. It is useful to examine these reasons, to understand what kinds of assumption the protocol makes about consistency of Frame Header field. For certain of these conditions retry is possible (i.e., the Frame can be sent again), whereas other conditions may require an implementation-dependent specific intervention, and Frame retry is not possible.

An F_RJT or P_RJT can be sent for a Data Frame for any of the defined reasons. If a similar error is detected on a Link Control Frame, a P_RJT can only be transmitted for the following reasons: (1) invalid D_ID or S_ID, (2) N_Port not available, temporarily or permanently, (3) Class not supported, or (4) Login required. Similarly, an N_Port can only reject a Link Control Frame and transmit a P_RJT using the "Unexpected ACK" reason code. For the other rejection reasons, if an N_Port detects an error in a Link Control

Figure 7.10
Frame reject reason codes.

Word 5, [31–24]	Word 5, bits [23–16]: Reason for rejection	Detailed Description
x'01' Retryable Error	x'01': Invalid D_ID	F_RJT:Fabric unable to locate N_Port P_RJT:N_Port doesn't recognize D_ID as own
	x'02': Invalid S_ID	F_RJT:S_ID mismatches Fabric-assigned value P_RJT:destination doesn't recognize S_ID as valid
	x'03': N_Port unavail. temporarily	F_RT - Valid address for online but functionally unavailable Port, e.g., performing Link Recovery
	x'04': N_Port unavail. permanently	F_RJT - Valid address for functionally unavailable Port, - may be offline, or powered down
	x'05': Unsupported Class	F_RJT, P_RJT - Class of service for rejected Frame is unsupported by Fabric or dest. N_Port
x'02' Non-retryable error	x'06': Delimiter usage error	F_RJT, P_RJT - SOF or EOF inappropriate for current state, e.g., SOFc1 during connection
	x'07':TYPE error	F_RJT, P_RJT- Type field of Frame is unsupported
	x'08': Invalid Link Control	P_RJT - Command specified in R_CTL Info. Cat. bits are invalid or unsupported as Link Control
	x'09': Invalid R_CTL	P_RJT - R_CTL field invalid or inconsistent
	x'0A': Invalid F_CTL	P_RJT: F_CTL field invalid or inconsistent
	x'0B': Invalid OX_ID	P_RJT: OX_ID field invalid or inconsistent
	x'0C': Invalid RX_ID	P_RJT: RX_ID field invalid or inconsistent
	x'0D': Invalid SEQ_ID	P_RJT: SEQ_ID field invalid or inconsistent
	x'0E': Invalid DF_CTL	P_RJT: DF_CTL field invalid
	x'0F': Invalid SEQ_CNT	P_RJT: SEQ_CNT field invalid or inconsistent. This RJT is not used to notify of missing Frames
	x'10': Invalid Param.	P_RJT - parameter field badly specified or invalid
	x'11': Exchange Error	P_RJT - error detected in Exchange (OX_ID) such as a Frame transmitted without Seq. Initiative
	x'12': Protocol Error	P_RJT - FC-2 signalling protocol error detected which is not covered by other codes.
	x'13': Length error	F_RJT, P_RJT - Frame has incorrect/invalid length
	x'14': Unexpctd ACK	P_RJT - ACK for unopened Seq. or Exchange.
x'01' Retryable	x'16': Login Required	F_RJT, P_RJT - Exchange initiated before Service Parameters were interchanged
	x'17': Excessive Seqs. attempted	P_RJT - N_Port initiated a Sequence which exceeded the capability of Recipient facilities
	x'18': Unable to establish Exchange	P_RJT - N_Port initiated a new Exchange which exceeded the capability of Responder facilities
x'02' Non-Retryable	x'19': Exp._Sec._Hdr unsupported	P_RJT - N_Port does not support the optional Expiration_Security_Header
	x'1A': Fabric path not available	F_RJT - Source and destination speeds don't match, or other error vs. multiple Fabric domains
x'01' or x'02	x'FF': Vendor Unique	F_RJT, P_RJT- Bits 7–0 of Word 5 specify additional vendor-dependent reason codes

Frame for a valid Exchange, it will initiate the Abort Sequence (ABTS) Protocol and will not transmit a reject Frame. If an error is detected in a Link Control Frame of an unidentified or invalid Exchange, the N_Port will discard the Frame, ignore the Link Control Frame error, and hope for the best.

If a Class 3 Frame satisfies a rejectable condition, the Frame is discarded, and no reject is sent. A reject Frame (either F_RJT or P_RJT) is not transmitted in response to another reject Frame (either F_RJT or P_RJT); the received RJT Frame is discarded, to keep both N_Ports from continuously transmitting RJT Frames to each other.

If a Frame within a Sequence is rejected, the entire Sequence is abnormally terminated or aborted. If the RJT Frame is delimited with an **EOFt** or **EOFdt**, the Port transmitting the RJT Frame has terminated the Sequence. A Port will only terminate the Sequence using a reject with an **EOFdt** delimiter in response to a connect-request (**SOFc1**). In Class 2 an N_Port will only terminate the Sequence on a reject in response to an interlocked Data Frame associated with X_ID assignment or reassignment (**SOFi2**). If the RJT Frame ends with an **EOFn**, the N_Port receiving the RJT Frame will perform the Abort Sequence protocol to abort the Sequences. Rejects will only be transmitted in response to valid Frames.

When forming an F_RJT or P_RJT, the Recipient will interchange the S_ID and D_ID fields, invert both the Exchange and Sequence context bits, and set the other bits as shown in Figure 7.8. The TYPE field is reserved. The SEQ_ID and SEQ_CNT fields are the same as for the Frame being rejected. The Parameter field is returned with action and reason codes for the rejection, as shown in Figure 7.10.

Notify (NTY) and End (END) Frames

The Notify (NTY) Frame is used by the Fabric to inform a destination N_Port that a Camp-On exists on its F_Port, and requests the destination N_Port to initiate the End Connection protocol with the existing connection, so that the Camped-On Connection can be initiated. It's only used for Fabrics that support Camp-On, and only sent to N_Ports that can intermix Class 1 and Class 2 traffic. The END Link Control command is used to (a) remove Dedicated Simplex Connections, (b) to remove outstanding Camp-On requests, and (c) to deactivate or remove Class 4 circuits (as denoted by the EOF delimiter).

Chapter 8

FC-2: Sequences and Exchanges

Introduction

At the level above Frames, data are grouped into two higher-level constructs, termed "Sequences" and "Exchanges." A Sequence is a set of one or more related Frames transmitted unidirectionally from one N_Port to another N_Port with corresponding Link Control Frames (ACKs, BSY, or RJT Frames) transmitted in response. An Exchange is a set of one or more related Sequences that may flow in the same or in opposite directions (but not simultaneously) between a pair of N_Ports.

This chapter describes in detail the rules for initiating, managing, and terminating Sequences and Exchanges. It also describes the format and handling of the Sequence Status Block and Exchange Status Block constructs, which are used to track the status of open and active Sequences and Exchanges at each N_Port of an Exchange.

Sequence Management

An N_Port which transmits a Sequence is termed the "Sequence Initiator" and the N_Port which receives the Sequence is termed the "Sequence Recipient." Error recovery is performed on Sequence boundaries, if required. If a Frame does not arrive intact, and the Exchange Error Policy requires error recovery, the entire Sequence will be retransmitted. Each Sequence is identified by an identifier value in every Frame, the SEQ_ID. The length of each Sequence is unbounded, since a Sequence can consist of an arbitrary number of Frames.

All Frames have **SOF** and **EOF** Ordered Sets as delimiters. A Sequence is initiated by a Frame with a **SOFi1**, **SOFi2**, **SOFi3**, or **SOFc1** delimiter. All intermediate Frames within a Sequence are transmitted with **SOFn1**, **SOFn2**, or **SOFn3** and **EOFn** delimiters. The Sequence is complete when an **EOFt**, **EOFdt**, or **EOFdti** has been transmitted or received for the appropriate SEQ_ID and all previous Data Frames and ACKs (if any) have been accounted for by the Initiator and Recipient, respectively.

The status of a Sequence is described using the words "active" and "open," which have somewhat different meanings depending on whether status is described at the Sequence Initiator or Recipient, and depending on the Service Class used for the Sequence. For the Sequence Initiator, a Sequence is open and active from the transmission of the Sequence's first Data Frame, and active until transmission of the Sequence's last Data Frame. In Class 1 and Class 2, the Sequence is open until (1) the ACK with a **EOFt** (or **EOFdt**, for Class 1) is received, (2) the Sequence is aborted by performing the ABTS protocol, or (3) the Sequence is abnormally terminated. In Class 3, the Sequence is open until either the deliverability is confirmed (e.g., by using the ABTS Basic Link Service Request) or an R_A_TOV timeout period has expired.

For the Sequence Recipient, a Sequence is active and open from the time any Data Frame is received. In Class 1 and 2 a Sequence is active and open until the **EOFt** (or **EOFdt**, for Class 1) is transmitted in the ACK to the last Data Frame, or there is an abnormal termination of the Sequence. In Class 3 a Sequence is active and open until all Data Frames up to the Frame containing **EOFt** have been received.

For both Sequence Initiator and Recipient, a Sequence is also considered "closed" (not open) when (1) a BA_ACC indicating closure is received in response to an ABTS Request, (2) the Sequence's Exchange is aborted using the ABTX Extended Link Service request, (3) the ACC reply to a Read Sequence Status or Read Exchange Status Extended Link request indicates either normal or abnormal completion, (4) a Primitive Sequence is received or transmitted during a Class 1 Dedicated Connection, or (5) a Logout Extended Link Service request is completed.

Sequence Initiation

An N_Port can initiate a new Sequence in an Exchange if (1) it holds the Sequence Initiative, (2) it has a unique SEQ_ID available, and (3) it will not exceed the limits on total concurrent Sequences, concurrent Sequences per Class of service, or open Sequences per Exchange, as established during N_Port Login.

Having an SEQ_ID "available" means that the N_Port can assign a SEQ_ID value for the Exchange which will be unique within the other fields of the Sequence Qualifier (S_ID, D_ID, OX_ID, and RX_ID). This means, for example, that the SEQ_ID cannot match the SEQ_ID of the preceding Sequence of the Exchange, and that if the N_Port can send M open Sequences per Exchange, then it must use at least M + 1 unique SEQ_ID values sequentially, to prevent overlap of outstanding SEQ_ID values. Any previous Sequences with the same SEQ_ID value must be closed, with the status known either through normal termination or by the Abort Sequence Protocol, described in the "Abort Sequence (ABTS) Command" section, on page 199, before a Sequence can be initiated. In Class 2, a SEQ_ID cannot be reused until all ACKs for the Sequence have been accounted for, either through ACK receipt or through R_A_TOV timeout.

Sequence Handling Validity

During the time that a Sequence is open, there are some specific rules that must be followed in the assignment of values to Frame Header fields and in the handling of Frame transmission within a Sequence. These rules apply both to the Data Frames and to Link Control Frames.

Sequence Qualifier: All the Frames in the Sequence should contain the S_ID, D_ID, OX_ID, RX_ID, and SEQ_ID values which are current for the Sequence. Note that the RX_ID value (and possibly the OX_ID value, with X_ID reassignment) can change during the course of the Sequence, as X_ID values are assigned or reassigned.

Unassigned X_ID: If an OX_ID or RX_ID field is not currently assigned for the Exchange, the field value must be set to x'FFFF.'

Exchange and Sequence Contexts: Exchange_- and Sequence_ Context (F_CTL bits 23 and 22) must be set to correctly indicate Exchange Originator/Responder and Sequence Initiator/Recipient.

SEQ_CNT management: Each Frame within a Sequence must contain a SEQ_CNT value, which follows the rules described in the "Sequence Count Management" section, on page 147 below.

Flow Control: The Frames must be transmitted following the end-to-end and buffer-to-buffer flow control rules described in Chapter 15.

Payload size: The size of the Data Frames cannot exceed either the limit specified by the Fabric or by the destination N_Port during Login. The Fabric has no limit on Class 1 Data Frame sizes but does have limits on connect-request, Class 2, and Class 3 Frames.

Class of service: Each Frame in the Sequence must have Frame delimiters for the same Class of service, since a Sequence cannot be transmitted in multiple Classes. This includes Frame delimiters for Link Control Frames.

Transmission delay: Each Data Frame must be transmitted within an E_D_TOV timeout period following transmission of the previous Frame, so that a Sequence timeout isn't detected.

If any of the rules are broken, either by the Sequence Initiator or Recipient, it is a protocol error. It is also a protocol error if the rules on ACK processing described in the "ACK Frames" section, on page 135 are broken.

If the Sequence Recipient detects the error, it will transmit a P_RJT with a reason code describing the problem. If the Sequence Initiator detects the error, it will abort the Sequence using the Abort Sequence Protocol described in the "Abort Sequence (ABTS) Command" section, on page 199. Acknowledgments to Data Frames within a Sequence are described in the "ACK Frames" section, on page 135.

Sequence Count Management

Each Data Frame within a Sequence is marked by a SEQ_CNT field, which is incremented every Frame to ensure sequentiality and detection of lost Frames. Link Control Frames echo these SEQ_CNT values, to match their corresponding Data Frames. If an acknowledgment corresponds to multiple Data Frames, as in ACK_N or ACK_0 usage, the SEQ_CNT value is set to the value of the highest SEQ_CNT Data Frame being acknowledged.

The basic rules for assignment of SEQ_CNT values to Data Frames were described in the "Sequence Count: SEQ_CNT" section, on page 128. These include assignment of the value x'0000' to the first Data Frame of an Exchange, with SEQ_CNT incremented by 1 for each subsequent Data Frame, and initiation of each new Sequence in the Exchange with either SEQ_CNT = x'0000' or a continuously increasing SEQ_CNT value. SEQ_CNT values generally assure uniqueness of Data Frames, but if the "Process with Infinite Buffers" Exchange Error Policy is used, with ACK_0 acknowledgment, it is legal to have non-unique SEQ_CNT values. The 15-bit limit on end-to-end credit, described in the "N_Port and F_Port Class Service Parameters" section, on page 171, ensures that SEQ_CNT values do not wrap within a Sequence before previous Frames have been acknowledged.

Normal Sequence Completion

When an N_Port acting as Sequence Initiator is done transmitting the data for a Sequence and desires to terminate it, it indicates that desire by setting End_Sequence (F_CTL bit 19) to 1 in the Frame Header of the last Data Frame. When End_Sequence is set to 1, a number of other F_CTL bits become meaningful that were not meaningful on previous Frames of the Sequence, as shown in Figure 7.8 and described in the "Frame Control: F_CTL" section, on page 120. These indications describe actions desired by the Sequence Initiator following termination of the Sequence.

On termination of a Sequence, a Sequence Initiator can do a number of things, including:

- indicating that the current Sequence is the last of the Exchange, by setting Last_Sequence (F_CTL bit 20) to 1;
- requesting that a Class 1 Connection be removed, by setting End_Connection (F_CTL bit 18) to 1;
- requesting a reply Sequence within the Exchange, on some systems, by setting Chained_Sequence (F_CTL bit 17) to 1;
- transferring Sequence Initiative to the other Port, by setting Sequence Ini-

tiative (F_CTL bit 16) to 1, or keeping Sequence Initiative, by setting it to 0;

- invalidating an X_ID value to free Exchange resources or doing an X_ID reassignment within an operation, on some systems, by setting X_ID Invalid (F_CTL bit 14) to 1;
- notifying the Sequence Recipient of how soon the next Sequence will be coming if Sequence Initiative is not transferred, by setting Continue Sequence Condition (F_CTL bits 7–6).

In addition to the assertion of the End_Sequence F_CTL bit, Sequence termination is indicated using the Frame delimiter Ordered Sets. In Class 1 and Class 2, the last Data Frame has End_Sequence = 1, and is terminated with the **EOFn** delimiter. The Sequence Recipient returns an ACK with End_Sequence = 1 and an **EOFt** or **EOFdt** delimiter to terminate the Sequence. In Class 3, the last Data Frame is terminated with the **EOFt** delimiter, to terminate the Sequence without acknowledgment.

Because Frames may be corrupted or lost during transmission, mere transmission of a Frame with End_Sequence set to 1 does not necessarily mean that the Sequence will be normally closed. The conditions which allow an N_Port to consider a Sequence to be complete and deliverable depend on whether the N_Port is the Sequence Initiator or Recipient and on the Class of service used for the Sequence. Conditions for Sequence completion are shown in Figure 8.1.

Figure 8.1
N_Port- and Class-dependent conditions for considering Sequences as complete.

	Completion condition vs. Class of Service		
	Class 1	**Class 2**	**Class 3**
Initiator	Final ACK received with **EOFt** or **EOFdt**	Final ACK received with **EOFt**	BA_ACC to ABTS Frame, or ACC reply to Read Exchange Status
Recipient	All Frames received, no Sequence errors, all ACKs transmitted	All Frames received, no Sequence errors, all ACKs transmitted	All Frames received, no Sequence errors, final Data Frame with **EOFt** received

In Class 2 transmission, where Frames transmitted over a Fabric may not arrive in order, Sequence completion is more complicated than for the other Classes. If Frames arrive at the Sequence Recipient in order, processing happens normally — the ACK to the last Data Frame has End_Sequence = 1 and an **EOFt** delimiter. If the last Data Frame transmitted arrives earlier than a Frame with a lower SEQ_CNT, the Recipient has two alternatives. It can withhold the ACK until all lower Frames have been accounted for, then transmit it normally, with End_Sequence = 1, a **EOFt** delimiter, and the History bit set to 0, indicating nothing worrisome in the Sequence history. Alternatively, it can send an ACK terminated by a **EOFn** delimiter, with

End_Sequence = 0, Continue_Sequence = b'00,' Sequence_Initiative = 1, and with the History bit set to 1, while remembering what the Continue_Sequence and Sequence_Initiative bits were in the highest SEQ_CNT Data Frame. When all Data Frames have been accounted for, the Recipient can send a final ACK with the lower SEQ_CNT of the last arriving Frame, with End_Sequence = 1, an **EOFt** delimiter, and the Continue_Sequence and Sequence_Initiative bits copied from the highest SEQ_CNT Data Frame. Of the two alternatives, the second is slightly more complicated, but it can allow the Initiator an extra Frame's worth of end-to-end Credit for the following Sequence, if Sequences are being streamed.

The Sequence Recipient confirms a requested transfer of Sequence Initiative by echoing the Sequence_Initiative bit from the last Data Frame in the **EOFt** ACK Frame. If the Initiator holds Sequence Initiative but sets Continue_Sequence to b'11' = "next Sequence is delayed," it will wait for the **EOFt** ACK before initiating the next Sequence.

When the Sequence is completed by each N_Port, the associated Sequence Status Block and Exchange Status Block are updated in each N_Port to show that the Sequence was completed and to show whether the Exchange Originator or Responder holds the Sequence Initiative. Sequence status in the Exchange Status Block is maintained until X + 2 Sequences have been completed (where X is the number of open Sequences per Exchange indicated during Login) or until the Exchange is terminated. Link facilities associated with the Sequence are released and available for other use. This frees up facilities for usage for other Sequences.

Exchange Management

An Exchange is composed of one or more unidirectional Sequences, initiated by either the Exchange Originator or the Exchange Responder. An Exchange may be (1) unidirectional, where only the Exchange Originator initiates Sequences, or (2) bidirectional, where both Exchange Originator and Responder can non-simultaneously initiate Sequences. The first Sequence of the Exchange is always transmitted by the Exchange Originator, since it originates the Exchange. Following the initial Sequence, usage of the Sequence Initiative bit (F_CTL bit 16) determines which N_Port may initiate the next Sequence for the Exchange. An N_Port can request Sequence Initiative for the Exchange using the "Request Sequence Initiative (RSI)" Extended Link Service Command. Each Exchange is associated with an Exchange Identifier for the Originator, termed the "OX_ID," and one for the Responder, termed the "RX_ID." Exchange management is handled independently of Class 1 Dedicated Connection management, so that an

Exchange may be performed across one or more Class 1 Connections and may contain one or more Class 1 Connections.

Exchange Origination and X_ID Assignment

Before any data communication Exchanges can be originated, an N_Port must complete the Fabric Login and the N_Port Login with the intended Responder of the Exchange, as described in Chapter 9, to determine Service Parameters for managing Exchanges and Sequences. Before Login only Extended Link Service Command Exchanges can be originated, to initialize the communication environment.

A new Exchange may be originated if (1) the originating N_Port has performed an N_Port Login with the intended Responder, (2) it has enough resources, including OX_ID values, available for use, and (3) it can Initiate a new Sequence, as described in the "Sequence Initiation" section, on page 145. The N_Port will originate the new Exchange by sending a Data Frame with First_Sequence (F_CTL bit 21) set to 1 in the Frame Header, an OX_ID of the originator's choosing, and the RX_ID field set to the "Unassigned" value, x'FFFF,' for assignment by the Responder.

At this point, there are two possibilities, depending on whether the Responder N_Port requires X_ID interlock, as specified during N_Port Login, shown in Figure 9.9. If the Responder N_Port requires X_ID interlock, then the Originator must delay sending any more Frames until the Responder returns an ACK Frame with the RX_ID field assigned. X_ID interlock does not apply to Class 3 transmission. The Originator will then include the assigned RX_ID and its own OX_ID in all subsequent Frames of the Exchange. X_ID interlock is required by default. If X_ID interlock is not required, as indicated during N_Port Login, the Originator can send the rest of the Frames for the first Sequence with RX_ID = x'FFFF' but must delay sending any more Sequences until at least one ACK for the first Sequence has been returned, assigning the RX_ID. In either case, the Exchange Originator must receive at least one ACK containing the Responder-assigned RX_ID value before initiating any new Class 1 or Class 2 Sequences. In Class 3, the RX_ID is not assigned unless Sequence Initiative is transferred and the Exchange Responder can assign an RX_ID in a Responder-initiated Sequence. If Sequence Initiative is never transferred, the RX_ID value will never be assigned.

In the first Frame of the Exchange, the Exchange Originator sets the Exchange Error Policy to be used for the Exchange, by setting the "Abort Sequence Condition" (F_CTL bits 5 and 4). There are four possible values, which are used for Exchanges with different Sequence deliverability and Sequence interdependency requirements, as described in the "Frame Control: F_CTL" section, on page 120, and in the "Exchange Error Policies for Class 1 and Class 2" section, on page 229.

On some systems where more complex record-keeping is required, like mainframes, an Association Header optional header included in each Sequence contains 64-bit Originator and Responder Operation_Associator values, for associating together data transmitted in multiple Exchanges (with possibly reused OX_ID and/or RX_ID values). In these cases, the OX_ID and RX_ID values alone don't assure uniqueness and may be reassigned or invalidated between Sequences. Tracking of Sequences in these cases is based on the Association Header values, as well as X_ID values, as described in the "Association_Header" section, on page 244.

Exchange Management

When an Exchange is originated by an N_Port, that N_Port assigns an Originator Exchange_ID (OX_ID) unique to that N_Port and builds an Exchange Status Block associated with the OX_ID to track the status of the Exchange and all Sequences associated with the Exchange while the Exchange is active. The value of OX_ID used is transparent to the Fabric and to the Exchange Responder, so the Originator can assign any value (other than x'FFFF,' which indicates "Unassigned") that simplifies record-keeping on the active Exchange.

When the destination N_Port, or Responder, receives the first Sequence of the Exchange, it assigns a Responder Exchange_ID (RX_ID) to the newly established Exchange (in Class 1 and Class 2) and builds an Exchange Status Block for tracking the Exchange. The RX_ID value provides the mechanism for tracking Sequences for multiple concurrent Exchanges from multiple S_IDs or from the same S_ID. The Exchange Responder has similar flexibility in assigning the RX_ID value, since the Originator only reflects the value in Frame Header fields, and assignment may be organized to provide efficient processing within the Responder N_Port. All Frames transmitted within the Exchange use Exchange_Context (F_CTL bit 23) to indicate whether the Frame was sent by the Exchange Originator (0) or Responder (1).

During the life of the Exchange, the Originator and Responder track the status of all the Sequences of the Exchange, accepting data from and delivering data to an FC-4 or ULP on a Sequence-by-Sequence basis as the Sequences are completed.

Sequences within a single Exchange with the same Initiator must be delivered to the FC-4 or ULP in the same order as transmitted, when they arrive valid and complete. There is no specification on sequentiality of delivery between different Exchanges, although there can be some interdependency between Exchanges for several of the Extended Link Service

Commands, as described in the "Extended Link Service Command Overview" section, on page 200.

Exchange Termination

An Exchange can be terminated by either the Exchange Originator or Exchange Recipient. To request termination of an Exchange, a Sequence Initiator can set Last_Sequence (F_CTL bit 20) to 1 in the last Data Frame of the Sequence. This bit can also be set in a sequential set of Data Frames including the last Frame of the Sequence, to give advance warning of the Exchange termination, as described in the "Frame Control: F_CTL" section, on page 120. The Exchange is normally terminated when the last Sequence is completed by the normal Sequence completion rules described in the "Normal Sequence Completion" section, on page 147. The OX_ID and RX_ID values are then available for immediate reuse.

An Exchange can be abnormally terminated by either the Exchange Originator or the Exchange Responder by (1) using the ABTS-LS Basic Link Service Request to terminate the Sequence and Exchange together, (2) using the ABTX Extended Link Service Request to terminate the Exchange (less commonly), or (3) using the LOGO Extended Link Service Request to Logout from the other N_Port of the Exchange. Abnormally terminating Exchanges is fairly disruptive. A Sequence Recipient can also cause a Sequence to be aborted by withholding an ACK to the last Sequence of the Exchange and relying on the Initiator to abort the Exchange following Sequence timeout. Even further, reception of an F_RJT or P_RJT Frame with a "Non-retryable Error" action code will cause the Sequence to be aborted and probably the Exchange as well. In all of these cases, the Ports may have to wait for an R_A_TOV period and use a Recovery_Qualifier timeout to ensure that all Frames for the Exchange are discarded before reusing any X_ID values or other Exchange resources.

Sequence and Exchange Status Blocks

While a Sequence or Exchange is active, both the transmitting and receiving Ports must perform record-keeping, to keep track of information such as which SEQ_CNT Frames have been sent or received, what Class(es) of service are used, and what OX_ID, RX_ID, or SEQ_ID values are being used.

The actual format and internal representation of the information used to track Sequence and Exchange information is implementation-dependent and need not be made available to other N_Ports. However, there may be cases where one N_Port does need to determine what information another N_Port has about a Sequence or Exchange, for example to determine if there is any discrepancy between the two sets of information, or to confirm completion of Class 3 Sequences.

Therefore, Fibre Channel provides mechanisms for allowing an N_Port to request that another N_Port return specific Sequence or Exchange information. These mechanisms involve Exchange of Extended Link Service requests, and are described in the "Read Exchange Status Block (RES)" section, on page 207, and in the "Read Sequence Status Block (RSS)" section, on page 209. Each N_Port must therefore, at a minimum, keep and maintain the information described in these requests. It must also be able to return the Sequence and Exchange information in the format shown in those sections. These groups of information are termed a "Sequence Status Block" and an "Exchange Status Block."

Sequence Status Block Rules

The Sequence Status Block is used to describe the overall implementation-independent information that must be maintained for each Sequence and returned by the Sequence Recipient as a reply to the "Request Sequence Status Block (RSS)" Extended Link Service Request described in the "Read Sequence Status Block (RSS)" section, on page 209.

In actual practice, an N_Port must keep more information than is described here, such as the timeout times for the Frames transmitted and the address(es) in Node memory where the Sequence data was transmitted from. The actual information recorded, and the format in which it is stored, is implementation-dependent. The Sequence Status Block describes the format in which the N_Port must make the implementation-independent information available to other N_Ports.

The format of a Sequence Status Block, along with the bit definitions for the S_STAT field, is shown in Figure 8.2.

The Sequence Status Block S_STAT bits must be updated as the status of the Sequence changes. If an ACK returns to the Sequence Initiator with the Abort Sequence Condition bits set to a non-zero value, that value is copied into bits 11–10 of S_STAT. If the Sequence becomes closed, bit 14 is set to 0. If the Sequence becomes inactive, bit 13 is set to 0. In either of these conditions, bit 12 indicates whether the Sequence was completed normally or abnormally by the Abort Sequence Protocol.

The Sequence Status Block must be maintained while the Sequence is open and active, as described in the "Sequence Management" section, on page 144. There is not a specification of when the Sequence Status Block resources may be reallocated to another Sequence. In practice, they will gen-

Figure 8.2
Sequence Status Block format.

Sequence Status Block

Word \ Bit	33222222 10987654	22221111 32109876	11111100 54321098	00000000 76543210
0	SEQ_ID	reserved	Lowest SEQ_CNT	
1	Highest SEQ_CNT		S_STAT	
2	Error SEQ_CNT		OX_ID	
3	RX_ID		reserved	

S_STAT Bit Definitions		
15: Sequence Context	0: Sequence Initiator	1: Sequence Recipient
14: Open	0: not open	1: open
13: Active	0: not active	1: active
12: Ending Condition	0: normal Ending Condition	1: abnormal Ending Condition
11-10: ACK, Abort Sequence Condition *	00: continue	01: Abort Sequence requested
	10: Stop Seq. requested	11: Abort w/Retrans. requested
9: ABTS protocol *	0: ABTS not completed	1: ABTS completed by Recipient
8: Retransmission performed *	0: Retransmission not completed	1: Retransmission completed by Recipient
7: Sequence timeout *	0: Sequence not timed out	1: Seq. timed out by Recipient
6: P_RJT transmitted	0: P_RJT not transmitted	1: P_RJT transmitted
5-4: Class	00: reserved	01: Class 1
	10: Class 2	11: Class 3
3: ACK **EOFt** transmitted	0: ACK **EOFt** not transmitted	1: ACK (**EOFt** or **EOFdt**) not transmitted
2-0: reserved		

Note: Bits marked by * clarify the reason for an abnormal ending condition, and are only valid when Bit 12 = 1.

erally be maintained until either the SEQ_ID value is reused or the Exchange is no longer open.

Exchange Status Block Rules

The Exchange Status Block is used to describe the overall implementation-independent information that must be maintained for each Exchange and returned as a reply to the "Request Exchange Status Block (RES)" Extended Link Service request, described in the "Read Exchange Status Block (RES)" section, on page 207.

As with the Sequence Status Block, in actual practice an N_Port must keep more information than is described here for each Exchange, and the format in which it is saved is implementation-dependent. The Exchange Status Block describes the format in which the N_Port must make the information available to other N_Ports.

The format of an Exchange Status Block, including the E_STAT field bit definitions, is shown in Figure 8.3.

The Exchange is active for the Exchange Originator and Responder after the first Frame of the Exchange is either transmitted or received, respectively. The Exchange remains open until (1) the last Sequence of the Exchange completes normally, (2) a timeout period of E_D_TOV has elapsed since the last Sequence of the Exchange was completed abnormally, or (3) the ABTX Extended Link Service request is completed normally. Once the Exchange is completed, the Exchange Status Block resources may be allocated to another Exchange.

Figure 8.3 Exchange Status Block format.

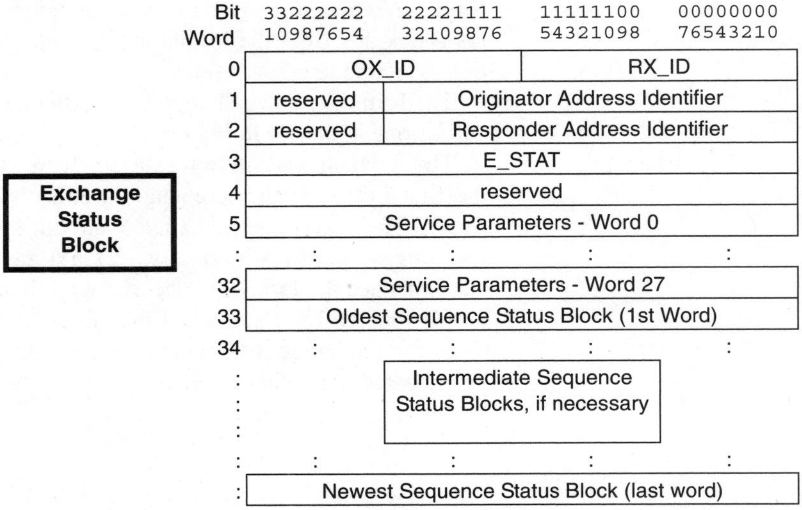

Word	Bits 31–24 (33222222 / 10987654)	Bits 23–16 (22221111 / 32109876)	Bits 15–8 (11111100 / 54321098)	Bits 7–0 (00000000 / 76543210)
0	OX_ID		RX_ID	
1	reserved	Originator Address Identifier		
2	reserved	Responder Address Identifier		
3	E_STAT			
4	reserved			
5	Service Parameters - Word 0			
:	:	:	:	:
32	Service Parameters - Word 27			
33	Oldest Sequence Status Block (1st Word)			
34	:	:	:	:
:	Intermediate Sequence Status Blocks, if necessary			
:	:	:	:	:
:	Newest Sequence Status Block (last word)			

Exchange Status Block

E_STAT Bit Definitions		
31: ESB Owner	0: Exchange Originator	1: Exchange Responder
30: Sequence Initiative holder	0: Other Port holds Initiative	1: This Port holds Initiative
29: Completion	0: open	1: complete
28: Ending Condition	0: normal	1: abnormal
27: Error type (valid only when bit 28=1)	0: Exchange aborted with ABTX	1: Exchange abnormally terminated
26: Recovery Qualifier	0: None	1: Active
25-24: Exchange Error Policy	00: Abort, discard multiple Sequences	01: Abort, discard a single Sequence
	10: Process with infinite buffering	11: Discard multiple Seqs. with immediate retransmission.
23: Originator X_ID invalid for the newest SSSB in the ESB	0: OX_ID valid	1: OX_ID invalid
22: Responder X_ID invalid for the newest SSB in the ESB	0: RX_ID valid	1: RX_ID invalid
21-0: reserved		

Chapter 9

Login and Logout Services

Introduction

Before meaningful data can be sent, an N_Port must determine a number of things about its operating environment. This includes such factors as whether the interconnect topology is Point-to-point, Fabric, or Arbitrated Loop; what other N_Ports are in the environment; what Classes of service the Fabric and N_Ports support; what Service Parameters are applicable to the various Ports and the various Classes of service; and what error recovery procedures will be used.

In order to determine these operating parameters, and to determine the characteristics and operating parameters of the network and the other Ports attached to it, an N_Port must perform Login Procedures. Prior to Login or following Logout, a default set of Service Parameters apply, describing the minimal service performance that an N_Port must provide. These default Service Parameters are shown in Figure 9.1. These minimal Service Parameters can be increased to higher-performance values on a F_Port and N_Port basis using the Login procedures between communicating Ports. The Login procedure is the method by which an N_Port determines these environmental characteristics.

Figure 9.1
Pre-Login Default Service Parameters.

Parameter	Value	Parameter	Value
Concurrent Sequences per Class	1	X_ID reassignment	prohibited
Total Concurrent Sequences	1	X_ID interlock	required
End-to-end Credit	1	Exchange Error Policy	Abort, discard multiple
Buffer-to-buffer Credit	1	Relative Offset	unused
Receive Data Field size, in bytes	128	Other optional features	unused
ACK Form	ACK_1		

Login and Logout Overview

Login is divided into two parts: "Fabric Login," which determines the network operating parameters, and "N_Port Login," which determines N_Port-to-N_Port operating parameters. For an N_Port's full Login, the Fabric Login is performed either once or twice, depending on whether the attached Fabric can assign Port identifiers; then an N_Port Login is performed with every N_Port in the environment with which the N_Port needs to communicate. Fabric and N_Port Logins are both accomplished using a similar procedure and are distinguished through the use of different destination identifiers (D_ID) and possibly different source identifiers (S_ID). The goal of both Fabric and N_Port Login procedures is to establish the Service Parameters that apply to further data communications.

Both Fabric and N_Port Login are carried out using Extended Link Service commands, which are described in detail in Chapter 11. The requesting N_Port transmits an Extended Link Service Request (either PLOGI or FLOGI), which is a Sequence in its own Exchange, with a particular header and Payload format identifying the service being requested. Under normal processing, the Recipient will ACK the Sequence and return another Sequence in the same Exchange, termed an "Extended Link Service Reply," which either accepts (ACC) the service request and returns acceptance information or rejects (LS_RJT) the request, with a Payload indicating the reason for the rejection. Fabric Login uses the "Fabric Login" (FLOGI) Extended Link Service command (see the "F_Port Login (FLOGI)" section, on page 206), N_Port Login uses the "N_Port Login" (PLOGI) Extended Link Service command (see the "N_Port Login (PLOGI)" section, on page 204) for the Requests, and both use the "Accept" (ACC) Extended Link Service Command for the reply.

Figure 9.2 shows the flow of Frames between the Initiator, Fabric, and Recipient for Fabric Login and N_Port Login, with and without a Fabric present. All traffic is shown using Class 2 service with one Frame per Sequence, although it would be possible to use other classes of service. The actual order of Classes tried for N_Port Login is: unbuffered Class 1, buffered Class 1, Class 2, Class 3, and finally, Class 4. Only the Classes supported by the Fabric (as determined during Fabric Login) are tried

Unless some type of implicit mechanism is used, an N_Port must in general perform both Fabric Login and N_Port Login before transmitting any meaningful data. In addition to these two types of Login, there is a third type of Login, termed "Process Login." Process Login lets multiple entities, termed "images" or "processes," share a single N_Port and establish communication pairings with N_Ports or with images behind other N_Ports. Process Login may be used, for example, to set up the operating environment between two SCSI-FCP processes or to allow multiple instances of a mainframe operating system to separate their communications over a shared N_Port.

Fabric Login and N_Port Login are described in the "Fabric Login" section, on page 161, and the "N_Port Login" section, on page 163, respectively, and Process Login and Logout are described in the "Overview of Process Login/Logout" section, on page 174.

Login procedures may not be necessary in some Fibre Channel installations. If, for example, a Point-to-point topology is built with N_Ports having known operating parameters, then the parameters are known implicitly, and the information that would normally be obtained through Login procedures

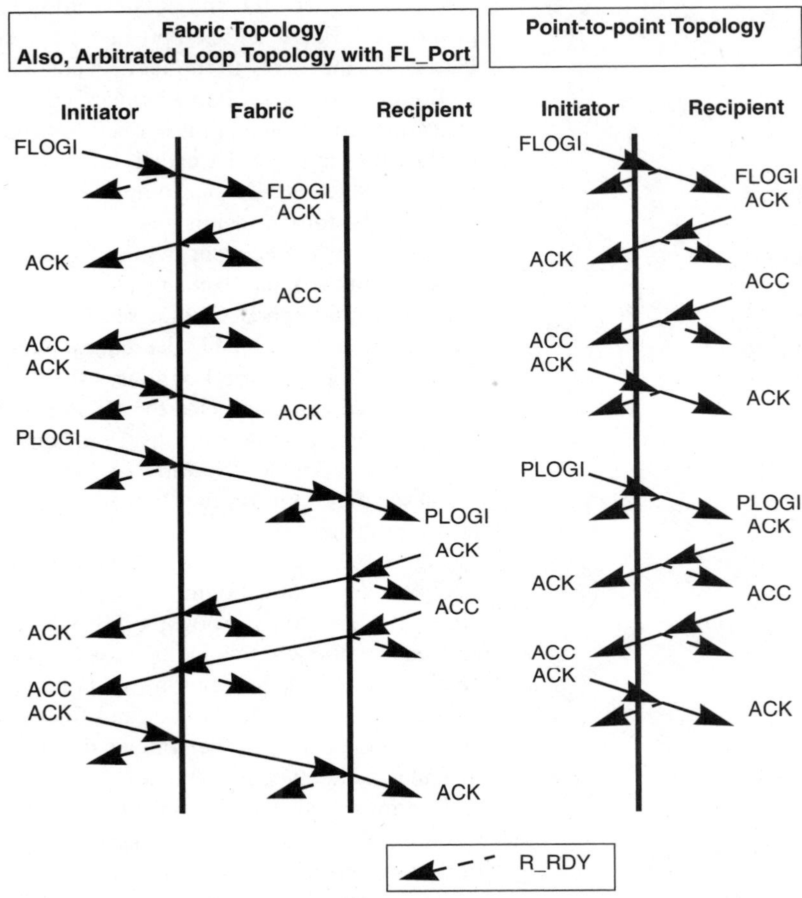

Figure 9.2
Frame flow during Fabric Login and N_Port Login.

can be initialized by the system's administrator. This implicit Login eliminates the need for the explicit Login procedures described here. Implicit Login procedures can apply to Fabric, N_Port, or Process Login.

Login between an N_Port and the Fabric or between two N_Ports is long-lived. The number of concurrent N_Ports with which an N_Port may be Logged in is a vendor-specific function of the N_Port facilities available.

Login with the Fabric is required for all N_Ports, regardless of the Class of service supported. Communication with other N_Ports cannot begin until the Fabric Login procedure is complete, either implicitly or explicitly. For an N_Port which supports Class 1 or Class 2 service, an N_Port is required to perform N_Port Login with each N_Port with which it intends to communicate. A Port may provide different Service Parameters to different N_Ports with which it is communicating, but generally the parameters will not be contradictory or conflicting between different Login partners.

The formats of the PLOGI, FLOGI, and ACC Frames are similar, as shown in Figure 9.3, although the fields within the Frames are filled in slightly differently in the three cases.

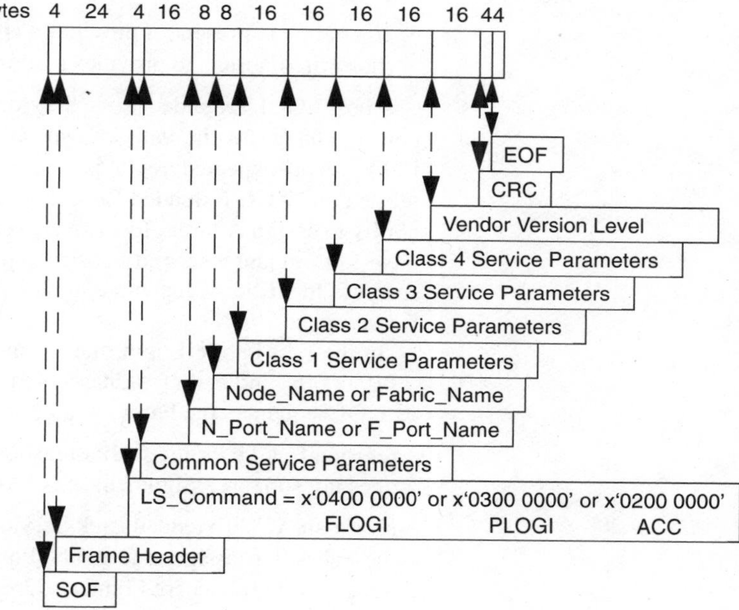

Figure 9.3
Format of the N_Port Login and Fabric Login requests and their Accept replies.

In addition to Fabric and N_Port Login, a pair of N_Ports may optionally follow a procedure for estimating an optimal end-to-end Credit value. If a source/destination pair uses too low a value of end-to-end Credit, the source may wait too long for ACKs to return before transmitting more Frames, wasting transmission bandwidth. To prevent this, a procedure, described in the "Procedure to Estimate End-to-End Credit" section, on page 176, can be used by a source N_Port to estimate the minimum end-to-end Credit needed to allow streaming transmission without transmission delay waiting for ACKs to be returned. This end-to-end Credit estimation procedure may be performed at any time, to ensure maximal bandwidth utilization.

Fabric Login

Fabric Login accomplishes three functions:

1. The Accept (ACC) to the FLOGI Extended Link Service Request contains

an indication of whether the ACC was generated by a Fabric or by an N_Port, showing whether the requesting N_Port is attached in a Fabric topology, or to a Point-to-point or an Arbitrated Loop topology.

2. If a Fabric is present, it will either assign an N_Port Identifier, if the Fabric supports the facility, or at least will confirm that an N_Port identifier chosen by the requesting N_Port is valid and unique.
3. If a Fabric is present, it provides a specific set of operating parameters for the entire Fabric and provides a buffer-to-buffer Credit value for the link.

The FLOGI Extended Link Service Request is sent as a Sequence in a new Exchange to the well-known Fabric F_Port address at D_ID = x'FF FFFE.' The expected reply is another Sequence in the same Exchange, termed an "ACC Extended Link Service Reply." Depending on the Fabric's ability to assign Address Identifiers, as indicated in the ACC, an N_Port may have to originate a second Fabric Login Exchange as confirmation of a test Address Identifier value and complete Fabric Login with an Address Identifier assigned.

In the first Fabric Login, the Sequence is set with the S_ID value set to x'00 0000,' indicating "Unassigned." Following reception of this first FLOGI Sequence, the Fabric will either

- return a F_RJT Frame with a reason code of "Invalid S_ID," if the Fabric does not support assignment of N_Port identifiers, or
- return an ACC Extended Link Service Reply addressed to the D_ID which the Fabric has assigned to the N_Port. Since the N_Port's D_ID is not yet assigned, it will accept Frames addressed to any D_ID value.

If the N_Port is attached in a Point-to-point topology with an N_Port, the N_Port will receive an ACC from the other N_Port, indicating by S_ID value and in the Payload that it is from an N_Port rather than an F_Port. Also, if the N_Port receives a FLOGI Sequence over the link, it knows that the attached Port is an N_Port, which is attempting to do its own Fabric Login. The N_Port will compare its own Port_Name with the Port_Name in the ACC. The N_Port with the higher Port_Name will transmit the PLOGI for N_Port Login, and the lower one will wait.

If the N_Port is attached to a Fabric which does not support assignment of N_Port identifiers, the N_Port will transmit a second FLOGI Extended Link Service Request. In this Request, the S_ID field is set to a value of the N_Port's choosing. This is a test S_ID, which the Fabric must confirm for validity and uniqueness. In response to this FLOGI Frame, the Fabric will either

- return a F_RJT Frame with a reason code of "Invalid S_ID," if the S_ID chosen is invalid, or
- return an ACC ELS Reply, confirming the N_Port's S_ID selection.

In the first of the above cases, where the Fabric has rejected the test S_ID value, the N_Port should either attempt another FLOGI with a different S_ID value or determine a valid and unique S_ID value by some other unspecified method.

At the end of this procedure, the N_Port has determined whether it is attached to an F_Port or is attached point-to-point to an N_Port. It has also determined a unique and valid N_Port identifier to use for further communications.

The Fabric operating parameters, if a Fabric is present, are included in the ACC Payload, with format as shown in Figure 9.3 and described in the "N_Port and F_Port Service Parameters" section, on page 166. If no Fabric is present, the N_Port ignores the parameter values in the ACC received and goes on to perform N_Port Login.

Not discussed above was the Class of service used for the Fabric Login. Fabric Login can be carried out in any Class of service, but since the Classes of service are optional, the Fabric may not support all Classes for Login. The N_Port, therefore, will attempt Fabric Login in Class 1, using a connect-request (**SOFc1**) delimiter for the first Frame of the Sequence and ensuring removal of the Connection when Fabric Login is over. If the Fabric cannot support Class 1, it will return a F_RJT with a reason code of "Class not supported," and the N_Port will attempt Fabric Login in Class 2 and then in Class 3 if necessary.

Class 1 communication normally requires that both N_Ports operate at the same speed, so N_Port Login should be performed in Class 1 if that Class is supported. Class 2 or 3 communication does not require that both N_Ports operate at the same speed. Link speed is not a Login parameter. A special type of Class 1, termed "Buffered Class 1," allows Class 1 operation at the lowest common port speed. This is described further in the "Buffered Class 1 Service" section, on page 187.

N_Port Login

Once the N_Port has finished Fabric Login, it must perform N_Port Login with each N_Port it intends to communicate with, unless the functions of N_Port Login have been carried out through implicit means (e.g., system administrator assignment). N_Port Login between N_Ports

- provides a specific set of operating parameters for communicating between N_Ports,
- initializes the destination end-to-end Credit, and
- in a Point-to-point topology, initializes the buffer-to-buffer Credit.

The operation of N_Port Login is similar to the operation of Fabric Login. The N_Port transmits a PLOGI Extended Link Service Request addressed to the D_ID of an N_Port with which it needs to communicate, and the addressed N_Port returns an ACC Reply. The contents of the PLOGI and ACC Sequences contain operating parameters for communication between the N_Ports. The Request/Reply operation lets the two N_Ports negotiate on operating parameters, such that they agree to communicate using the most functional parameters which are supported by both N_Ports. A single PLOGI Request/ACC Reply Exchange establishes communication between the two N_Ports. A second N_Port Login Exchange with the Originator and Responder interchanged is not required unless the other N_Port wished to alter the existing Login parameters.

N_Port Login is complete when each N_Port has received the Service Parameters for every other N_Port on the network with which it needs to communicate. The method for determining the IDs of all other Ports on the Fabric is unspecified. If a configuration change in the topology occurs, such as an N_Port being removed or added, N_Ports may have to reinitiate N_Port Login to prevent collisions in address identifiers or Port_Names.

Normally, an N_Port that is logged in will perform a Logout before doing a new N_Port Login. However, if an N_Port receives or transmits a PLOGI Sequence during normal operations, it will behave as if it had performed a Logout with the remote N_Port of the PLOGI.

The procedure for performing N_Port Login varies slightly, depending on whether the N_Port is connected directly to another N_Port in a Point-to-point topology or attached to a Fabric. If the N_Port is attached directly to another N_Port, which it determined during Fabric Login, then there are two differences from the Fabric-attached situation. First, each N_Port knows the worldwide Port_Name of the other N_Port. Second, neither N_Port has been assigned an address identifier by the Fabric. It is the responsibility of the N_Port with the higher priority to initiate PLOGI and to assign address identifiers for both N_Ports.

In a Point-to-point topology, therefore, the N_Port with the lower Port_Name waits for a PLOGI from the attached N_Port, and the N_Port with the high Port_Name sends a PLOGI with S_ID and D_ID values set to non-zero values to be used for further communications. The N_Port with the lower Port_Name will accept the D_ID value in the PLOGI Frame and use it as its own.

In a Fabric topology, the N_Ports have unique and valid N_Port address identifiers before N_Port Login, which are known to the Fabric. The N_Ports can then determine from the Fabric which other N_Port address identifiers are available, and can then begin N_Port Login to the necessary destination N_Ports. Since the N_Ports do not know each other's Port_Names at this point, there is no relative priority, and an N_Port which sends a PLOGI to another N_Port may receive a previously transmitted

PLOGI from that N_Port before receiving the ACC in return. If this happens, the N_Port responds as though its PLOGI had never been transmitted — no special processing is required, since the Login carried out in each direction will yield identical agreed-upon Service Parameters.

Logout

The destination Logout procedures provide a method for canceling or removing service between two N_Ports. This allows any resources devoted to maintaining service, such as memory space reserved for storing Service Parameters, to be released for other uses. There are both explicit and implicit methods for performing Logout with another N_Port. There are no methods for performing Logout with a Fabric, since the Fabric has no major resource dedicated to an N_Port which could usefully be released, except possibly for a small routing table entry.

Explicit Logout is accomplished using a Logout (LOGO) Extended Link Service Request, with an accompanying ACC Extended Link Service Reply. Before an N_Port initiates Logout, it should terminate any active Sequences. Otherwise, the Sequences would be in an indeterminate state, and further operations would be unpredictable. An N_Port that receives a LOGO Frame while another Sequence is active may reject the Logout Request by transmitting an LS_RJT Sequence with the "Unable to perform command requested" reason code. After Logout, the Login Service parameters revert to the default values shown in Figure 9.1, unless some type of implicit Login Parameters apply. Figure 9.4 shows the format of the LOGO and corresponding ACC Extended Link Service Commands. The N_Port follows the format shown in Figure 9.4.

Implicit Logout occurs when an N_Port receives or transmits the NOS (Not Operational) or OLS (Offline) Primitive Sequence in the normal three consecutive transmission words. This implicit Logout affects the Fabric Login parameters, and an N_Port must perform another Fabric Login before transmitting or receiving any other communications. If there are any configuration changes, the N_Port must explicitly log out and perform N_Port Login with each of the N_Ports with which it intends to communicate before accepting or transmitting any further communications. Implicit Logout with the NOS or OLS Primitive Sequences is a very disruptive procedure, since it may imply significant changes in the Fabric on re-Login, so the N_Port may have to assume that a complete Fabric and N_Port Login is required, including waiting for the R_A_TOV timeout period before continuing to ensure that there are no operations pending which might disrupt the new Login.

Figure 9.4 Payload format of Logout (LOGO) Command and corresponding Accept (ACC).

Logout (LOGO)

Word \ Bit	33222222 10987654	22221111 32109876	11111100 54321098	00000000 76543210
0 (LS_Command code)	0000 0101	0000 0000	0000 0000	0000 0000
1	0000 0000	N_Port Identifier 0BBB BBBB BBBB BBBB BBBB BBBB		
2	N_Port_Name - Most Significant Word nnnn nnnn nnnn nnnn nnnn nnnn nnnn nnnn			
3	N_Port_Name - Least Significant Word nnnn nnnn nnnn nnnn nnnn nnnn nnnn nnnn			

Accept (ACC) for LOGO

Word \ Bit	33222222 10987654	22221111 32109876	11111100 54321098	00000000 76543210
0	0000 0010	0000 0000	0000 0000	0000 0000

N_Port and F_Port Service Parameters

The formats of the FLOGI, PLOGI, and corresponding ACC Extended Link Service commands are shown in Figure 9.3. The formats and usages of the individual fields within these commands are detailed in the following sections. The Common Service Parameters, described in the "Common Service Parameters" section, on page 166, describe Service Parameters that are independent of Class of service. The (F/N)_Port_Name or Port_Name and Node_Name or Fabric_Name fields provide 64-bit names that are guaranteed to be unique worldwide for every piece of Fibre Channel hardware. The Class 1, 2, and 3 Service Parameters, described in the "N_Port and F_Port Class Service Parameters" section, on page 171, describe service parameters that depend on the Class of service used for communications.

Common Service Parameters

The Common Service Parameters field describes parameters which are common across all Classes supported by the F_Port or N_Port. The format of the Common Service Parameter fields is shown for both FLOGI and PLOGI Frames and their corresponding ACC ELS Replies in Figure 9.5. The definitions of the bit fields for the Common Service Parameter fields are shown in Figure 9.6.

Figure 9.5
LS_Command codes and Common Service Parameters for FLOGI and PLOGI Commands.

FLOGI Common Service Parameters

Word	Bit 33222222 10987654	22221111 32109876	11111100 54321098	00000000 76543210
0 (LS_Command code)	0000 0100	0000 0000	0000 0000	0000 0000
1	FC-PH Version HHHH HHHH LLLL LLLL		Buffer-to-buffer Credit 0BBB BBBB BBBB BBBB	
2	Common Features rrrN rrMB rDrr rrrr		B-to-B recv data field size 0000 FFFF FFFF FFFF	
3	R_A_TOV (only meaningful in ACC from Fabric) tttt tttt tttt tttt tttttttt tttttttt			
4	E_D_TOV (only meaningful in ACC from Fabric) tttt tttt tttt tttt tttttttt tttttttt			

PLOGI Common Service Parameters

Word	Bit 33222222 10987654	22221111 32109876	11111100 54321098	00000000 76543210
0 (LS_Command code)	0000 0011	0000 0000	0000 0000	0000 0000
1	FC-PH Version COVN Arrr rDrr rrrr		Buffer-to-buffer Credit (Point-to-point) 0BBB BBBB BBBB BBBB	
2	Total concurrent Seqs. 0000 TTTT TTTT TTTT		Relative offset by Info Cat. FFFF FFFF FFFF FFFF	
3	E_D_TOV (only meaningful in point-to-point topology) tttt tttt tttt tttt tttttttt tttttttt			

The usages of the various fields outlined in the figure are as follows:

LS_Command code: contains a 1-byte code indicating the ELS command plus 3 reserved bytes.

FC-PH version: contains a code indicating which version of FC-PH the Initiator supports. There are only very minor differences between the recent versions.

Buffer-to-buffer Credit: indicates the number of connectionless buffers available at the transmitting Port. This field may be different in the Request (PLOGI or FLOGI) and the Reply (ACC), since different Ports may have different numbers of buffers.

Common features: defines support for some particular features. Bit 31 indicates support for Continuously Increasing Offset, both as Sequence Initiator and Sequence Recipient, in the Information Categories with support indicated in the "Relative Offset by Information Category" field. Bit 30 indicates support for Random Relative Offset. Bit 29 indicates whether

Figure 9.6
Bit field definitions for FLOGI and PLOGI Common Service Parameters

FC-PH Version	
x'06': FC-PH 4.0	x'10': FC-PH-2
x'07': FC-PH 4.1	x'20': FC-PH-3
x'08': FC-PH 4.2	
x'09': FC-PH 4.3	

Buffer-to-buffer Credit
Number of buffers for Class1 connect-request, Class 2, and Class 3 Frames

B-to-B receive data field size
Maximum size for Class 1 connect-request, Class 2, and Class 3 Frames. Valid Range: 128-2112

Common Features			
Bit Definition	Meaning		Validity
	0	1	
31: C — Continuously Increasing Offset	Not supported	Supported	PLOGI & ACC
30: O — Random Relative Offset	Not supported	Supported	PLOGI & ACC
29: V — Valid Vendor Version Level	Not supported	Supported	all
28: N — N_Port/F_Port	N_Port	F_Port	all
27: A — Alternate BB_Credit Management	FC-PH method	Alternate	all
26: reserved			
25: M — Multicast	Not supported	Supported	Fabric's FLOGI ACC
24: B — Broadcast	Not supported	Supported	Fabric's FLOGI ACC
23: reserved			
22: D — Dedicated Simplex	Not supported	Supported	PLOGI & ACC
22 - 16: reserved			

PLOGI Common Service Parameters	
Total Concurrent Sequences	Maximum total concurrent Sequence for all Classes of service
Relative Offset by Information Category	Bit position map of Relative offset support onto 16 Information Categories
R_A_TOV	R_A_TOV value in milliseconds x'0001 D4C0' (120 sec.) is the default
E_D_TOV	E_D_TOV value in milliseconds x'0000 2710' (10 sec.) is the default

the Vendor Version Level field at the end of the Sequence Payload contains valid information. Bit 28 indicates whether the Sequence transmitter is an N_Port or an F_Port. Bit 27 indicates usage of the Alternate Buffer-to-Buffer Credit Management method described in the "Alternate Buffer-to-Buffer Flow Control" section, on page 271. Bit 22 indicates whether Dedicated Simplex, as described in the "Dedicated Simplex" section, on page 187, is supported by the N_Port.

Buffer-to-buffer receive data field size: specifies the largest Data_Field size that can be supported with buffering of Class 1 connect-request, and Class 2 and Class 3 Frames. The size cannot be smaller than 128 bytes, although a value of at least 256 bytes is "recommended."

Total concurrent Sequences: specifies the total number of Sequences that can be supported in all Classes.

Relative Offset by Information Category: is a 16-bit field specifying whether either Continuously Increasing or Random Relative Offset fields in Frame Headers are supported by the N_Port, for both Sequence Initiation and Reception. Each bit indicates Relative Offset support for one of the 16 Information Categories, with one bit mapped to each Category.

R_A_TOV: specifies the Resource_Allocation_Timeout Value to use for timing out Frames in a Fabric topology, as described in the "R_A_TOV" section, on page 220.

E_D_TOV: specifies the Error_Detect_Timeout value to use between the two N_Ports in a point-to-point topology or over the Fabric in a Fabric topology, as described in the "E_D_TOV" section, on page 219.

The timeout values in the ACC for the PLOGI will be greater than or equal to the value in the PLOGI, since the two N_Ports must agree on the longer of the two supported timeout values. The E_D_TOV value in the ACC will be the value used by each N_Port. The R_A_TOV value is set to twice the E_D_TOV in a Point-to-point topology.

N_Port_Name or F_Port_Name

The N_Port_Name or F_Port_Name field is an 8-byte or 16-byte field which uniquely identifies an N_Port or F_Port for purposes, such as diagnostics, which may be independent and unrelated to Fibre Channel network addressing. Each (N/F)_Port must have an (N/F)_Port_Name which is unique within the address domain of the Fabric. In general, these names will actually be unique worldwide. Bits 63 to 60 specify the format of the name; the rest of the bits specify the name. The possible formats of the name are shown in Figure 9.7. The usage of these names is further detailed in the "Network_Header" section, on page 244.

Figure 9.7
Formats of worldwide Port_Names and Node_Names.

NAA ID Bits 63–60	Network Address Authority	Network	Name Scope	Name_Identifier Format (bits 59–0)	
				Bits 59–48	Bits 47–0
x'0'	ignored	–	–	–	
x'1'	IEEE Std. 802.1A ULA	Hetero-geneous	WWN	0 (12 bits)	IEEE address
x'2'	IEEE Extended	FC Networks	FCN	N_Port id in Node	Node IEEE address
				F_Port in Switch	Switch IEEE address
x'3'	Locally assigned	FC Networks	FCN	Fabric unique	
x'4'	IP	Hetero-geneous	WWN	zero(59–32), IP address (31–0)	
x'5'	IEEE Registered		WWN	24-bit IEEE company_id, 36-bit vendor-specified identifiers (VSID)	
x'6'	IEEE registered Extended		WWN	24-bit IEEE company_id, 36-bit vendor-specified identifiers (VSID), plus an extra 64-bit vendor specified identifier extension	
x'7'–x'B'	reserved	–	–	–	
x'C'	Reclaimed from CCITT				
x'D'	reserved	–	–	–	
x'E'	Reclaimed from CCITT				
x'F'	reserved	–	–	–	

All Name_Identifiers apply to N_Ports, Nodes, F_Ports, and switchs, except:
IEEE Extended names don't apply to Nodes or Fabrics (redundant with IEEE Name_Identifiers).
For Name Scope: FCN = Fibre Channel Network, WWN = WorldWide Name

Node_Name or Fabric_Name

The Node_Name or Fabric_Name field specifies an 8-byte field which uniquely identifies the Node or Fabric, similar to the Port_Name fields. The possible Node_Name formats are the same as the (N/F)_Port_Name formats shown in Figure 9.7.

N_Port and F_Port Class Service Parameters

The Class Service Parameter fields describe parameters relating to the Classes of service supported by the N_Port or F_Port participating in the Login. As shown in Figure 9.8, each of the three Classes has four words in the FLOGI, PLOGI, or ACC command, specifying parameters for the Class of service. The bit field definitions for the Class Service Parameter fields are shown in Figure 9.9.

Figure 9.8
F_Port and N_Port Class Service Parameters.

F_Port Class Service Parameters

FLOGI, ACC Words
Class 1: 9–12
Class 2:13–16
Class 3: 17–20

Bit / Word	33222222 10987654	22221111 32109876	11111100 54321098	00000000 76543210
9,13,17	Service Options VISS QDCB EEEE EEEE		Initiator Control FLOGI: DDDD DDDD DDDD DDDD ACC: not meaningful	
10,14,18	Recipient Control FLOGI: CCCC CCCC CCCC CCCC ACC: Reserved for Fabric use		Receive Data Field Size FLOGI: rrrr NNNN NNNN NNNN ACC: not meaningful	
11,15,19	Concurrent Sequences: FLOGI: rrrrrrrr LLLLLLLL ACC: not meaningful		N_Port End-to-end Credit 0MMM MMMM MMMM MMMM ACC: not meaningful	
12,16,20	FLOGI: Open Sequences per Exchange: rrrr rrrr xxxx xxxx ACC: CR_TOV tttt tttt tttt tttt		FLOGI: Reserved rrrr rrrr rrrr rrrr tttt tttt tttt tttt	

N_Port Class Service Parameters

FLOGI, PLOGI, ACC Words
Class 1: 9–12
Class 2:13–16
Class 3: 17–20

Bit / Word	33222222 10987654	22221111 32109876	11111100 54321098	00000000 76543210
9,13,17	Service Options VISS QDCB 0000 0000		Initiator Control XXPP ZNGC 0000 0000	
10,14,18	Recipient Control ZNXL LrSS CCCrr rrrr		Receive data field size 0000 NNNN NNNN NNNN	
11,15,19	Concurrent Sequences rrrr rrrr LLLL LLLL		N_Port End-to-end Credit 0NNN NNNN NNNN NNNN	
12,16,20	Open Sequences per Exch. rrrr rrrr xxxx xxxx		Class 6 Multicast RX_ID xxxx xxxx xxxx xxxx	

Figure 9.9
Bit field definitions for Class Service Parameters.

Service Options				
Bit	**Definition**	**Class**	**Usage**	
31: V	Class Validity	all	0: invalid	1: valid
30: I	Intermix Mode	1	0: not supported	1: supported
29–28: SS	Stacked Connect-requests	1F	00: no request 10: request transparent	01: request lock-down 11: request both
27: D	Sequential Delivery	2F,3F	0: out-of-order	1: sequential delivery requested

Initiator Control				
Bit	**Definition**	**Class**	**Usage**	
15–14: Xr	X_ID Reassignment	1P,2P	00: not supported 10: reserved	01: supported 11: required
13–12: Pa	Initial Process Associator	1P, 2P, 3P	00: not supported 10: reserved	01: supported 11: required
11: K	ACK_0 as Initiator	1P, 2P	0: ACK_0 incapable	1: ACK_0 capable
10: N	ACK_N as Initiator	1P, 2P	0: ACK_N incapable	1: ACK_N capable

Recipient Control				
Bit	**Definition**	**Class**	**Usage**	
31: K	ACK_0 as Recipient	1P, 2P	0: ACK_0 incapable	1: ACK_N capable
30: N	ACK_N as Recipient	1P, 2P	0: ACK_N incapable	1: ACK_N capable
29: I	X_ID Interlock	1P, 2P	0: not required	1: required
28–27: Ep	Error policy supported- discard or process	1P, 2P, 3P	00: only discard 10: both supported	01: reserved 11: reserved
25–24: Cs	Categories per Sequences	1P, 2P, 3P	00: 1 Cat. / Seq. 10: reserved	01: 2 Cat. / Seq. 11: more than 2
19–16: r	reserved for Fabric-specific uses	1F, 2F, 3F	Fabric-specific	

Notes:
(1P, 2P, 3P) = bit validity in N_Port Class (1, 2, 3) Service Parameters. (1F, 2F, 3F) = bit validity in Fabric Class (1, 2, 3) Service Parameters. 1 = bit validity for both Fabric and N_Port Class 1 Service Parameters.

The usages of the various fields outlined in the figure are as follows:

Service Options: specify general support for Class options. Bit 31 of the first word in each set of Class Service parameters specifies whether the Port sending the Sequence supports the Class of service. If this bit is set to

0, the entire set of 16 bytes for the Class service parameters should be ignored. Bit 30 indicates whether the Intermix function for Class 1 service is supported or Requested, and is only valid in the Class 1 Service Parameters for both PLOGI (and ACC) and FLOGI (and ACC). Intermix must be supported by both the N_Port and the Fabric (if present) or the other N_Port for it to be functional. Bit 29 indicates whether the N_Port requests and the Fabric grants support for stacked connect-requests, as described in the "Stacked Connect-Requests" section, on page 257. Bit 28 indicates whether the N_Port requests and the Fabric grants support for Sequential delivery of Class 2 and Class 3 Frames.

Initiator Control: specifies which protocols, policies, or functions the Port supports when acting as Sequence Initiator. The definitions of the bits in the PLOGI and ACC ELS Commands are shown in Figure 9.8.

Recipient Control: specifies which protocols, policies, or functions the Port supports when acting as Sequence Recipient. The definitions of the bits in the PLOGI and ACC ELS commands are shown in Figure 9.8.

Receive Data Field Size: contains a binary value which specifies the largest data field size for a Frame that can be received by the N_Port transmitting the service parameters. Valid values are between 128 and 2,112, with at least 256 being recommended. This field only applies to Class 1 Frames — connect-request, Class 2, and Class 3 Frame size limits are specified in the Common Service Parameters field.

Concurrent Sequences: specifies the number of concurrent Sequences which can be open, specified separately for each Class.

N_Port End-to-end Credit: indicates the maximum possible number of outstanding Class 1 and Class 2 Data Frames. Bit 15 is reserved, placing a 15-bit limit on the maximum allowable value. This prevents a difficulty from SEQ_CNT wrapping, since it allows an unambiguous determination of whether a received 16-bit SEQ_CNT value is higher or lower than an expected SEQ_CNT value.

Open Sequences per Exchange: specifies the number of concurrently open Sequences allowed per Exchange. Each N_Port must be able to track two more Sequence Status Blocks than the number specified here, to be sure of accurately tracking status for all Sequences.

Vendor Version Level

Vendor Version Level is a 16-byte field which specifies version levels for an N_Port which deviates in a vendor-specific manner from the FC-PH version levels specified in the Common Service Parameters. This field only contains valid information if word 1, bit 29 in the Common Service Parameters is 1.

The first 2 bytes specify a code assigned to a specific vendor. The next 6 bytes specify one or more vendor-specific versions supported by the N_Port supplying these Login parameters. The levels may be encoded or bit-assigned. The last 8 bytes are reserved for vendor-specific use.

Overview of Process Login/Logout

There are some cases where the addresses and data grouping functions provided in Fibre Channel are not flexible enough for some required operations. As an example, if a Fibre Channel N_Port is implementing the SCSI protocol for a system with multiple SCSI Targets, as a way of multiplexing function over the Fibre Channel hardware, there must be some way for the incoming Information Units to be directed to one of the SCSI Targets behind the N_Port. Another layer of addressing is required, beyond the addressing to the N_Port address identifier.

Similarly, the data grouping provided by Fibre Channel may not be flexible enough for some operations. On mainframe-type systems, where a channel operation may initiate operations covering multiple Exchanges, there must be a way of grouping the Exchanges together through some operation identifier with a more general scope than the Exchange identifier.

Both of these extra levels of addressing, or grouping, are carried in the optional Association Header, which has the format shown in the "Association_Header" section, on page 244. If included in a Frame, the Association Header has fields which may contain Originator and/or Responder Process Associators, which address multiple "images" behind a single N_Port, and Originator and Responder Operation Associators, which can provide association of multiple Exchanges into "Operations."

The Operation Association function is left for later discussion in the "Association_Header" section, on page 244, but the Process Associator function is related to determining the operating environment and the independent entities which may transmit or receive data, so it is included here.

Figure 9.10 shows the organization of "images" behind N_Ports. Each image is described as a group of one or more related processes, which operate as independent senders or receivers of data behind an N_Port, and can be addressed using the Association Header. Each image can send data to or receive data from another N_Port, or an image behind another N_Port, once they have interchanged Service Parameters. The Service Parameters used by one image may be different from those used by a different image at the same N_Port.

The "Process Login/Logout" procedures are defined as procedures for an image to exchange Service Parameters with another image or with an N_Port. Process Login therefore operates as a third type of Login which is

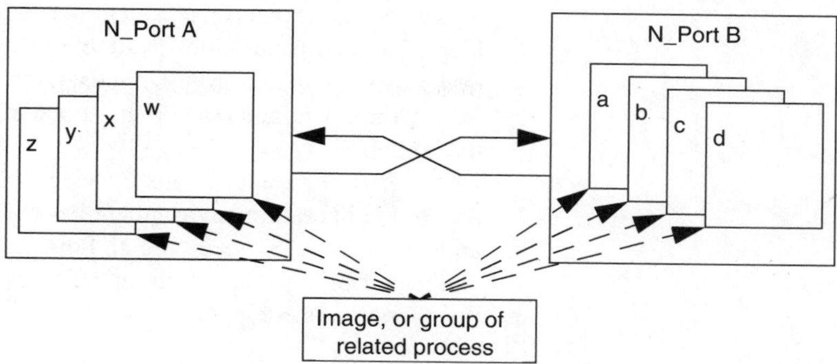

Figure 9.10 Multiple images are addressable behind a single Port using Process Associators.

executed following Fabric and N_Port Login on those networks which implement this optional layer of addressing.

Process Login and Logout are carried out using two Extended Link Service commands: Process Login (PRLI) and Process Logout (PRLO). The PRLI command allows for one or more images at one N_Port to be related to corresponding images at another N_Port as an "image pair." In addition, the PRLI allows one or more FC-4 capabilities to be announced by the initiating N_Port to the Recipient N_Port. The Recipient FCP_Port indicates its acceptance or rejection of the modes of operation in its response to the PRLI command.

Support for the actual PRLI Extended Link Service Request is optional since implicit Login can (and probably will) be established by configuration conventions outside the Fibre Channel mechanisms.

Each PRLI and its corresponding ACC ELS Reply includes a Payload of one or more four-word "service parameter pages," each of which establishes a single image pair. An image pair is a combination of two images behind two different N_Ports which can initiate communications with each other. Service Parameters are indicated for each image pair separately. The image pair is identified by a "Originator Process Associator" and "Responder Process Associator," which are assigned during Process Login and are included in the Association Header in the first Frame of each new Sequence directed between the images in the pair.

The actual Service Parameters exchanged are specified in the specifications for the FC-4 which is using the image pair functionality. The service parameter page specifies up to 32 bits for usage by the FC-4 layer in determining the ULP protocol-dependent Service Parameters supported. As an example, in the SCSI-FCP FC-4 document, Service Parameter bits are included specifying whether the image can act as a SCSI Target or Initiator, whether commands and data can be mixed in the same Information Unit, and whether the XFER_RDY Information Unit is disabled. These are all SCSI-FCP-related functions, and are transparent to the Fibre Channel levels.

The use of a Binding PRLI page requires that the Originator of the PRLI Exchange have precise and detailed knowledge of the process associator identifiers and capabilities available in the PRLI Responder. This information may be obtained from Directory Services, implicitly from configuration information obtained outside the scope of FC, or by performing an Informative PRLI.

The PRLO Extended Link Service Command is used to request invalidation of the operating environment between an image at the initiating N_Port and an image at the receiving N_Port. As with PRLI, the PRLO needs to be confirmed by response with an ACC Extended Link Service Reply before the operation is complete.

Procedure to Estimate End-to-End Credit

The allocation by an N_Port of end-to-end Credit to different destination N_Ports can be a major factor determining overall N_Port performance in streaming data. This is particularly important when Sequences are being streamed to multiple destinations simultaneously, with multiplexing on a Frame-by-Frame basis. In this case, the end-to-end Credit allocated to each N_Port interacts with end-to-end Credit values allocated to different N_Ports.

The end-to-end Credit value allocated to a particular source should be large enough that Frames can be continuously transmitted, i.e., the end-to-end Credit value is high enough that the source N_Port will not send out EE_Credit number of Frames before acknowledgments return for earlier Frames.

On the other hand, setting the end-to-end Credit higher than the value required to allow continuous streaming of Frames does not increase performance any further, and may decrease performance if the allocation of resources to provide this higher-than-required EE_Credit_Count diminishes the resources available for receiving Frames from another source N_Port.

For optimizing overall system performance, it is therefore useful to determine the minimum end-to-end Credit value that lets Frames be continuously transmitted and to not set the end-to-end Credit value any higher than that value. The procedure described here allows estimation of this best end-to-end Credit value. This procedure is performed after Fabric and N_Port Login, and can be part of normal optimization of transmission performance. It is included here because it provides determination of the operating environment characteristics.

This procedure uses three optional Extended Link Service commands, as described in Chapter 11. These include the Establish Streaming (ESTS) (see the "Establish Streaming (ESTS)" section, on page 209), Estimate Credit (ESTC) (see the "Estimate Credit (ESTC)" section, on page 210), and Advise Credit (ADVC) ELS commands (see the "Advise Credit (ADVC)" section, on page 210). Since these commands are related, an N_Port would either support all three or none of them. If they are not supported, and an N_Port receives one of them from a different N_Port, it will send the Link Service Reject (LS_RJT) Extended Link Service Reply Sequence, with a reason code of "Command not supported" (x'0B'), as described in the "Link Service Reject (LS_RJT)" section, on page 203.

The general procedure is that the source N_Port transmits an "Establish Streaming" (ESTS) Sequence to set up the procedure, and the return of an ACC Sequence indicates that the destination N_Port is ready to begin receiving Frames. The source N_Port then streams "Establish Credit" (ESTC) Frames continuously until it receives an acknowledgment from the destination N_Port. The number of Frames transmitted before the ACK returns determines the best end-to-end Credit value to use with that particular destination N_Port. The "Advise Credit" (ADVC) Sequence is then sent, to notify the destination N_Port of the value found. The destination N_Port can then do a new N_Port Login to advertise the new end-to-end Credit value. The series of Sequences and operations used in the procedure is shown in Figure 9.11.

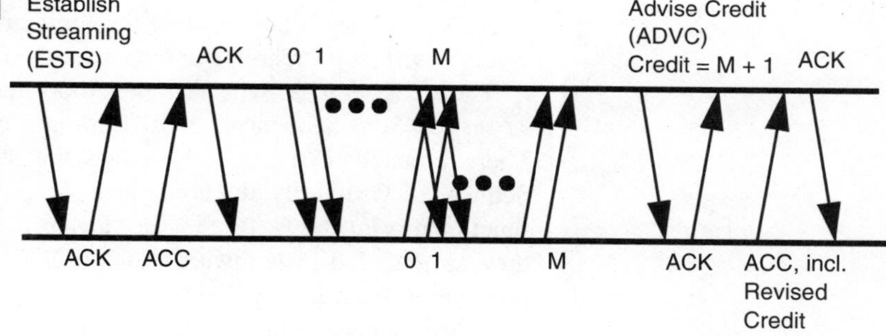

Figure 9.11 Frames and Sequences transmitted in procedure for establishing streaming credit.

The procedure is separately applicable to both Class 1 and Class 2 Sequences. Since Class 1 and Class 2 Sequences flow through Fabric and N_Ports differently, and since Class 2 Sequences are also affected by buffer-to-buffer Credit, the values obtained for the two Classes may differ substantially. To perform this procedure for Class 1 or Class 1 Intermix, a Class 1 connection must have been established before the procedure is begun.

The value found in one iteration of this procedure may not be the true optimal value, since performance can depend heavily on what other traffic is traversing the network, as well as on a number of other factors. It may therefore be worthwhile to perform the procedure several times and to use a value in the ADVC Sequence based on the results from several iterations of the procedure.

Procedure Details

The Establish Streaming (ESTS) Sequence is used to provide the destination N_Port with an opportunity to communicate the maximum end-to-end Credit it can provide, to assure streaming capability during the streaming phase of the procedure. This is a temporary allocation, termed "streaming credit" (L).

This Sequence is initiated as a new Exchange by the Exchange's Originator. The **SOF** delimiter identifies the Class of service for which the source N_Port is requesting streaming credit. The destination replies with an ACC Frame, with the same format as the N_Port Login. The Payload contains the streaming credit value (L) allocated in the end-to-end Credit field of the Class service parameters field. All other fields are ignored by the source N_Port (Sequence Recipient). This establishes an end-to-end Credit value which is large enough that the source can transmit a large number of Frames relative to the single Frame transmission time.

Immediately following return of the ACC to the ESTS, the source N_Port will stream Estimate Credit (ESTC) Frames consecutively, starting from SEQ_CNT = x'0000,' until it receives the first ACK (ACK_1 or ACK_N) from the destination N_Port. If the source does not receive an ACK (ACK_1 or ACK_N) after it has reached the limit imposed by the streaming credit value, it will stop streaming and wait for the first ACK to be received.

The ACK returned will have the Abort Sequence F_CTL bits (bits 5 and 4) set to "Stop Sequence" (b'10'), to stop the streaming of Data Frames. There is another reason for setting the Abort Sequence bits to "Stop Sequence." Ordinarily, the upper level (in this case, the link service support functionality in the N_Port) is not aware of the reception of ACKS, since they are handled in hardware. For the Estimate Credit (ESTC) Sequence, however, it must be notified when the first ACK is received. The setting of the Abort Sequence bits to "Stop Sequence (b'10') under normal operation would normally indicate that the Sequence Recipient has run out of upper level buffer space for the incoming data. The "Stop Sequence" indication in an ACK Frame must therefore be surfaced to the upper level. Using the "Stop Sequence" mechanism here assures that the Link Service support functionality upper level receives the notification of ACK receipt that it would not ordinarily receive.

If the highest SEQ_CNT transmitted by the source N_Port at the time it receives the first ACK is M, the number of outstanding Frames (i.e., credit estimated for continuous streaming) will equal M + 1. If ACK is received within the streaming credit limit (L > M), this value of M + 1 represents the minimum credit required to utilize the maximum bandwidth of the system. If the ACK is received after reaching the streaming credit limit (L), this value is less than the optimal credit limit credit required to utilize the maximum system bandwidth.

The source N_Port will follow all the normal rules in closing the Sequence, by sending the last Data Frame of the Sequence and waiting for the corresponding ACK to be received.

The size of the Payload of the ESTC Frame is not fixed and can be chosen by the N_Port transmitting the ESTC Frames to determine a size-dependent value for the end-to-end Credit. The Payload may contain any valid data bytes, and the SEQ_CNT will follow the normal rules for Sequence transmission.

After the Estimate Credit (ESTC) Sequence has been terminated by the source N_Port, the Advise Credit (ADVC) Sequence is transmitted to advise the destination N_Port of the streaming credit value M + 1 determined by the source N_Port. The destination N_Port will reply using an ACC Sequence, with a revised end-to-end Credit value in its Payload. This value may be different from the value sent by the source N_Port, based on the destination N_Port's buffering scheme, buffer management, buffer availability, and N_Port processing time. The value in the ACC is the final value to be used by the source N_Port for revised end-to-end Credit.

Since the maximum Frame size can be unequal in forward and reverse directions between two N_Ports, the estimate credit may be performed separately for each direction of transfer. Credit modification applies only to the direction of the transfer of the ESTC frames.

An alternative to using this procedure is for a source to simply accept an initially estimated end-to-end Credit value that may be too small. If, at a later time, data transfers are unable to stream continuously, the source may arbitrarily request an increase in the end-to-end Credit by using an ADVC Extended Link Service Command Sequence and see if the destination grants the additional end-to-end Credit in the ACC Frame.

Chapter 10

Classes of Service

Introduction

A significant problem in designing data communications systems, especially systems that are targeted toward transmitting many types of data or message traffic, is that different types of traffic patterns require different kinds of network performance. For example, some kinds of traffic require long-term data transfer between pairs of Nodes, while others require a single Port to send small pieces of data to multiple destination Nodes in turn. A network architecture which is optimized for one kind of traffic pattern may not perform well at all on a different traffic pattern, since efficient flow control and routing control algorithms depend on the traffic patterns of the data transmission.

To address these issues, the Fibre Channel architecture allows the N_Port sending data to give the network help in determining what kind of performance is needed. This is done by specifying that the data is to be sent in one of three Classes of service. The Class of service used for transmission determines what type of flow control will be used for the Sequence and what resources are required to handle transmission.

Following are the general characteristics of the various Classes of service:

Class 1: provides Dedicated Connections, where a Sequence Initiator requests that all the links and related resources on the route to the Sequence Recipient are guaranteed for the duration of a communication period, possibly including multiple Sequences and Exchanges, until either the Initiator or Recipient requests that the Connection be removed.

Class 2: provides a multiplexed service, which allows routing and flow control on a Frame-by-Frame basis with acknowledgment of successful delivery. No resources are guaranteed beyond the boundaries of each Frame, and Frames may be interleaved from multiple Sequences and multiple N_Ports over the links and at the N_Ports.

Class 3: is a Datagram type of service, where the Fabric and N_Ports provide no guarantee or indication of successful delivery, and any management of Sequence integrity is handled by a higher-level function.

Class 4: provides Dedicated Connections that guarantee Quality of Service Parameters at a fractional portion of the link capabilities. The originator and the Fabirc negotiate Dedicated Connections with guaranteed minimum and maximum bandwidth, and maximum Frame delivery latency, for a total of up to 254 connections per N_Port, with each obtaining a guaranteed fraction of the transmission resources.

Class 6: provides a Unidirectional Dedicated Connection between an originator an a multicast group. Responses from all members of the group are aggregated by a multicast server to provide a single response to the originator, resulting in a reliable multicast service across a Fabric.

Class 1 service can be compared to a telephone network, where an Initiator dials a telephone number to request a Connection that either party can remove by hanging up, and during the Connection there is no interference from any other traffic on the network. Class 2 service is more like a packet-switching network, where multiple Frames traverse the network between multiple sources and destinations and compete for limited network bandwidth and buffering resources. The network must do much more ongoing management in Class 2 than in Class 1, where the Frame management and flow control is performed by the communicating N_Ports. Class 3 provides no notification of Frame delivery and provides no direct knowledge to the source N_Port of whether Frames arrive at their destinations or not. This can provide an increase in performance for transmission under ULPs that provide their own flow control and acknowledgment and also can simplify management for traffic patterns such as broadcast, where a single transmitted Data Frame might generate a number of Link Response Frames.

In addition to the three Classes of service, there is an "Intermix" facility, which combines the advantages of Class 1 with Class 2 and Class 3 service. Support for Intermix allows Class 2 or Class 3 Frames to be interleaved between Class 1 Frames during a Class 1 Dedicated Connection, when not all of the bandwidth is being used. This allows better link utilization when two N_Ports have established a Class 1 Dedicated Connection but don't have enough data to transmit to keep all the links completely busy in both directions. In a Fabric environment, this can happen quite often, so Intermix support in the Fabric and in N_Ports is generally well worth the extra resources required to support it.

Class 4 addresses the desire for quality of service guarantees for transmission across a network at something less than the capabilities provided with Class 1, and between multiple source/destination pairs. Class 4 assumes Intermix capabilities for the un-reserved bandwidth of the network.

Class 6 is aimed primarily at providing efficient replication of data, which is especially useful in high-reliability installations which need to failover capabilities in case of failure of systems or loss of data. Doing the Frame replication at the Fabric level, rather than the N_Port level, improves bandwidth utilization by at least 2 times. To alleviate the originating N_Port from having to manage response from all destinations in the multicast group, the Fabric aggregates responses together appropriately to give the service of an acknowledged multicast capability.

The choice of which Class of service to use for a particular Sequence or group of Sequences can be made at the FC-2 level or by the ULP through the FC-4 level, determined by the amount of data in the Sequence and by any other knowledge that is available about Fabric performance or network topology.

Support for the various Classes of service is not required by either the N_Ports or by the Fabric and is indicated in the Class Service Parameters during Fabric or N_Port Login. All Ports are required to recognize all the types of Frame delimiters, to know if they have received Frames in a Class of service that they don't support. If a Port receives a Frame with an **SOF** delimiter for a Class of service which it doesn't support, it will discard the Frame and issue a P_RJT with the "Class not supported" reason code for Class 1 and Class 2 Frames. Following Class rules, a Class 3 Frame is discarded in this case with no indication returned.

Figure 10.1 shows an example of the flow of Data Frames, ACK Frames, and R_RDY flow control signals for N_Port Login over a Fabric, to demonstrate the differences between the three Classes. Following are detailed descriptions of operations under each of the Classes of service.

Class 1 Service

Class 1 service is particularly useful on Fabric topologies, where establishment of Dedicated Connection can allow routing to be done once per Connection, rather than once per Frame, and can allow the source and destination N_Ports to handle Link Control Frame generation and management without Fabric involvement, simplifying Fabric operations and possibly freeing up Fabric resources for handling other traffic.

Under Class 1 service, a Sequence Initiator begins establishing a Connection by sending a connect-request Frame. A connect-request is a normal Data Frame, which may or may not contain Payload data, that has a **SOFc1** Frame delimiter. It is only semi-useful to think of a connect-request as the first Frame of a Class 1 transmission, since it is handled entirely differently from every other Frame of the Class. Although it is allowed to share a SEQ_ID value, SEQ_CNT association, and Sequence Payload data with later Frames transmitted with **SOFi1** or **SOFn1** delimiters, a connect-request actually flows through the Fabric much more like a Class 2 Frame than like a Class 1 Frame. It's best to think of the connect-request as a single-Frame message to the network and the destination to set up a connection for Sequence transmission that might contain a Frame's worth of the Sequence's Payload for efficiency.

When the Fabric receives a connect-request indicating a particular destination N_Port D_ID, it begins establishing a Connection with that N_Port, ensuring that later Frames in the Connection can be routed with minimal Fabric-level control or buffering. When the destination N_Port receives the connect-request, it performs whatever setup functions are necessary to establish the Connection and transmits an ACK back to the source. The Fabric finishes establishing the Connection while routing the ACK Frame to the Sequence Initiator. When the Initiator receives the ACK, the Connection is fully established, and the two N_Ports control management of all Frames,

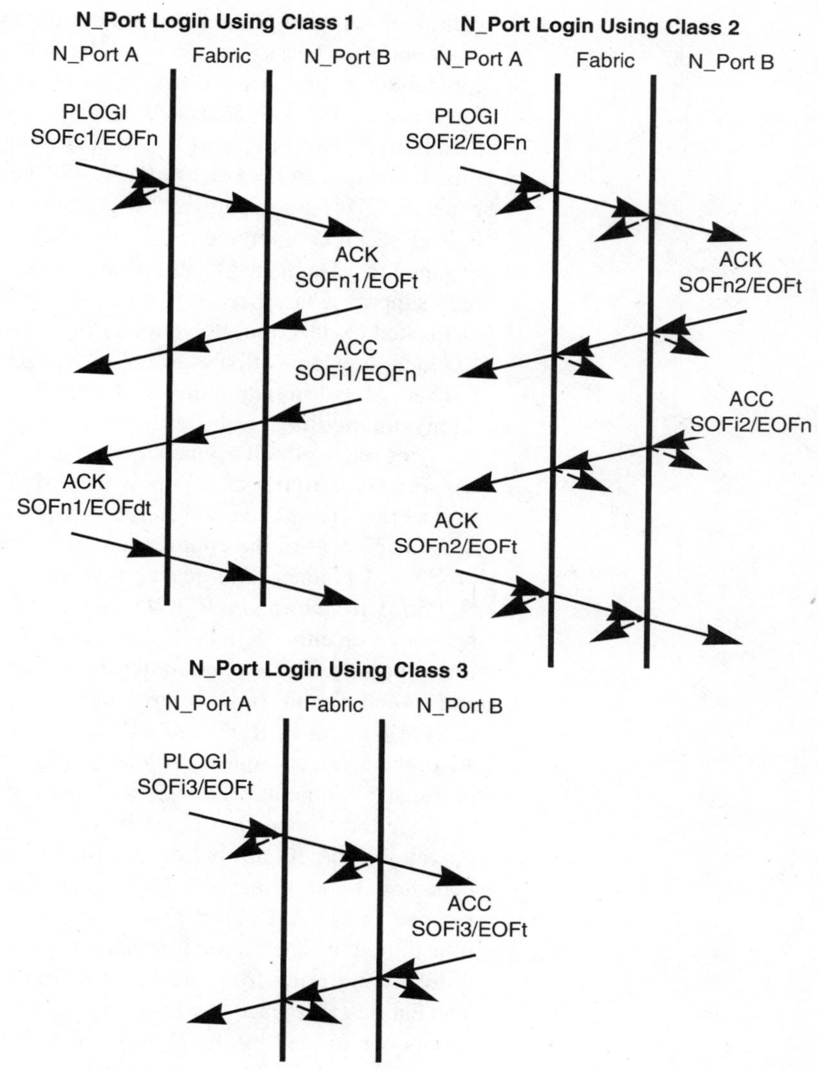

Figure 10.1
Examples of N_Port Login over a Fabric using the three Classes of service.

Sequences, and Exchanges transmitted until either N_Port removes the connection.

Frame delimiters are used to manage establishment and removal of Dedicated Connections. The connect-request Frame which requests establishment of the Connection uses an **SOFc1** delimiter. A Connection can be removed by either N_Port by terminating a Frame with an **EOFdt** or **EOFdti** delimiter. All Sequences within the Connection except the Sequence which contains the connect-request Frame are begun with Data Frames using the

SOFi1 delimiter and terminated with Link Control Frames using the **EOFt** delimiter. All other Class 1 Frames use the **SOFn1** and **EOFn** delimiters

Support for Class 1 service must be indicated both by the Fabric, as indicated in the Fabric Login Class Service Parameters, and by both Initiator and Recipient N_Ports, as indicated in the N_Port Login Class Service Parameters. If an N_Port or F_Port which doesn't support Class 1 service receives a Frame with an **SOFc1** delimiter, it will issue a RJT Frame with a reason code of "Class not supported." It is a serious protocol error if an N_Port or F_Port receives a Frame with an **SOFi1** or **SOFn1** delimiter while not engaged in a Dedicated Connection. It is also a serious error if a Port which only supports Class 1 receives a Class 2 or Class 3 Frame while engaged in a Dedicated Connection. Behavior in these cases is unpredictable.

Once the Connection is established, each N_Port can originate multiple Exchanges and initiate multiple Sequences within the Connection. There is no maximum time for a Connection to be established. When the Connection does get removed, all Sequences and Exchanges should be complete, since any Frames arriving at a Port with **SOFi1** or **SOFn1** delimiters after the Connection is removed will cause a serious protocol error.

Management of the connect-request Frame is similar to the management of Class 2 Frames. This means that the Fabric, if present, and destination N_Port will return an R_RDY signal for the connect-request and may respond with either F_RJT, F_BSY, P_RJT, or P_BSY Link Control Frames, if appropriate. The connect-request operates under both buffer-to-buffer and end-to-end Credit flow control management, so the source N_Port will receive both an R_RDY and an ACK back for the connect-request Frame. Also, the connect-request Payload size is limited by both Fabric and N_Port buffer size limits, like Class 2 and Class 3 Frames.

Once the Connection is established, however, the Fabric is assumed to operate essentially as a wire. This implies a number of things about Class 1 behavior. First is that the Fabric cannot do any reordering, so all Data Frames and Link Control Frames arrive in the same order that they were transmitted in. The Fabric is assumed to eliminate all buffering, or at least to hide any buffering from the N_Ports, in order to guarantee full bandwidth. The Fabric can't place any limits on Frame size, so maximum Class 1 Frame size is determined by the N_Ports at N_Port Login. Also, the Fabric can't generate any F_RJT or F_BSY Frames for any Data Frames inside a Class 1 Connection. No R_RDY Frames are generated for any Data Frames or Link Control Frames, and all flow control is controlled by the end-to-end Credit mechanism. The destination N_Port must be able to receive the Frames and cannot issue a P_BSY, but it can issue a P_RJT, if appropriate.

All three forms of ACK can be used for Class 1 Frames. The rules about not mixing ACK forms within a Sequence still apply, so the destination N_Port using ACK_N must issue an ACK_N with a count of 1 for the connect-request. An N_Port using ACK_0 may either issue two ACK_0 acknowledgments per Sequence, since one must be returned for the connect-request to establish the Connection, or may use an ACK_1 or ACK_N with count of 1.

If a Fabric receives a Class 2 or Class 3 Frame destined to either of the N_Ports in a Dedicated Connection, and either the Fabric or destination N_Port does not support Intermix, it will either discard it (Class 3) or discard it and return an F_BSY with the "N_Port Busy" reason code.

Variations on Class 1 Service

Dedicated Simplex. Dedicated simplex lets a single N_Port simultaneously establish a Class 1 connection to one other N_Port as Connection Initiator and a separate Class 1 connection to another N_Port as Connection Recipient. This function is established in an attempt to more efficiently use both fibers on an N_Port, since there are few applications which can simultaneously exchange in both directions between two Nodes. It requires defining the reserved byte in word 1, bits 31 to 24, as a CS_CTL field, indicating "Class Control," with a bit defined in the **SOFc1** Frame Header requesting the Class 1 connection as a dedicated simplex connection. The connection will be unidirectional, and both the Initiator and Recipient can simultaneously establish dedicated simplex connections with other N_Ports in the opposite function. Even though this is a Class 1 Connection, there may not be a Class 1 Connection for the return path, so ACKs must be returned in Class 2, and the N_Ports and switch elements must be able to support Intermix.

Buffered Class 1 Service. Buffered Class 1 service is implemented to allow establishment of Class 1 dedicated connections between N_Ports of different speeds. Buffered Class 1 is an mechanism to allow an N_Port operating at, for example, 531.25 Mbps to exchange data with one operating at 1,062.5 Gbps. The two N_Ports will establish Class 1 connections at the highest common bandwidth (531.25 Mbps in this case) and the Fabric will buffer and control the flow of the data transmitted at the higher speed while it is received at the lower speed.

Class 2 Service

In Class 2 service, the network and destination N_Port provide connectionless service with notification of delivery or non-delivery between the two N_Ports. Each Frame is routed independently through the Fabric, so one N_Port can transmit consecutive Frames to multiple destinations and can receive consecutive Frames from multiple sources.

The Fabric routes Frames based on the D_ID field, using information on physical location of the D_ID destination Port obtained during Fabric Login. When the destination N_Port receives the Frame, it normally returns an ACK, which is also routed back through the Fabric, if present, based on the D_ID field of the ACK.

The bandwidth available between any two N_Ports is dependent on how many other N_Ports on the Fabric are sending data and on Fabric utilization. There is no requirement that a Fabric be non-blocking, so that, for example, Class 2 traffic between N_Ports A and B may have a detrimental effect on traffic between N_Ports C and D.

In Class 2 service, the source N_Port transmits Frames using either the **SOFi2** or **SOFn2** and either the **EOFn** or **EOFt** delimiters. The allowable size of Class 2 Frames is limited by the smaller of the buffering capacity of the Fabric and the destination N_Port.

Class 2 uses both buffer-to-buffer and end-to-end flow control mechanisms, so that every Frame will generate an R_RDY from the attached Port and will be acknowledged with an ACK Link Control Frame. All forms of ACK are valid, and either the Fabric or N_Port can return a RJT or BSY if necessary.

The Fabric is not required to deliver either Data Frames or Link Control Frames in order. This can allow better link utilization on a multi-stage switched Fabric network, as Frames for a single Sequence can be simultaneously sent on different paths.

An N_Port which supports Class 2 must be able to support reordering of out-of-order Frames based on the SEQ_CNT value. The standard does not specify how far out of order a Sequence of Frames can go but only specifies that a Frame must be delivered within an E_D_TOV period after the Frame with the previous SEQ_CNT value. With a default E_D_TOV value of 10 s, and a transmission time for a 2,048-byte Frame of roughly 20 μs per Frame at full speed, a Frame could conceivably be delayed in the Fabric until almost 500,000 Frames have passed, and then be delivered intact. The end-to-end Credit mechanism does not limit how far out of order Frame delivery may go, since it is based on the number of Frames sent and acknowledged, not on a window of SEQ_CNT values. In practice, keeping track of Frames this far out of sequence is not feasible, and vendors will have to determine an out-of-order window, with the window size depending on Fabric characteristics, for determining when the Frames have been really lost or only delayed.

An N_Port can determine when an out-of-order condition has occurred because a Frame has been lost, and not just delayed, by using the timeout period. If an indication of Frame delivery (positive or negative) has not been received at an N_Port within the E_D_TOV time period, then either the Data Frame or its corresponding Link Response Frame has been lost.

Both the Fabric and the destination N_Port for Class 2 Data Frames can issue BSY or RJT Frames. If the Frame is busied, the BSY is routed back to the source of the Frame, which may optionally retransmit the busied Frame a vendor-dependent number of times. If the source N_Port keeps receiving BSY Frames in response to a Data Frame beyond its retransmission limit, it must assume that the Frame can never be delivered and must begin error recovery procedures.

In a Point-to-point or Arbitrated Loop topology, the behavior of Class 1 and Class 2 service is very similar, with the exception that there is no equivalent to a connect-request in Class 2, and the Frame delimiters and Frame size limits must be appropriate to the Class of service used.

Class 3 Service

Class 3 service provides a connectionless service without notification of delivery between N_Ports. The transmission and routing of Class 3 Frames is the same as for Class 2 Frames. However, when a Frame arrives at its destination N_Port, the N_Port does not return an ACK, indicating successful delivery, or a BSY or RJT indicating unsuccessful delivery.

This kind of service can be useful in some cases for ULPs which do their own error detection and recovery or which don't need reliable data transmission. Class 3 is particularly useful in implementing multicast transmission, where a single Frame is delivered to multiple destinations, since it circumvents the difficulty of dealing with multiple, possibly different, responses found in Class 6 multicast. This is normally only possible on switched fabrics, as described in the "Multicast" section, on page 308. On Arbitrated loops, a similar function can be achieved by Frame replication, as described in the "Primitive Signals and Sequences for AL" section, on page 280. Another good example of Class 3 usage is for initialization of Arbitrated Loops, as described in the "Loop Initialization" section, on page 284, where no response notification of delivery is required.

In Class 3 service, the source N_Port transmits Frames with either **EOFi3** or **SOFn3** and either **EOFn** or **EOFt** delimiters. The Fabric routes each Frame based on the D_ID value. The Fabric is not required to guarantee in-order delivery, and the destination N_Port which supports Class 3 must support reordering of Frames based on SEQ_CNT. The same conditions on determining whether Frames are lost or just delayed in the Fabric apply to Class 3 Sequences as were described for Class 2 Sequences.

Flow control under Class 3 uses only the buffer-to-buffer Credit mechanism, with an R_RDY Ordered Set being returned across a link for each Frame when facilities are available for receiving the next Frame.

If the Fabric or destination N_Port are busy on receiving a Class 3 Frame or detect an error that would cause a RJT to be generated for a Class 2 or Class 1 Frame, they will discard the Class 3 Frame without issuing a BSY or a RJT Frame or otherwise notifying the source N_Port.

Even though there is no notification of successful delivery, there are several ways for a Sequence Initiator to determine which Frames of a Sequence have been successfully delivered. Two high-level methods are by transmission of the Read Exchange Status (RES) and Read Sequence Status (RSS) Extended Link Service commands, both of which indicate whether or not complete Sequences have been received. For more detail, the BA_ACC Reply to the Abort Sequence (ABTS) Basic Link Service command can include the SEQ_ID of the last complete delivered Sequence and the SEQ_CNT of the last delivered Frame of that Sequence, as described in detail in the "Abort Sequence (ABTS) Command" section, on page 199.

Performance and Reliability with Class 3 Service

It is a very strong design objective of the Fibre Channel architecture that it not deliver corrupted data to a ULP without an indication that it is corrupted. This requirement, along with a limitation in the size of the SEQ_ID field and uncertainty in data delivery time through the Fabric, can cause some serious performance implications for Class 3 transmission.

Recall that the SEQ_ID field is 8 bits long. This implies that only 256 Sequences can be outstanding in a particular Exchange at once, since if the SEQ_ID field is reused within an Exchange, then the two Sequences will have the same S_ID, D_ID, OX_ID, RX_ID, and SEQ_ID values, and two Frames with equal SEQ_CNT values could be mistaken for each other.

Consider the following series of occurrences under Class 3 transmission. A two-Frame Sequence is transmitted with SEQ_ID = 3. The first Frame is delivered intact, but the second Frame is discarded, due to congestion or a Frame Header error. The Sequence Initiator gets no indication, as is normal under Class 3. 255 more Sequences are transmitted, and another Sequence is transmitted, also with SEQ_ID = 3. The first Frame of this Sequence is discarded or delayed, and the second Frame arrives at the destination. The delivered Frame from the last Sequence is not distinguishable from the discard Frame from the first Sequence by the X_ID or SEQ_ID fields, and if the SEQ_CNT fields happen to match, the Recipient would have no indication that they are not from the same Sequence. The data error would not be caught at the Fibre Channel level, and the corrupted data from the two Sequences would be delivered to the ULP with the indications for a single intact Sequence.

In Class 1 and Class 2, the Sequences become closed when ACKs for all the Frames have returned. It will in general then be unusual for more than a few Sequences to be outstanding. In Class 3, however, since no ACKs are returned, Sequences are normally outstanding until the R_A_TOV time period has expired, when all Frames either must be delivered or discarded. The default value for the R_A_TOV timeout value is 120 s, which would be enough time to send over 2,400,000 Sequences of length 4,096 bytes. Since only 256 Sequences could actually be sent, only 256/2,400,000, or about 0.01%, of the available bandwidth could be used for Sequences of this size. This is clearly a problem.

There are several ways to get around this limitation. The most obvious is to not use Class 3 service for data which must be delivered reliably. Alternatively, the ABTS Basic Link Service Request could be used to ensure that Sequences are closed — this is slower than the ACKs used in Class 1 and Class 3 but is much faster than waiting for an R_A_TOV period. Finally, if the ULP provides adequate mechanisms for detecting data corruption, then it may not matter, for that particular ULP, whether the Fibre Channel levels deliver corrupted data.

Clearly, neither of these solutions is very pretty. The choice of what to do in this situation will be up to the vendors and will depend on exactly what transmission characteristics are required.

Intermix

Intermix is an option to Class 1 service that allows interleaving of Class 2 and Class 3 Frames during an established Class 1 Dedicated Connection. This facility is included to allow better utilization of links when a source N_Port is not sending or receiving data quickly enough to fully utilize the link capacity but does not want to remove a Class 1 Connection due to the overhead required to reestablish it. It can also be useful to minimize the latency of Class 2 and Class 3 traffic, which doesn't have to be delayed until the removal of a Class 1 Connection.

Under Intermix, the Fabric must still guarantee full link bandwidth to the Class 1 traffic. It can, however, allow any unused bandwidth at either the transmitting or receiving end to be used for delivery of Class 2 and Class 3 Frames. Effectively, this means that the Fabric must assure that Class 1 Frames have priority over Class 2 or Class 3 Frames, so that Class 2 and 3 Frames can be delayed and buffered if there is not enough available bandwidth. Clearly, support for Intermix will require some amount of vendor-dependent buffering capability in the Fabric for Frames which are delayed because of the interleaving. If there is not enough buffering space, the Fabric might discard, BSY, or RJT a Class 2 or Class 3 Frame even if Intermix is supported. A Fabric may even truncate a Class 2 or Class 3 Frame and terminate it with an **EOFa** delimiter to guarantee Class 1 bandwidth if a Class 1 Frame arrives at the Fabric to be delivered while a Class 2 or Class 3 Frame is being Intermixed.

There is a feature in the Frame transmission protocol which can let Intermix be used even over links that are being driven at the full link speed by the Sequence Initiator. The standard specifies that an Initiator must allow six Primitive Signals between each Frame, to allow clock synchronization and delimiter processing time. However, at a Recipient, only two Primitive Signals are required between Frames. The Fabric therefore has the option of removing up to four words between Frames and interleaving part of a Class 2 or Class 3 Frame into the resulting space. By doing this repeatedly, with enough buffering in the Fabric, it is possible to continuously insert a small number of Class 2 or Class 3 Frames, even over links that are being driven at the maximum possible speed.

For Intermix to be supported between two N_Ports, both N_Ports and the Fabric must support Intermix, as indicated in the Class Service parameters during Fabric and N_Port Login. This allows a source N_Port to send Class 2 or Class 3 Frames during a Class 1 Connection, either to the Connection partner or to a third N_Port. It also allows a third Port to interleave Class 2 and Class 3 Frames to either of the two N_Ports in the Connection.

When not all the elements support Intermix, processing of Frames can get fairly complex. If, for example, a Fabric that supports Intermix receives a Frame from a third N_Port destined for an N_Port in the Connection that doesn't support Intermix, it may optionally buffer the Frame for a time, hoping for the Connection to be removed, before discarding the Frame and (for Class 2) returning an "N_Port Busy" F_BSY Frame. The same behavior can be followed if the Fabric does not support Intermix.

If a Frame involved in a Dedicated Connection that supports Intermix attempts to interleave Class 2 or Class 3 Frames over a Fabric that doesn't support Intermix, it is a violation of the Login Service Parameters, and the Frame delivery is unpredictable.

Class 4 — Fractional

Class 4 is a newer connection oriented service which works to provide better quality of service guarantees for bandwidth and latency than Class 2 or Class 3 allow, while providing more flexibility than Class 1 allows.

Roughly, Class 4 provides a Class 1 type of dedicated connection service, but instead of all the bandwidth of a particular port being allocated to one other N_Port on the Fabric, the bandwidth (in each direction) is divided among up to 254 Virtual Circuit (VC) connections to other N_Ports on the fabric. A separate Quality of Service Facilitator, at the well-known address x'FF FFF9' is used to negotiate, manage, and maintain the quality of service for each VC, and assure consistency among all the VCs set up across the full Fabric to all ports.

A Class 4 circuit includes one Virtual Circuit (VC) in each direction, with a set of Quality of Service parameters for each VC. These Quality of Service parameters include guaranteed transmission and reception bandwidths and/or guaranteed maximum latencies in each direction across the Fabric.

To set up a Class 4 circuit, the circuit initiator (CTI) sends a Quality of Service Request (QoSR) Extended LinkService command to the QoS Facilitator (QoSF). The QoSF makes sure that the Fabric has the available transmission resources to satisfy the requested QoS parameters, then forwards the request to the circuit recipient (CTR). If the Fabric and the recipient can both provide the requested QoS, the QoS Request is accepted, and the transmission can start in both directions. If the requested QoS parameters cannot be met, the request is rejected with an

Class 4 has the concept that the VCs can be in a "Dormant" state, with the VC set up at the N_Ports and through the Fabric, but with no data flowing, or a "Live" state, where data is actively flowing.

In Class 4, the Fabric manages N_Port transmission by using a buffer-to-buffer flow control mechanism similar to the R_RDY mechanism. However, in Class 4, separate flow control is used for each Virtual Circuit that the N_Por manages, so a new Primitive Signal, called VC_RDY, is used. VC_RDY resembles R_RDY, but it contains a Virtual Circuit identifier byte in the Primitive Signal, indicating which VC is being given the buffer-to-buffer credit.

Each VC_RDY indicates to the N_Port that a single Class 4 frame is needed from the N_Port if it wishes to maintain the requested bandwidth. However, the Fabric is expected to make any unused bandwidth available for other Live Class 4 circuits, and for Class 2 or Class 3 frames, so the VC_RDY does allow other frames to be sent from the N_Port.

There are some scalability difficulties associated with Class 4 service, since the Fabric must negotiate resource allocation across each of 254 possible VCs on each N_Port. Also, Class 4 on Loops, either private or public, is not clearly defined. However, Quality of Service is an increasingly important aspect of network design and management, so there may be places where Class 4 is very useful, particularly where Class 1 doesn't provide enough flexibility in network resource management.

Class 6 — Uni-Directional Dedicated Connection

Class 6 is a reliable multicast service, where a single Initiator N_Port sends Sequences to a D_ID which has previously been allocated to a multicast group, and the Fabric replicates the Frames and distributes them to all the Sequence Recipients in that multicast group. When the Recipient N_Ports respond to the data frames, the Fabric recognizes the multicast ID in the S_ID frame header field of the responses, and aggregates the ACK information together to present a single response to the Sequence Initiator.

The result of this is a reliable, acknowledged multicast service which appears (except for the more complex registration and setup) to each N_Port as a Class 1 dedicated connection. The main difference for normal Class 1 Dedicated connections, as far as the N_Ports are concerned, is that it's a Simplex connection — data Frames can only go from the sequence initiator to the multiplexed recipients.

There are a few interesting features of this approach.

- The registration of a multicast group, and aliasing of a set of N_Port IDs behind a MG_ID uses the Extended Link Service commands exchanged with an Alias Server at ID x'FF FFF8'. (GAID, FACT, NACT, etc.), and must be done before communications can begin.
- ACK aggregation is simple if all the Recipients return the same response at the same time. If they don't, the Fabric has to buffer responses, and aggregate them them together. Since the responses include ACK_0, ACK_1, ACK_N, P_BSY, or P_RJT link control frames, and Link Reset or Link Reset Response Primitive Sequences, returned at various times, the Fabric operations required to keep consistency between both ends of the Dedicated Connection can get fairly complex.
- The Exchange_ID rules require that a RX_ID values be unique across all Exchanges on a single N_Port. However, a single normal RX_ID value would not normally be uniquely available across all N_Ports in a multicast group. To assure this uniqueness, an extra field is added to the Class Service Parameters for N_Port Login, shown in Figure 9.8, to allow the N_Port and the Multicast Server to negotiate a unique RX_ID. This may also help the Fabric identify frames as part of a Class 6 Dedicated Connection.

These various complexities, and the difficulties of implementing Class 6 on Loop topologies, will probably limit the usage of Class 6 for some time.

Chapter 11

Link Services

Introduction

In addition to normal data transmission procedures, any communications protocol requires procedures for performing link level and setup operations such as controlling link status, polling other Ports for verification of the status of various operations, and for aborting operations already begun, in case of errors. Under Fibre Channel, these operations are provided for under the heading "Link Services."

There are three types of Link Services:

- Basic Link Services, which are distinguished by being single Frame commands that can either be sent inside an existing Sequence or inside a newly initiated one;
- Extended Link Services, which are transmitted within their own Exchange, as independent operations separate from any ongoing communication; and
- FC-4 Link Services, which follow the same procedures used for Basic and Extended Link Services in performing FC-4 level operations.

The Link Service Frames and Sequences are composed of Link Data Frames and operate according to the normal R_RDY Primitive Signal and the Link Control Frame (ACK, RJT, and BSY) response rules used for Data Frames. Basic Link Service Frames cannot contain optional headers, so the DF_CTL Frame Header field is meaningless on these Frames. The Extended Link Service and FC-4 Link Data Frames may contain only the Expiration/Security optional header. Link Service Frames can be sent in any supported Class of service.

Basic Link Service Commands

The Basic Link Service commands are shown in Figure 11.1. All Basic Link Service commands must be supported by all N_Ports. Five Basic Link Service commands are defined: three Basic Link Service Request (BLS Requests); "No Operation" (NOP), "Abort Sequence" (ABTS), and "Remove Connection"; and two Basic Link Service Replies (BLS Replies); "Basic Accept" (BA_ACC), and "Basic Reject" (BA_RJT). As shown in Figure 7.2, the Basic Link Service Frames are distinguished from other Frames by the values in bits 31–24 of the Frame Header R_CTL field.

The BLS Requests are transmitted as single Frames without any payload, possibly as part of an already-open Sequence. The BA_ACC and BA_RJT Frames are transmitted as replies to the ABTS BLS request and have a small Payload indicating Reply parameters to the ABTS request.

Figure 11.1
Basic Link Service commands.

Encoded Value, Word 0 [27–24]	Description	Abbrev.	Request/ Reply	Payload
x'0'	No Operation	NOP	Request	None
x'1'	Abort Sequence	ABTS	Request	None
x'2'	Remove Connection	RMC	Request	None
x'4'	Basic Accept	BA_ACC	Reply	dependent on service request
x'5'	Basic Reject	BA_RJT	Reply	BA_RJT reason and explanation

No Operation (NOP)

The No Operation BLS Request contains no data field and has no meaning in itself. However, the Frame Header F_CTL field and the Frame delimiters are examined by the Fabric and the Recipient N_Port, so actions are taken based on these elements of the Frame.

For example, an NOP Frame sent with **SOFc1** and **EOFn** delimiters can be used to initiate a Class 1 Connection. An NOP Frame with End_Sequence F_CTL bit set can be used to normally terminate a Sequence without sending a data field Payload. It could also be used to keep the Sequence Recipient from timing out a Sequence when there is no data available for longer than an E_D_TOV period or to remove a Class 1 Connection, when there is no data to include in a Data Frame.

Beyond the fact that the NOP has no Data Field, it acts as a normal Data Frame, with all attendant control and signaling properties. Clearly, this can be a very useful and widely used Frame. The OX_ID, RX_ID, SEQ_ID, S_ID and D_ID fields are set to be appropriate to the Sequence in which the NOP is being sent.

Remove Connection (RMC)

The Remove Connection BLS Request can be used to request immediate removal of a Class 1 Connection. This overrides the normal termination method and abnormally terminates all open Sequences on the Connection. An ACK Frame with an **EOFt** delimiter is returned, and recovery on all open Sequences must follow by normal means. RMC is sent as part of an ongoing Sequence or as a separate Sequence, with the S_ID, D_ID, OX_ID,

RX_ID, and SEQ_ID values set to match those of the Exchange and Sequence in which the RMC is being transmitted. Naturally, End_Connection (F_CTL bit 18) is set to 1 in the Frame Header.

Basic Accept (BA_ACC)

The Basic Accept (BA_ACC) is a single Frame Basic Link Service reply Sequence that notifies the transmitter that a Basic Link Service command has been completed. The normal procedures for transferring Sequence Initiatives apply. The OX_ID and RX_ID are set to match the Exchange in which the ABTS Frame was transmitted. The SEQ_ID are assigned following the normal rules for SEQ_ID assignment.

The only current use of the BA_ACC Basic Link Service command is as a reply to the ABTS command, where it has the Payload format shown in Figure 11.2.

Figure 11.2
Format of the Basic Accept to the ABTS Basic Link Service command.

Basic Accept (BA_ACC) to ABTS

Word \ Bit	33222222 10987654	22221111 32109876	11111100 54321098	00000000 76543210
0	validity	SEQ_ID	reserved	
1	OX_ID		RX_ID	
2	Low SEQ_CNT		High SEQ_CNT	

Notes:
The "Validity field indicates the validity of the SEQ_ID field: x'80' = valid, x'00' = invalid.
The SEQ_ID field, if valid, is the SEQ_ID of the last Sequence deliverable to the ULP.
Usage of the Low and High SEQ_CNT fields is described in the ABTS definition.

Basic Reject (BA_RJT)

Basic Reject is used to notify the Initiator of a BLS Request that the request has been rejected. The 4-byte Payload for BA_RJT, which includes reject reason codes and reason explanation codes, is shown in Figure 11.3. These reasons may be specific to a specific BLS Request, although BA_RJT is only sent in reply to the ABTS BLS Request. The OX_ID and RX_ID match the Exchange in which the BLS Request was transmitted. The SEQ_ID is assigned following the normal rules for SEQ_ID assignment.

Figure 11.3
Basic Reject BLS Response format, BA_RJT reason codes, and reason explanation codes.

Bit / Word	33222222 10987654	22221111 32109876	11111100 54321098	00000000 76543210
Basic Reject (BA_RJT) — 0	reserved	Reason Code	Explanation Code	Vendor Unique

Reason Code	Description
x'01'	Invalid Command code - The command code in the Sequence being rejected is invalid
x'03'	Logical Error - The request identified by the command code is invalid or logically inconsistent for the conditions present
x'05'	Logical Busy - The basic link service is logically busy and unable to process the request at this time.
x'07'	Protocol error - An error has been detected which violates FC-2 protocol rules, which is not specified by other error codes.
x'09'	Unable to perform command request - The recipient of a link service command is unable to perform the request at this time,
x'FF'	Vendor unique error - May be used by vendors to specify additional reason codes.
others	reserved

Explanation code	Description	Applicable Commands
x'00'	No additional explanation	ABTS
x'03'	Invalid OX_ID - RX_ID combination	ABTS
x'05'	Sequence Aborted, no Sequence information provided	ABTS
others	reserved	reserved

The only current usage of the BA_RJT Basic Link Service command is as a reply to the ABTS command, where it has the Payload format shown in Figure 11.3.

Abort Sequence (ABTS) Command

The ABTS Basic Link Service Frame can be used to abort one or more Sequences, to determine the status at the Recipient of an outstanding Sequence, and to abort an entire Exchange. The Initiator of a Sequence with a questionable or indeterminate completion status can sent the ABTS command to the Sequence Recipient to determine the status and further processing required. The Recipient will build a BA_ACC Frame in the format shown in Figure 11.2. The settings in the field indicate what Sequences have been completed and gives a window of Frames to discard on receipt. This window is active for longer than the R_A_TOV timeout period to ensure that no Frames are active from the first Sequence transmission when the recovery procedure is finished and further Sequences will be transmitted. The full operation of the Abort Sequence protocol for aborting no Sequences, aborting all the incomplete Sequences of an Exchange, or aborting an entire Exchange is described in the "Sequence Recovery" section, on page 233.

Extended Link Service Command Overview

Extended Link Service commands (ELS commands) are used by an N_Port to solicit a destination Port (either F_Port or N_Port) to perform some link-level function or service. These commands can be used either for determining environmental characteristics, for determining the status of some operation or action at another Port, for performing Login or Logout functions, or for aborting existing Exchanges.

Extended Link Service commands are split into two categories: "Extended Link Service Requests" (ELS Requests) and "Extended Link Service Replies" (ELS Replies). Both Requests and Replies are termed "Extended Link Service commands." The terminology used here slightly modifies the terminology used in the Fibre Channel standards documents, for clarification. The Recipient of an ELS Request may either accept the Request, accept it with qualifications (in some specific cases), or reject the Request entirely. The response is generally indicated in an ELS Reply that is returned to the Request Initiator. The defined Extended Link Service commands, both ELS Replies and ELS Requests, are shown in Figure 11.4. All of the commands are ELS Requests except "Link Service Reject" (LS_RJT) and "Accept" (ACC), which are ELS Replies.

Each command is composed of a single Sequence of one or more Frames, with an LS_Command code (Extended Link Service command code) being specified in the first Payload byte of the first Frame. R_CTL, Information Category, and TYPE field bits for the Extended Link Service Requests and Replies are shown in Figure 7.2. Each Extended Link Service Request is combined with its corresponding Reply in a single Exchange, with the exception of the ESTS, ESTC, and ADVC Sequences. The ACC Reply Sequence to all ELS Requests (except ESTS, ESTC, and ADVC) will terminate the Exchange by setting the Last_Sequence F_CTL bit (bit 20) to 1 on its last Frame. Basically, all other normal rules for Frame, Sequence, and Exchange management apply to Extended Link Service Frames, Sequences, and Exchanges.

Figure 11.4 Extended Link Service commands, with LS_Command codes and usage requirements.

Cmd. Code Word 0 [31-24]	Description	Abbrev.	N_Port sup-port	Exchanges	Reply Required ?
x'01'	Link Service Reject (1)	LS_RJT	req'd	same as rqst	(is reply)
x'02'	Accept (1)	ACC	req'd	same as rqst	(is reply)
x'03'	N_Port Login	PLOGI	req'd	single	yes
x'04'	F_Port Login	FLOGI	req'd	single	yes
x'05'	Logout	LOGO	req'd	single	yes
x'06'	Abort Exchange	ABTX	opt.	single	yes
x'07'	Read Connection Status	RCS	opt.	single	yes
x'08'	Read Exchange Status Block	RES	opt.	single	yes
x'09'	Read Sequence Status Block	RSS	opt.	single	yes
x'0A'	Request Sequence Initiative	RSI	opt.	single	yes
x'0B	Establish Streaming	ESTS	opt.	single, multiple	yes
x'0C'	Estimate Credit	ESTC	opt.	single, multiple	no
x'0D'	Advise Credit	ADVC	opt.	single, multiple	yes
x'0E'	Read Timeout Value	RTV	opt.	single	yes
x'0F'	Read Link Error Status Block	RLS	opt.	single	yes
x'10'	Echo	ECHO	opt.	single	yes
x'11'	Test	TEST	opt.	single	no
x'12'	Reinstate Recovery Qualifier	RRQ	req'd(2)	single	yes
x'20'	Process Login	PRLI	opt.	single	yes
x'21'	Process Logout	PRLO	opt.	single	yes
x'22'	State Change Notification	SCN	opt.	single	no
x'23'	Test Process Login State	TPLS	opt.	single	yes
x'24'	Third Party Process Logout	TPRLO	opt.	single	
x'30'	Get Alias_ID	GAID	opt.	single	yes
x'31'	Fabric Activate Alias_ID	FACT	opt.	single	yes
x'32'	Fabric Deactivate Alias_ID	FDACT	opt.	single	yes
x'33'	N_Port Activate Alias_ID	NACT	opt.	single	yes
x'34'	N_Port Deactivate Alias_ID	NDACT	opt.	single	yes
x'40'	Quality of Service Request	QoSR	opt.		
x'41'	Read Virtual Circuit Status	RVCS	opt.		
x'50'	Discover_N_Port Service Parm	PDISC	opt.		
x'51'	Discover F_Port Service Parm	FDISC	opt.		
x'52'	Discover Address	ASICS	opt.		
x'53'	Report Node Capability	RNC	opt.		
others	reserved	-	-	-	-

(1): LS_RJT and ACC are transmitted as Information Category "Solicited Control" (R_CTL[27–24] = x'3'). All others are "Unsolicited Control" (R_CTL[27–24] = x'2').
(2): RRQ is only required for N_Ports supporting Class 2 and Class 3 service.

A Sequence Initiator will transmit an ELS Request Sequence to solicit the destination N_Port to perform a particular link-level service. The last Data Frame of the Request will transfer the Sequence Initiative to allow an ELS Reply to be transmitted in response. Only the ESTC (Estimate Credit) and TEST (Test) requests do not require replies.

Figure 11.4 shows the required and optional Extended Link Service commands. Perhaps surprisingly, an N_Port is not required to be able to initiate the PLOGI Extended Link Service command. This allows simpler N_Port implementation on peripheral devices, such as disk drives, which must be very inexpensive, and can depend on another N_Port on the network to generate PLOGI. However, if such an N_Port receives the PLOGI LS_Command, the N_Port must be able to respond properly.

The Originator of the ELS Request will assign a SEQ_ID and an OX_ID when the command Sequence and Exchange are originated. The Responder will assign an RX_ID value when the command Sequence is received. The value x'FFFF' may be used for both the OX_ID and RX_ID if only one Exchange is open with the Originator, so that uniqueness is assured.

There are two ELS command Sequences in which transmission of the Sequence to a remote N_Port can collide with reception of the identical command from the same N_Port. These are the ABTX (Abort Exchange) and the RSI (Request Sequence Initiative) ELS commands. Clearly, if both Originator and Responder of an Exchange request Sequence Initiative, there is a collision, and operations at the Initiator and Recipient of the ABTX Sequence should be different. To circumvent this problem, the Originator of the referenced Exchange takes priority and sends an LS_RJT ELS Reply with a reason code of "Command already in Progress," so that the Responder replies to the Originator's Request.

Types of Extended Link Service Commands

Following are detailed descriptions of the Extended Link Service Requests and Replies. The Extended Link Service commands are grouped by relevant function, since the various commands are used in distinctly different ways. These include ELS Replies; ELS Requests for Login, Logout, and Aborting Exchanges; Requests for Status Determination and Initiative Requests; Requests for Streaming Credit Determination; Requests for Miscellaneous Functions; Requests related to Alias_IDs; and Requests related to Class 4 Service. Requests related to Process Login, Logout, and Status are described in the "Overview of Process Login/Logout" section, on page 174.

Extended Link Service Replies

Two Extended Link Service commands are used as ELS Replies: Link Service Reject (LS_RJT) and Accept (ACC). These ELS Replies indicate a response to the ELS Requests. The format of LS_RJT is fixed, but the format of the ACC Reply depends on which ELS Request is being accepted.

Link Service Reject (LS_RJT)

The Link Service Reject ELS Reply notifies the Initiator of an ELS Request Sequence that the Request has been rejected, and provides codes indicating the reason for the rejection and a short explanation, as shown in Figure 11.5 and also in Figure 11.6. The reason and explanation codes are listed here in detail since understanding how operations work is helped by understanding what happens when they don't work as expected. The LS_RJT explanations are mostly useful for just a subset of the Link Service Requests — the newer explanation codes relate to N_Port ID aliasing, and Class 4 service.

Accept (ACC)

The Accept (ACC) Extended Link Service Reply notifies the Initiator of an ELS Request that the request has been accepted. ACC is a valid reply Sequence to all ELS Requests except ESTC and TEST. The ACC Payload following the LS_Command code of x'02' is completely dependent on which ELS Request is being replied to and is described for each ELS Request in the relevant section.

ELS Requests: Login, Logout, and Abort Exchange

The two ELS Requests used for Login, PLOGI, and FLOGI are discussed in detail in Chapter 9. The LOGO ELS Request is used to release resources associated with a Login. The ABTX ELS Request could be used by either an Exchange Originator or Responder to abort an open Exchange, although much of the same function is provided by support for the ABTS-LS procedure described in the "Abort Sequence (ABTS) Command" section, on page 199.

Figure 11.5
Link Service Reject (LS_RJT) format and reason codes.

Link Service Reject (LS_RJT)

Word	Bit 33222222 10987654	Bit 22221111 32109876	Bit 11111100 54321098	Bit 00000000 76543210
0	reserved	Reason Code	Explanation Code	Vendor Unique

Reason Code	Description
x'01'	Invalid LS_Command code - the LS_Command code in the Sequence being rejected is invalid or not supported.
x'03'	Logical Error - The request identified by the LS_Command code and payload content is invalid or logically inconsistent for the conditions present.
x'05'	Logical Busy - The link service is logically busy and unable to process the request at this time.
x'07'	Protocol error - An error has been detected which violates extended link service protocol rules and which is not specified by other error codes.
x'09'	Unable to perform command request - The recipient of an extended link service command is unable to perform the request at the time.
x'11'	Command not supported - The recipient of a link service command does not support the command requested.
x'FF'	Vendor unique error - May be used by vendors to specify additional reason codes
others	reserved

N_Port Login (PLOGI)

The N_Port Login ELS Request notifies the Recipient of the Initiating N_Ports requested N_Port Common and Class Service Parameters and requests an ACC in reply to complete N_Port Login. Transmission of the PLOGI Sequence and return of the ACC reply constitute a negotiation on Service Parameters such that both N_Ports can agree on shared value to be able to operate together. This N_Port Login procedure is described in detail in the "N_Port Login" section, on page 163.

The format of the PLOGI and its corresponding ACC are shown in detail in Figure 9.3, and described further in the "N_Port and F_Port Service Parameters" section, on page 166.

The PLOGI can be directed to any other N_Port on the network or to a Fibre Channel service (Management server, Time server, etc.), using the appropriate well-known Address Identifiers listed in the "Address Identifi-

Figure 11.6
Link Service Reject (LS_RJT) explanation codes.

Code	Description	Applicable commands
x'00'	No additional explanation	all (1)
x'01'	Service Parameter error - Options	FLOGI, PLOGI
x'03'	Service Parameter error - Initiator Control	FLOGI, PLOGI
x'05'	Service Parameter error - Recipient Control	FLOGI, PLOGI
x'07'	Service Param. error - recv. data field size	FLOGI, PLOGI
x'09'	Service Parameter error - concurrent Seqs.	FLOGI, PLOGI
x'0B'	Service Parameter error - credit	ADVC, FLOGI, PLOGI
x'0D'	Invalid N_Port/F_Port Name	FLOGI, PLOGI
x'0E'	Invalid Node/Fabric Name	FLOGI, PLOGI
x'0F'	Invalid Common Service Parameters	FLOGI, PLOGI
x'11'	Invalid Association Header	ABTX, RES, RRQ, RSI
x'13'	Association Header Required	ABTX, RES, RRQ, RSI
x'15'	Invalid Originator S_ID	ABTX, RES, RRQ, RSI, RSS
x'17'	Invalid OX_ID - RX_ID combination	ABTX, RES, RRQ, RSI, RSS
x'19'	Command (request) already in progress	ABTX, PLOGI, RSI
x'1F'	Invalid N_Port Identifier	RCS, RLS
x'21'	Invalid SEQ_ID	RSS
x'23'	Attempt to abort invalid Exchange	ABTX
x'25'	Attempt to abort inactive Exchange	ABTX
x'27'	Recovery Qualifier required	ABTX
x'29'	Insufficient resources to support Login	FLOGI, PLOGI
x'2A'	Unable to supply requested data	ADVC, ESTS, RCS, RES, RLS, RSS, RTV
x'2C'	Requests not supported	all but FLOGI, PLOGI, LOGO
x'2D'	Invalid Payload Length	FLOGI, PLOGI
x'30'	No Alias IDs available for this Alias Type	GAID
x'31'	Can't activate Alias ID (no resources available)	FACT, NACT
x'32'	Alias ID can't be activated (invalid ID)	FACT, NACT
x'33'	Alias ID can't be deactivated (doesn't exist)	FDACT, NDACT
x'34'	Can't deactivate Alias ID (resource problem)	FDACT, NDACT
x'35'	Service Parameter conflict	NACT
x'36'	Invalid Alias_Token	GAID
x'37'	Unsupported Alias Token	NACT
x'38'	Alias Group cannot be formed	GAID
x'40'	QoS Parameter Error	QoSR
x'41'	VC_ID Not Found	QoSR
x'42'	Insufficient Resources for Class 4 circuit	QoSR
Others	reserved	

(1) all = ABTX, ADVC, ESTS, FLOGI, PLOGI, LOGO, RCS, RES, RLS, RSS, RTV, RSI, PRLI, PRLO, TPLS, TPRLO, GAID, FACT, FDACT, NACT, NDACT, QoSR, RVCS, PDISC, FDISC, ADISC, RNC

ers: S_ID and D_ID" section, on page 115. When an N_Port receives a PLOGI from another N_Port, any active or open Sequences with the N_Port performing the re-Login will be abnormally terminated — the PLOGI restarts the communication.

F_Port Login (FLOGI)

The F_Port Login ELS Request is initiated by an N_Port and directed toward the entity at the well-known Fabric F_Port Identifier x'FF FFFE,' to initiate the Fabric Login procedure. The procedure establishes operating parameters for communicating with a Fabric, if present. This Fabric Login procedure is described in detail in the "Fabric Login" section, on page 161.

The format of the FLOGI and its corresponding ACC are shown in detail in Figure 9.3 and are described further in the "N_Port and F_Port Service Parameters" section, on page 166.

Logout (LOGO)

The Logout (LOGO) ELS Request is used to request invalidation of the service parameters and Port_Name which have been saved by an N_Port, freeing those resources. This provides a means by which an N_Port may command Logout, or removal of service, with another N_Port. The Payload of the LOGO ELS Request contains the N_Port Identifier and N_Port_Name, as shown in Figure 9.4, and the usage of the command, with its corresponding reply, are described in the "Logout" section, on page 165.

Aside from allowing non-disruptive disconnection from the network, this ELS command can be used to allow an N_Port to change its own N_Port identifier. Any open Exchange at the Initiator and Recipient N_Ports which uses the existing N_Port identifier will be abnormally terminated on reception or transmission of the LOGO ELS Request.

Abort Exchange (ABTX)

The Abort Exchange ELS Request is established as a way to request abnormal termination of an open Exchange. The Abort Exchange ELS Request is transmitted in a separate Exchange to abort an existing Exchange. It contains in the Payload the OX_ID and RX_ID values for the Exchange to be aborted, along with possibly an Association Header, if the Recipient requires it, a source N_Port Identifier value, and a flag to indicate whether a Recovery_Qualifier is required.

This command has been largely superceded by the ABTS-LS procedure, which allows aborting Exchanges with easier record-keeping and less overhead, so it is not detailed here.

ELS Requests: Status Determination and Initiative Request

Three Extended Link Service Requests are defined for determining the status of Connections, Exchanges, and Sequences at remote Nodes. Included also in this section is the ELS Request for Sequence Initiative on an existing Exchange. The format of these ELS Requests is shown in Figure 11.7, as are those for the ACC Payloads for the RCS and RSI Requests. The Payloads of the Replies to the RES and RSS Requests are shown in Figure 8.2 and Figure 8.3 as Exchange and Sequence Status Blocks.

Read Connection Status (RCS)

The Read Connection Status ELS Request is used to request the Fabric controller to return the current Dedicated Connection status for the N_Port specified in the RCS Sequence Payload. The RCS command provides the means by which an N_Port may interrogate the Fabric for the Connection status of other N_Ports within the Fabric. The Cnct-Stat field shown in Figure 11.7 contains 4 bits, describing whether a connect-request was delivered or stacked, whether a Connection is established, and whether the Connection supports Intermix. The N_Port identified in the ACC, if valid, is associated with the N_Port specified in the RCS Request.

Read Exchange Status Block (RES)

The Read Exchange Status Block (RES) ELS Request is used to request the Exchange status for the Exchange originated by the S_ID specified in the RES. Full identifier information is included for the Exchange in the Request, including OX_ID, RX_ID, and Association Header, if required, although the destination N_Port may not require them all to access the Status Block. The ACC to the RES ELS Request returns the Exchange Status Block information in the format shown in Figure 8.3. The Remote N_Port will naturally keep more information on the Exchange, but the other information is not accessible to another N_Port through standard means.

Figure 11.7
Payloads for the RCS, RES RSS, and RSI Requests.

Read Connection Status (RCS)

Word \ Bit	33222222 10987654	22221111 32109876	11111100 54321098	00000000 76543210
0	0000 0111	0000 0000	0000 0000	0000 0000
1	0000 0000	N_Port Identifier		

ACC for RCS

Word \ Bit	33222222 10987654	22221111 32109876	11111100 54321098	00000000 76543210
0	0000 0010	0000 0000	0000 0000	0000 0000
1	Cnct-Stat	Identifier of connected N_Port		

Read Exchange Status Block (RES)

Word \ Bit	33222222 10987654	22221111 32109876	11111100 54321098	00000000 76543210
0	0000 1000	0000 0000	0000 0000	0000 0000
1	0000 0000	Originator S_ID		
2	OX_ID		RX_ID	
3-10	Association Header : :			

Read Sequence Status Block (RSS)

Word \ Bit	33222222 10987654	22221111 32109876	11111100 54321098	00000000 76543210
0	0000 1001	0000 0000	0000 0000	0000 0000
1	SEQ_ID	Originator S_ID		
2	OX_ID		RX_ID	

Request Sequence Initiative (RSI)

Word \ Bit	33222222 10987654	22221111 32109876	11111100 54321098	00000000 76543210
0	0000 1010	0000 0000	0000 0000	0000 0000
1	0000 0000	Originator S_ID		
2	OX_ID		RX_ID	
3-10	Association Header : :			

ACC for RSI

Word \ Bit	33222222 10987654	22221111 32109876	11111100 54321098	00000000 76543210
0	0000 0010	0000 0000	0000 0000	0000 0000

Read Sequence Status Block (RSS)

The Read Sequence Status Block (RSS) ELS Request is used to request the Sequence status for the Sequence specified by the fields in the RSS. Full information is included for the Sequence, including Association Header, if valid, although the destination N_Port may not use them all in accessing the Status Block. The ACC to the RSS ELS Request returns the Sequence Status Block information in the format shown in Figure 8.2. The Remote N_Port will naturally keep more information on the Sequence, but the other information is not accessible to another N_Port through standard means.

Request Sequence Initiative (RSI)

The Request Sequence Initiative ELS Request is used to request that the Sequence Initiative be passed to Initiator of the RSI. The RSI is sent in a separate Exchange so as to not disturb the Exchange for which Sequence Initiative is being requested. This is not a disruptive request — if a Sequence is active, the Initiator will terminate the Sequence normally, using the End_Sequence and Sequence_Initiative F_CTL bits. If no Sequences are active, the previous Initiator will initiate a new single-Frame Sequence containing only the NOP Basic Link Service command, with the End_Sequence and Sequence_Initiative F_CTL bits set, to transfer Sequence Initiative. The Payload of RSI and the ACC for RSI are shown in Figure 11.7.

ELS Requests: Credit Determination

The Establish Streaming (ESTS), Estimate Credit (ESTC), and Advise Credit (ADVC) ELS Requests are used to estimate the most efficient value of end-to-end Credit to use between a particular Initiator/Recipient N_Port pair. The procedure for establishing credit is described in detail in the "Procedure to Estimate End-to-End Credit" section, on page 176.

Establish Streaming (ESTS)

The Establish Streaming ELS Request is used to request a temporary allocation of credit, known as streaming credit, large enough to perform continuous streaming of Data Frames. The Payload of this Request is a single word containing the LS_Command for ESTS, x'0B.' The ACC reply is in the form of the PLOGI and FLOGI Requests, as shown in Figure 9.3. The end-to-end Credit in the ACC is set to a large value for the Class of service requested in the ESTS.

Estimate Credit (ESTC)

The Estimate Credit Link Service command is used to estimate the minimum credit required to achieve the maximum bandwidth for a given distance between an N_Port pair. The ESTC Link Service command is unusual in having a variable Frame size determined by Login with the destination N_Port. The Payload of the Frame is arbitrary and will generally be ignored at the Recipient. There is no Reply, as described in the "Procedure to Estimate End-to-End Credit" section, on page 176.

Advise Credit (ADVC)

The ADVC Link Service command is used to request the destination N_Port to change the end-to-end Credit to the value included in the Payload. In the procedure to estimate end-to-end Credit, this will presumably be the value determined to be the minimum required for streaming, as described in the "Procedure to Estimate End-to-End Credit" section, on page 176. This Request can be used outside of the procedure, to request a new value of the end-to-end Credit with the destination N_Port acting as Sequence Recipient. The ACC reply indicates the actual revised end-to-end Credit, which might or might not be different from the previous value.

ELS Requests: Miscellaneous Functions

This section describes several ELS Requests which are used independently and do not depend on or relate to other ELS Requests. These include "Read Timeout Value," which is used to determine the E_D_TOV; R_A_TOV, "Read Link Error Status Block," which reads the Status Block described in the "Link Error Status Block Rules" section, on page 240; "Echo," which can be used to test connectivity by requesting an echo; "Test," which can test connectivity by sending ignored data; and "Reinstate Recovery Qualifier," which is used to clear status after aborting an Exchange or set of Sequences.

Figure 11.8 Payloads for the RTV, RLS, ECHO, TEST, and RRQ ELS Requests and Replies.

Read Timeout Value (RTV)

Word \ Bit	33222222 10987654	22221111 32109876	11111100 54321098	00000000 76543210
0	0000 0111	0000 0000	0000 0000	0000 0000

ACC for RTV

Word \ Bit	33222222 10987654	22221111 32109876	11111100 54321098	00000000 76543210
0	0000 0010	0000 0000	0000 0000	0000 0000
1	0000 0000	Originator S_ID		

Echo (ECHO) ACC is identical

Word \ Bit	33222222 10987654	22221111 32109876	11111100 54321098	00000000 76543210
0	0001 0000	0000 0000	0000 0000	0000 0000
1-N	Payload to be returned : :			

Test (TEST)

Word \ Bit	33222222 10987654	22221111 32109876	11111100 54321098	00000000 76543210
0	0001 0001	0000 0000	0000 0000	0000 0000
1-N	Payload :			

Reinstate Recovery Qualifier (RRQ)

Word \ Bit	33222222 10987654	22221111 32109876	11111100 54321098	00000000 76543210
0	0001 0010	0000 0000	0000 0000	0000 0000
1	0000 0000	Originator S_ID		
2	OX_ID		RX_ID	
3-10	Association Header :			

ACC for RRQ

Word \ Bit	33222222 10987654	22221111 32109876	11111100 54321098	00000000 76543210
0	0000 0010	0000 0000	0000 0000	0000 0000

Read Timeout Value (RTV)

The Read Timeout Value ELS command requests an N_Port or F_Port to return the Resource_Allocation_Timeout Value and the Error_Detect_Timeout value for the network as a count of 1-ms. increments in the ACC reply Sequence. Generally, the destination of this ELS Request would be the well-known Fabric F_Port identifier at x'FF FFFE,' although another N_Port could return the value as well. The usage of the E_D_TOV and R_A_TOV parameters is discussed in Chapter 12. The format of the ELS Request and ACC Reply is shown in Figure 11.8.

Read Link Error Status Block (RLS)

The Read Link Error Status Block ELS command requests an N_Port or F_Port to return an ACC containing information on long-term link integrity, as described in the "Link Error Status Block Rules" section, on page 240. Ports which don't track this optional information will LS_RJT this command. The format for the ACC returned, less the LS_Command code of the ACC, is shown in Figure 12.1.

Echo (ECHO)

The Echo ELS command consists of a single Frame of arbitrary data, which the Initiator is requesting that the Recipient return in the ACC reply Payload. This command provides a means to test connectivity with a loop-back diagnostic function. Other than the limitations on Frame size for the Class of service used for the Request and Reply, and the requirement of transmitting valid data, the contents of the Payload are arbitrary.

Test (TEST)

The Test ELS command consists of a single Frame of arbitrary data, which the Initiator is requesting to transmit to the Recipient. There is no ELS Reply. This command can be used, for example, to test connectivity and to provide system loading for performance diagnostics. Other than the limitations on Frame size for the Class of service used for the Request, and the requirement of transmitting valid data, the contents of the Payload are arbitrary.

Reinstate Recovery Qualifier (RRQ)

The Reinstate Recovery Qualifier ELS command is used to notify the destination N_Port that the Recovery Qualifier is available for reuse, which is to say that the R_A_TOV period has expired, all the Frames in the invalid range have been discarded, and the Recovery Qualifier can be cleared. This Request can only be sent by either the Originator or the Responder of the Exchange identified by the OX_ID and RX_ID values of the Recovery Qualifier. It is sent in a separate Exchange, of course, which just contains ELS Request and ELS Reply Sequences. Resources associated with the aborted Exchange or Sequences should be available for use directly following transmission of the ACC Reply. Both the Originator and Responder must ensure that the OX_ID and RX_ID pair in the Recovery Qualifier match the OX_ID and RX_ID pair in the Reinstate Recovery Qualifier Request, or there would be a serious protocol error. The Payload formats for the RRQ Request and its ACC Reply are shown in Figure 11.8.

ELS Requests for Alias IDs

It is a generally useful function to be able to have a single N_Port known by several different IDs. This allows packets with different D_ID values to be routed to the same Port, but to be handled differentlyonce they get there.

In Fibre Channel, the currently-defined services that use Aliases are multicast service (either unreliable Class 3, or reliable Class 6, or hunt groups. In either case, the Fabric must contain an Alias Server, which handles Extended Link Service Requests related to Alias_IDs.

Get Alias_ID (GAID)

This Extended Link Service Request is used to set up an Alias_ID on a switched Fabric for a list of 1 or more N_Ports. The Alias Server (at address x'FF FFF8' sends the GAID to the Fabric Controller (at address x'FF FFFD'). The payload consists primarily of a list of 1 or more N_Port IDs, an Alias_Token which identifies the Alias Group, and a 80-byte list of Service Parameters for the group. The Accept ELS returned contains an Alias_ID which can work across all of the N_Ports in the GAID payload with the requested Service Parameters.

Fabric Activate and Deactivate Alias_ID (FACT, FDACT)

The FACT and FDACT ELS Requests are used to activate and deactivate the Alias_ID for a set of N_Ports. This ELS Request also contains the list of N_Port IDs used in the GAID request, along with the Alias_ID returned in the ACC for that GAID request. Once the FACT is acknowledged with an ACC, the Alias_ID can be used to address the entire group until the FDACT is sent and accepted.

N_Port Activate and Deactivate Alias_ID (NACT, NDACT)

The NACT is sent to a particular N_Port by an Alias Server to tell it to assign the passed alias address identifier as an alias. The NACT contains the 80-byte set of Service Parameters to be used for the Alias_ID — these don't need to have the same values as the Service Parameters used by the N_Port in its original Login.

ELS Requests for Class 4 Service

Since Class 4 service, as described in the "Class 4 — Fractional" section, on page 192, requires specific guarantees from the Fabric for minimum/maximum bandwidth, and maximum frame delivery latency, it is necessary for the Class 4 circuits to be set up, using Extended Link Services, through a negotiation between the Class 4 circuit initiator and the Fabric.

Quality of Service Request (QoSR)

The QoSR ELS Request requests setup of a Class 4 circuit with a particular service level. The QoSR includes fields to specify the requested Maximum and Minimum Bandwidth, and the Maximum delay in microseconds from the Initiator to the Recipient, as well as the maximum data field size. The maximum data field size is important since it may be difficult to guarantee the QoS parameters for frames larger than this. The ACC Frame indicates the QoS parameters that the Fabric is able to currently guarantee for the circuit.

Read Virtual Circuit Status (RVCS)

The RVCS ELS Request is sent from the requesting N_Port to the Quality of Service Facilitator, and contains an N_Port identifier in the payload. The QoSF returns an ACC containing a list of Virtual Circuit identifiers and Address identifiers for each Class 4 circuit with the designated N_Port.

FC-4 Link Services

The actual commands used for FC-4 Link Services are specified separately for each of the different FC-4 ULP interfaces, and will be dependent on the protocol being supported. The FC-4 Link Service mechanism provides a set of guidelines for specifying how link-level services should be performed within an FC-4, to ensure that they resemble the Basic and Extended Link Services supported directly within Fibre Channel. The Frame Header fields for FC-4 Link Service Frames are shown in Figure 7.2.

FC-4 Link Services are assumed to follow the request-response model of used for the majority of Extended Link Services. A source Port (either N_Port or F_Port) initiates a Sequence of one or more Frames, using the "Unsolicited Control" Information Category, requesting the destination Port (again either N_Port or F_Port) to perform a function or service in support of the FC-4 protocol. The Recipient responds with a reply Sequence in the same Exchange, using the "Solicited Control" Information Category. The normal rules for Sequences and Exchange management would apply. In addition, the following rules further restrict Exchanges used for FC-4 Link Services:

- The Frame Header Routing, Information Category, and TYPE bits are set as shown in Figure 7.2.
- FC-4 Link Service Exchanges can only be originated following N_Port Login and must use the "Abort, Discard multiple Sequences" Exchange Error Policy.
- An Exchange error is detected if the reply does not return within a time period of twice R_A_TOV after transfer of Sequence Initiative.
- If the Exchange Originator detects an Exchange error, it will abort the Exchange and retry the FC-4 Link Service Request in a different Exchange.
- If the Sequence Initiator receives an ACK with the Abort Sequence Condition bits set "Abort Sequence, perform ABTS" (b'01'), it will abort the Sequence and retry it only once after the BA_ACC is received.

These rules allow Link Service operations specified at the FC-4 level to operate within the normal Sequence, Exchange, and timeout guidelines that are supported for the Basic and Extended Link Services.

Chapter 12

Error Detection and Recovery

Introduction

There are two fundamental levels of error detection in Fibre Channel. Link integrity focuses on the inherent quality of the received transmission signal. Sequence integrity focuses on the integrity of the received data, within a Frame, within a Sequence, and between Sequences in an Exchange. Error recovery procedures must be defined to operate at both of these levels.

When the integrity of the link is in question, a hierarchy of Primitive Sequences are used to reestablish link integrity, as described in the "Link Recovery Protocols" section, on page 104. When Primitive Sequence protocols are finished, or when data transmission shows errors with overall link integrity, data recovery on a Sequence-by-Sequence or Exchange basis may be required.

A separate type of error recovery must be implemented to recovery from end-to-end or buffer-to-buffer Credit errors that can occur as a result of transmission errors in Primitive Signals, Data Frames, and Link Control Frames. Credit recovery is handled separately from Sequence recovery, although when a sequence is aborted, any end-to-end Credit for outstanding ACK Frames of the Sequence should be recovered. End-to-end Credit loss is usually only detectable when either a receive buffer overflows (for too high a credit value) or (more commonly) when data transmission suffers or stops altogether due to lost credit. Recovery of credit using the Link Credit Reset Primitive Sequence is described in the "Link Credit Reset (LCR) Frame" section, on page 134, and in the "End-to-End Flow Control" section, on page 265 and the "Buffer-to-Buffer Flow Control" section, on page 270, discussing flow control.

Beyond the link recovery mechanisms described in the "Link Recovery Protocols" section, on page 104 and the credit recovery mechanisms described in the "Link Credit Reset (LCR) Frame" section, on page 134 and Chapter 15, error recovery processing happens at the Sequence and higher levels. This chapter describes mechanisms for detecting and recovering from transmission errors at the Sequence level.

A Sequence provides the basis for ensuring the integrity of the block of data transmitted and received. Exchange management ensures that Sequences are delivered in the manner specified by the Exchange Error Policy, described in the "Exchange and Sequence Integrity" section, on page 228. Each Frame within a Sequence is tracked on the basis of X_ID (RX_ID and/or OX_ID), SEQ_ID, and the SEQ_CNT within the Sequence. Each Frame is verified for validity during reception as described in the "Frame Reception and Frame Validity" section, on page 226, and errors in any Frame of a Sequence will generate recovery procedures for the entire Sequence.

The basic method of detecting lost Frames and transmission errors is through timeouts. When an expected action or operation has not occurred within a timeout period, the N_Port takes action to determine the reason for the failure and to correct the transmission. If an error has occurred, the normal recovery mechanism is through the use of the ABTS Basic Link Service Request and/or the ABTX Extended Link Service Request and related operations. In some cases a Sequence Recipient can also request Sequence retransmission. This chapter therefore describes the Fibre Channel timeout periods in detail and then covers mechanisms for recovering from Sequence errors.

Timeout Periods

Three timeout periods are specified at the Fibre Channel level. There is some implementation-dependent flexibility in the actual values used, but the timeout values used should be roughly equal over all devices on the network.

R_T_TOV

The Receiver_Transmitter_Timeout value (R_T_TOV) is a very short timeout value used by the receiver logic to detect loss of synchronization between transmitters and receivers. The value for R_T_TOV is 100 ms. When a receiver goes into the Loss of Synchronization state, as described in the "Transmitter and Receiver States" section, on page 85, for more than the R_T_TOV, then a Link Failure is detected, and the Link Failure protocol described in the "Link Recovery Protocols" section, on page 104 can be initiated to attempt to resynchronize the transmitter and receiver at the byte and word levels.

E_D_TOV

A short timeout period is known as the Error_Detect_Timeout value (E_D_TOV). The E_D_TOV is intended to represent the maximum round-trip time that an operation could require. The actual value will depend on the processing time of the N_Ports on the network, the transmission time through the Fabric, and time to do Fabric processing, such as for a connect-request.

The E_D_TOV is specified as a timeout value for communications between two N_Ports. In practice, a single N_Port probably will not maintain different E_D_TOV values for each N_Port it is logged in with. The value for E_D_TOV is negotiated between the communicating N_Ports at N_Port Login time. A field in the Common Service Parameters of the PLOGI Sequence and its ACC Reply Sequence specifies the value to be used. The Sequence Initiator and Recipient use the higher of the values requested in the PLOGI and the ACC. A default value of 10 s would be encoded in units of 1 ms as x'0000 2710.' The actual value used will strongly depend on system characteristics such as the time for an N_Port to process a Frame. The value used must also have a specified relationship to the R_A_TOV value, which depends on the maximum time that a Frame could possibly be delayed within a Frame before being delivered.

In most cases, an N_Port waits for the E_D_TOV timeout period before performing some recovery action. However, there are three cases when E_D_TOV is used as the maximum time before an action must happen. (1) Consecutive Data Frames within a single Sequence must be separated by less than E_D_TOV, (2) transmission of an ACK Frame must occur less than E_D_TOV after receipt of its last corresponding Frame, and (3) retransmission of a Class 2 Data Frame must happen sooner than an E_D_TOV timeout period after receipt of a F_BSY or P_BSY.

R_A_TOV

The Resource_Allocation_Timeout value (R_A_TOV) is used to time out operations that depend on the maximum possible time that a Frame could be delayed in a Fabric and still be delivered. An example of this is for determining when to send a Reinstate Recovery Qualifier (RRQ) Extended Link Service Request to recover from using the ABTS ELS Request to recover from Sequence errors, as described in the "Abort Sequence Protocol for Sequence Recovery" section, on page 234 and the "Reinstate Recovery Qualifier (RRQ)" section, on page 213. The Initiator of the RRQ ELS Request must ensure that all Frames in the Recovery_Qualifier range have either been delivered or discarded, so it must wait the R_A_TOV period before initiating the Sequence.

The value of R_A_TOV is negotiated to be the larger of two requested values at Login time, similar to the E_D_TOV. The value used depends on the topology. In a networked topology, the ACC returned from the F_Port following the FLOGI ELS Request specifies the R_A_TOV to be used. In a point-to-point topology, twice the E_D_TOV value is used. An N_Port may determine another N_Port's value for R_A_TOV via the Read Timeout Value (RTV) Extended Link Service request. R_A_TOV values may differ to different N_Ports and between Classes of service.

This facility means that a Fabric must be able to either ensure delivery of a Frame within the maximum delivery time used to determine R_A_TOV or ensure that the Frame is discarded. The R_A_TOV value must be based on the maximum possible time over all paths joining source and destination N_Ports.

CR_TOV

The "CR_TOV" (Connection Request timeout value) is used to qualify the length of time during which connect-requests can be active. A connect request should be either (a) responded to, or (b) discarded, before the expiration of this timeout period.

This is particularly important for Fabrics which implement either Stacked ConnectRequests or Camp-On, where a connect-request may be active in the Fabric until a previously-existing Connection is terminated.

FC-AL Loop Timeout

When a Port that contains the functions required for operation on an Arbitrated Loop is powered on, it transmits a Primitive Sequence that is specific to the Arbitrated Loop and monitors the incoming fiber for similar Arbitrated Loop Primitive Sequences. This operation is described in detail in Chapter 16. If it does not receive any Primitive Sequences used for Arbitrated Loop within two FC-AL Loop Timeout periods, it knows that it is connected to a network with Ports that don't implement the Arbitrated Loop functionality, and it will revert to "OLD-PORT" behavior, without using Arbitrated Loop procedures.

Usage of Timeouts

Fibre Channel recognizes four different kinds of timeouts. Handling of three of these is fairly simple. These simpler timeout mechanisms involve the following procedures:

Link Failure timeout: This timeout is the most serious and is triggered when a receiver has detected a Loss of Signal or a timeout during the Link Reset Protocol or is in the Loss of Synchronization state, unable to establish byte or word synchronization for more than the R_T_TOV period. An N_Port detecting this timeout performs the Link Failure Protocol, as described in the "Link Recovery Protocols" section, on page 104.

Link timeout: A link timeout is detected when all transmission has stopped for some reason and all transmitted Frames have been outstanding for longer than the E_D_TOV timeout period. The actual reason may depend on the Class of service for the Frames being transmitted. Upon detection of a link timeout, the N_Port performs the Link Reset protocol, as described in the "Link Recovery Protocols" section, on page 104.

OLS Transmit timeout: This timeout, described in the "Online to Offline Protocol" section, on page 105, is invoked when the N_Port is in the process of going offline, to ensure that data is not sent to an offline Port.

The most complicated of the timeout-related operations is the Sequence timeout, described below.

Sequence Timeout

Sequence timeouts are the basic mechanism for detecting errors within a Sequence. During the course of a Sequence, both the Initiator and the Recipient recognize "expected events" that would normally occur if the Sequence was progressing correctly. If an expected event doesn't occur within a timeout period, the Port which detects the timeout will expect that the event will never occur, and can begin error recovery operations.

There are other methods for detecting Frame errors within a Sequence, which can be implemented as performance enhancements in order to detect an error sooner than the timeout period. Sequence timeouts are applicable to all Exchange Error Policies, but the expected events, and the recovery procedures, depend on the Class of service used for the transmission.

Class 1 and 2 Sequence Timeout. In Class 1 and Class 2 transmission, both the Sequence Initiator and the Sequence Recipient use a timer facility with the E_D_TOV timeout period between expected events. The expected event for the Initiator following Data Frame transmission is a Link Control Frame received in response (preferably an ACK). The expected event for the Recipient following reception of a Data Frame for a Sequence which is active and incomplete is reception of another Data Frame. Other events halting Sequence transmission, such as receipt of the Link Credit Reset (LCR) Frame, the Abort Sequence (ABTS) Basic Link Service Request, or the Abort Exchange (ABTX) Extended Link Service request will also stop the Sequence timer. When a Sequence Recipient receives the last Data Frame transmitted for the Sequence, it will verify that all Frames have been received before transmitting the final ACK (**EOFt** or **EOFdt**) for the Sequence and stopping tracking of the Sequence timer.

If the E_D_TOV timeout period expires for an expected event at either the Initiator or the Recipient before the Sequence is complete, a Sequence timeout is detected. A Sequence timeout results in either (a) aborting the Sequence by the Initiator, using the Abort Sequence protocol, (b) abnormal termination of the Sequence by the Recipient, or (c) aborting the Exchange by either the Sequence Initiator or Recipient.

If a Sequence Initiator detects a Sequence timeout, it will begin processing for the Abort Sequence protocol by transmitting the ABTS Basic Link Service command, as described in the "Abort Sequence Protocol for Sequence Recovery" section, on page 234. This notifies the Recipient that the Initiator detected a Sequence timeout error. The ABTS Frame can also be used to abort the Exchange, using the "ABTS-LS" error recovery procedure described in the "Example 3: Aborting an Entire Exchange" section, on page 237.

If a Sequence Recipient detects a Sequence timeout, it posts the detected condition in the Exchange Status Block (and Sequence Status Block) associated with the Sequence and waits for the Sequence Initiator to initiate recovery (e.g., via ABTS). All Data Frames for this Sequence that are subsequently received, if any, are acknowledged by returning an ACK with Abort Sequence Condition F_CTL bits set to either "Abort Sequence, perform ABTS" (b'01') or "Immediate Sequence retransmission requested" (b'11'), depending on the Exchange Error Policy. If no other Frames in the Sequence are transmitted, the Initiator will detect a Sequence timeout for the Frame which the Recipient did not acknowledge, and will initiate the Abort Sequence protocol.

In Class 2, if a Sequence has been aborted using the ABTS Frame and the Sequence Recipient supplies the Recovery_Qualifier (OX_ID, RX_ID, and a SEQ_CNT range of low and high SEQ_CNT values), the Sequence Initiator will not transmit any Data Frames within that range within an R_A_TOV timeout period. Both the Sequence Initiator and Sequence Recipient discard any Frames having SEQ_CNT values within the range. After R_A_TOV has expired, the Sequence Initiator will use a Reinstate Recovery Qualifier (RRQ) ELS Request to indicate that all Frames within the SEQ_CNT range have either been received or discarded, and the Recovery_Qualifier is no longer needed. This procedure is discussed in more detail in the "Abort Sequence Protocol for Sequence Recovery" section, on page 234.

Class 3 Sequence Timeout. In Class 3, there is no expected event following Frame transmission for the Initiator. For the Recipient, the expected event following Data Frame reception is reception of another Data Frame for the same Sequence within the E_D_TOV timeout period. The Sequence timer is stopped and reset following reception of each Data Frame. The Sequence timer will also be stopped by reception of the ABTS BLS Request.

The Recipient stops tracking the Sequence timeout when the last Data Frame and all previous Data Frames have been received.

Under normal operations, there is no direct Class 3 facility to let a Sequence Initiator know if a Sequence was received intact. This can have serious consequences, if not accounted for. For example, a Sequence error occurring under the "Abort, discard multiple Sequences" Exchange Error Policy would cause abnormal termination of the Sequence, and the Recipient would then discard all Frames for every other Sequence in the Exchange, forever.

There are several mechanisms to allow an Initiator to determine Sequence completion status. Transmission of the ABTS BLS Frame by the Sequence Initiator causes return of a BA_ACC which indicates the SEQ_ID of the latest complete Frame delivered and the range of invalid SEQ_CNT values received following. Also, in bidirectional Exchanges, it is possible to infer proper Sequence delivery through acknowledgment mechanisms sent by the ULP.

Usage of the Link Credit Reset Frame. In Class 2 it is possible to lose end-to-end Credit as a result of one or more Sequence timeouts. This doesn't happen in Class 1 since Data Frame and Link Control Frame transmission is sequential. The LCR Link Command Frame can be transmitted by an N_Port which detects a loss of end-to-end Credit (the Sequence Initiator) to the Sequence Recipient, to reset the end-to-end Credit to the Login value. The Initiator of the LCR Frame can then perform normal recovery for the Sequence(s) that timed out. When an N_Port receives an LCR Frame, it will discard the data in its buffers (of all Classes) associated with the S_ID of the LCR Frame and abnormally terminate any open Sequences associated with the discarded Frames.

F_BSY may be returned by the Fabric if it is unable to deliver the LCR Frame. A RJT may also be returned if either the S_ID or D_ID is invalid or an invalid delimiter is used.

Link Error Detection and Recovery

Link errors occur when the basic integrity of the link is in question. There are three levels of Link Errors detected. The first level is a Primitive Sequence Protocol Error, which occurs when a Port in the active state receives a Link Reset Response (LRR) Primitive Sequence. In response to this error, the Port will abnormally terminate any Class 1 Dedicated Connections, follow the Link Reset protocol described in the "Link Recovery Protocols" section, on page 104 to reset the link, and begin recovery of any abnormally terminated Sequences. The second level of error is when code violations occur, indicating that the received signal is not being converted into valid transmission character codes. The third level of link error detection is when there is question about the basic operability of the link. Some indications of this are detections of Loss of Signal, Loss of Synchronization to transmission words or bytes, reception of the Not_Operational (NOP) Primitive Sequence, or an R_T_TOV timeout during a Link Reset protocol.

In these cases, the Port will attempt to recover the link by implementing one of the three layers of link error recovery described in the "Link Recovery Protocols" section, on page 104. These protocols involve transmission of a layered hierarchy of Primitive Sequences, for establishing the operating state of the link and of the Port on the other side of the link. When these Primitive Sequence protocols are finished, the link should be up and operational, with synchronized transmitters and receivers sending and receiving Idle Primitive Signals, ready to begin transmitting data and control signals to recover higher level operations if necessary and go on doing communications.

Link Recovery: Secondary Effects

If the recovery action from an error involves invoking one of the Link Recovery Primitive Sequence protocols, active Sequences may be adversely affected. The exact effects will depend on the Class of service used for the Sequence. In Class 1, transmission or reception of a Primitive Sequence causes immediate removal of the Connection and reset of the end-to-end and buffer-to-buffer Credit values to the F_Port (if present) and remote N_Port. This causes immediate abnormal termination of any Class 1 Sequences, so that no more Class 1 Frames for the Connection will be sent or received. Note that this does not necessarily abnormally terminate the entire Exchange.

In Class 2 and 3, Primitive Sequence transmission or reception is not as disruptive, although any Class 2 or 3 Frames received during a Primitive Sequence Protocol will be discarded.

When the Primitive Sequence Protocol processing is done, end-to-end and buffer-to-buffer Credit values have been reset, and both Ports are in the active state and transmitting Idles, recovery on a Sequence-by-Sequence basis of the abnormally terminated Sequences can begin, under FC-4 and ULP control.

Frame Reception and Frame Validity

Assurance of Sequence integrity requires that all Frames within the Sequence are received intact, with no Frames missing. There are some rules specified concerning what constitutes a Frame and concerning the validity of a received Frame.

At a receiver, a Frame always starts at a **SOF** delimiter and ends when any other Ordered Set is detected. A link failure such as Loss of Synchronization or Loss of Signal can also terminate a Frame. Any data bytes received outside the scope of the **SOF** and **EOF** delimiters are discarded as not being part of any Frame.

For the Frame to be valid, there must be no code violations within the Frame, the Frame must be terminated by an **EOFt**, **EOFn,** or **EOFdt** delimiter; must have a correct CRC error detection coding for the Frame contents; and must be a multiple of 4 bytes long. The length between the **SOF** and **EOF** delimiters must be within the allowable size specified in the N_Port Login service Class parameters, and if the length is exceeded, the Port must consider the payload invalid, although the header may be intact. A Frame terminated with **EOFa** is assumed to have an invalid Frame Header and Payload, and is discarded, while a Frame terminated with **EOFni** or **EOFdti** has an invalid Payload but may have a valid header.

If an N_Port detects an invalid Frame, the Frame is discarded, and the Link Error Status Block fields for Invalid Transmission Word and/or Invalid CRC are updated (see the "Link Error Status Block Rules" section, on page 240). Since there is no error detection mechanism for Frame Headers separate from the Frame payload, it is not known whether any of the Frame Header fields are valid. This means, for example, that a P_RJT cannot be sent for an invalid Frame, since it may be sent to the wrong destination or sent with the wrong SEQ_ID.

Detection of Missing Frames

Aside from causing Frame corruption, transmission errors can also cause Frames to be lost in transit. This occurs when, for example, a transmission error causes corruption in the D_ID field or the **SOF** delimiter. In addition to the mechanisms for detecting invalid Frames, N_Ports must incorporate mechanisms to detect Frames which are lost or missing.

The basic mechanism for detecting lost Frames is the SEQ_CNT value. Since Frames are transmitted with continuously increasing SEQ_CNT values, if a particular SEQ_CNT is missing in a stream of received Frames, then that Frame was lost in transmission. The mechanism is complicated by the fact that some Fabrics may not guarantee that Class 2 and Class 3 Frames arrive in the transmitted order, so Frames may not arrive at a destination in order of continuously increasing SEQ_CNT values. In addition, even on a Fabric that guarantees ordering, Frames may not be received in order if a Class 2 or Class 3 Frame is busied or rejected and a later Frame is not. Missing Frame detection in these cases depends on each Frame being delivered within the E_D_TOV timeout period following the previous Frame.

The mechanisms for detecting missing Frames based on SEQ_CNT depend on having continuously increasing SEQ_CNT values. In Fibre Channel, this condition applies within a single Sequence or across multiple streamed Sequences of an Exchange with continuously increasing SEQ_CNT values. The SEQ_CNT mechanism would not work, for example, for detecting when a full Sequence is lost if continuously increasing SEQ_CNT values are not used, as described in the "Sequence Count: SEQ_CNT" section, on page 128. Error detection in this case is through the Sequence timeout mechanism described in the "Sequence Timeout" section, on page 222.

In all Classes, if a Data Frame with a particular SEQ_CNT value is delivered, then the Data Frame with the next SEQ_CNT value should either be delivered (1) next (with in-order transmission and no BSY or RJT of the Frame) or should at least be delivered (2) within an E_D_TOV timeout period (in all cases). If neither of these conditions is met, then the Frame is missing, and the Sequence must be considered invalid.

Reliance on the E_D_TOV period for detecting missing Frames when in-order delivery is not guaranteed can be fairly complex and inefficient in some implementation, since it requires independent timeout checking facilities for each Active Sequence, and recovery cannot begin until after the E_D_TOV period has expired. To simplify receiver design and increase recovery performance, implementations will sometimes implement a "missing Frame window" mechanism, which limits the number of outstanding Frames possible at once. If the missing Frame window is of size W, then the source will ensure that the acknowledgment for Frame N is received, for example, before transmitting Frame N + W. That way, the receiver knows that if it has received Frames up to SEQ_CNT = M, then it should only receive Frames up to SEQ_CNT = M + W. If it receives the Frame with SEQ_CNT = M + W + 1, and it has received and acknowledged only up to SEQ_CNT = M, then it knows that a Frame within the window is missing. The value for W should be larger than the end-to-end Credit, for efficiency, and will depend somewhat on Fabric characteristics. This is clearly an implementation-dependent feature, which can improve recovery performance.

If a missing or invalid Frame is detected within a Sequence, the Sequence is considered to be invalid, and error recovery must be performed on the Sequence.

Exchange and Sequence Integrity

Proper delivery of an intact Exchange depends on two factors. The first is delivery of intact Sequences — all Data Frames in each Sequence must be intact. The second is proper management of the relationship between Sequences. This section and the following sections describe the requirements for assuring proper Sequence integrity and Exchange integrity.

In Class 3, since there are no ACKs or other Link Control Frames, only the Sequence Recipient is aware of any invalid or missing Frames. The Recipient does not take any actions besides notifying the FC-4 or upper level for recovery at the FC-4 level, although the Sequence Initiator can use the ABTS ELS Request to find out the delivery status of Frames.

Proper integrity management of a Sequence entails correct (1) Sequence initiation, (2) ordered delivery of intact Frames by SEQ_CNT within a Sequence to the ULP, (3) Sequence completion, and (4) ordering between Sequences with continuously increasing SEQ_CNT. These factors are all detailed in the "Sequence Management" section, on page 144.

Sequence errors are detected in three ways, including (1) detection of a missing or invalid Frames (see the "Frame Reception and Frame Validity" section, on page 226), (2) detection of a Sequence timeout (see the "Sequence Timeout" section, on page 222), or (3) detection of a rejectable condition within a Frame (discussed in the "F_RJT and P_RJT Frames" section, on page 140 on RJT Frames).

The effect of Sequence errors and the actions taken in response to them depends on the Exchange Error Policy. The discarding of Sequences, the delivery order of Sequences, and the recovery policies for Sequences are specified in the Exchange Error Policy, which is identified in the Abort Sequence Condition bits (F_CTL bits 5 and 4) in the first Frame of the first Sequence of the Exchange. The four Exchange Error policies are discussed in the "Exchange Error Policies for Class 1 and Class 2" section, on page 229.

Detection of Class 1 or Class 2 Sequence errors by the Recipient is conveyed back to the Sequence Initiator in the Abort Sequence Condition bits (F_CTL bits 5 and 4) in an ACK Frame or by a P_RJT Frame. If the Recipient cannot transmit an ACK or P_RJT Frame, for example because the Data Frame was corrupted, then either the Sequence Initiator or Sequence Recipient or both will detect a Sequence timeout.

On systems which use X_ID invalidation and association of Exchanges using Operation Associators within the Association Header optional header field to allow X_ID reassignment during an operation, there may be certain periods in an Exchange when one or both X_ID fields may be unassigned (x'FFFF'). This is discussed in the "Association_Header" section, on page 244. If an X_ID is unassigned, special error recovery for both the Sequence

Initiator and Recipient may be required to abnormally terminate or recover an Exchange.

Exchange Error Policies for Class 1 and Class 2

There are two fundamental Exchange Error Policies: "Discard" and "Process." There are three types of discard policies, and one type of process policy. A discard policy means that a Sequence is delivered in its entirety, with all Frames intact, or not at all. The process policy allows an incomplete Sequence to be deliverable to the ULP in some cases if the first and last Frames are valid. The Process Error Policy can be used for applications such as video transmission, which don't require every Frame to be intact and don't benefit from retransmission. Following are discussions of the three different Discard Policies and the Process Exchange Error policy.

Rules Common to All Discard Policies

In all the discard policies, a Frame error will cause that Frame and all subsequent Frames in the Sequence to be discarded. These Frame errors include detection of any invalid or missing Frames as described in the "Frame Reception and Frame Validity" section, on page 226, conditions which cause a P_RJT, as described in the "F_RJT and P_RJT Frames" section, on page 140, or an internal malfunction. The Sequence Recipient will record the type of error in the Sequence Status Block, and will generate ACKs for the invalid Frame and any subsequent Frames in the same Sequence with the Abort Sequence Condition set to "Abort Sequence, perform ABTS" (F_CTL bits 5-4 = b'01'). If the Sequence Initiator receives this type of ACK, detects an internal error, or detects a missing Link Response Frame by SEQ_CNT, and the Sequence has not already been terminated by receipt of an ACK with an **EOFt** or **EOFdt** delimiter, it will begin aborting the Sequence by transmitting an Abort Sequence (ABTS) Basic Link Service command. The full Abort Sequence protocol is described in the "Abort Sequence Protocol for Sequence Recovery" section, on page 234.

In the case of Sequence timeouts, the Sequence Initiator will abort the Sequence using the Abort Sequence (ABTS) Basic Link Service Request. Since sending the ABTS requires end-to-end Credit, it might have to perform the Link Reset Protocol to reset credit values by transmitting the Link Reset (LR) Primitive Sequence. It may have to determine the Sequence Status on the Sequence Recipient by sending and waiting for a reply to either the Read Sequence Status Block (RSS) or the Read Exchange Status Block (RES) ELS Request, as described in the "ELS Requests: Status Determination and Initiative Request" section, on page 207, if the status is uncertain.

If a Sequence Recipient wishes to stop receiving a Sequence due to some condition at the FC-4 level, it can return a "Stop Sequence" condition by returning an ACK with the Abort Sequence Condition (F_CTL bits 5-4) set to "Stop Sequence" (b'10'). Since this is a less drastic recovery action, carried out in cooperation with the FC-4 level, the Frames already received may not necessarily have to be discarded. The Initiator will interrupt and normally terminate the Sequence, if it is still active, and will notify the FC-4. This facility is described further in the "Stop Sequence Protocol Overview" section, on page 239.

Abort, Discard Multiple Sequences Error Policy

The "Abort, Discard Multiple Sequences" Error Policy requires that for a Sequence to be deliverable all Data Frames must be accounted for and received intact and any previous Sequences from the same Initiator in the same Exchange must also be deliverable. These policies are used when the ordering of Sequence delivery is important to the FC-4. Sequences must be delivered to the FC-4 in the same order as transmitted. This Exchange Error Policy is the default policy used before N_Port Login, is applied to all Classes of service, and must be supported by all N_Ports.

In Class 1, if a Sequence Recipient detects a missing Frame error or an internal malfunction for a Sequence within an Exchange which requested "Abort, discard multiple Sequences" Exchange Error Policy, it will request that the Sequence be aborted by setting the Abort Sequence Condition F_CTL bits to "Abort Sequence, perform ABTS" (b'01') on the ACK for the Frame during which the error was detected. Any further Frames received in the same Sequence or in later streamed Sequences will generate ACKs with the same Abort Sequence Condition field. If a Data Frame of the Sequence is received with the End_Sequence, End_Connection, Sequence_Initiative, or Invalidate_X_ID bits set to 1, the Recipient will ignore them and set them to 0 in the ACK Frame, to prevent any further independent processing before the Abort Sequence protocol is performed.

In Class 2, most processing following an error is identical to that for a error in a Class 1 Sequence error, as described above. However, the last ACK for the Sequence in error will do normal F_CTL bit processing and delimiter generation as described under normal Sequence completion.

Discard Multiple with Immediate Retransmission

The "Discard multiple Sequences with immediate retransmission" is a special case of the "Abort, discard multiple Sequences" Error Policy which can only apply to Exchanges where all Sequences are in Class 1 and which allows a Sequence Recipient to request immediate Sequence retransmission under the guidance of the FC-4 level. If the Sequence Recipient is unable to support the "Discard multiple Sequence with retransmission" Exchange Error Policy for an Exchange with an N_Port which requested it, it will follow the rules for "Abort, Discard multiple Sequences."

In this Exchange Error Policy, if a Sequence Recipient detects an invalid or missing Frame error in a Class 1 transmission, or detects an internal malfunction for a Sequence, it can request that the Sequence be aborted and immediately retransmitted by setting the Abort Sequence Condition F_CTL bits to "Immediate Sequence retransmission requested" (b'11') on the ACK for the Frame during which the error was detected.

For errors detected other than a missing Frame, the Abort Sequence Condition F_CTL bits are transmitted for any subsequent ACKs transmitted. The Sequence Recipient may continue to transmit ACKs for subsequent Frames of the Sequence and any subsequent streamed Sequences until it receives a new Sequence (**SOFi1**) with the Retransmission bit (F_CTL bit 9) set to 1, or an ABTS Frame is received. If an ACK is transmitted for the last Data Frame of the Sequence, the End_Sequence, End_Connection, Sequence_Initiative, and Invalidate_X_ID F_CTL bits on the Data Frame are ignored, and those bits are set to 0 in the ACK Frame, with bits 5 and 4 set to "Immediate Sequence retransmission requested" (b'11').

If a later Data Frame of the Sequence is received with the End_Sequence, End_Connection, Sequence_Initiative, or Invalidate X_ID bits set to 1, the Recipient will ignore them and set them to 0 in the ACK Frame, to prevent any further independent processing before the Abort Sequence protocol is performed.

If a Sequence Initiator receives an ACK with the Abort Sequence Condition F_CTL bits set to "Immediate Sequence retransmission requested" (b'11'), it will begin retransmission of the first non-deliverable Sequence by starting a new Sequence and setting the Retransmission bit (F_CTL bit 9) to 1 until it has received at least one ACK indicating that the retransmitted Sequence has been successfully received. If the Sequence Initiator is unable to determine the correct Sequence boundary to begin retransmission, it can either transmit the ABTS BLS Request or the RES ELS Request to find out the current status. The detailed rules for Class 1 Sequence retransmission are described in the "Sequence Retransmission for Class 1 Recovery" section, on page 238.

Abort, Discard a Single Sequence Error Policy

Under the "Abort, discard a single Sequence" Exchange Error Policy, Sequences are independently deliverable. If all Frames of a Sequence are accounted for and received intact, it is deliverable, regardless of the status of previous Sequences in the Exchange. This policy is useful if the Payload of the Sequences delivered contains sufficient FC-4 or upper level information to process the Sequence independently of other Sequences within the Exchange. Sequences are still delivered to the FC-4 for the ULP on a Sequence-by-Sequence basis in the same order as received.

Processing of a Class 1 or Class 2 Sequence received following an error is the same as in the "Discard Multiple Sequences" Exchange Error Policies, with the Abort Sequence Condition F_CTL bits set to "Abort Sequence, perform ABTS" (b'01') to notify the Initiator of the error. However, when a new Sequence is initiated, ACK processing goes back to normal for the new Sequence and for all later Sequences. Clearly, if an error occurs in a Sequence, and the Initiating side decides to retransmit the data, the retransmitted data (in a new Sequence) may arrive at the destination later than Sequences which were not corrupted, so the ULP must be able to handle out-of-order Sequences.

Process with Infinite Buffering Error Policy

The "Process with infinite buffering" Error Policy is intended for applications such as video in which loss of a single Frame may have minimal or no effect on the Sequence being delivered. Under this Exchange Error Policy, a Sequence does not have to be complete to be deliverable, and there is no dependency on previous Sequences. It may sometimes even allow the usage of the Payload of some invalid Frames. Frames are still delivered to the FC-4 for the ULP in the same order as transmitted.

In the process policy, as opposed to the three discard policies, the Recipient will ignore errors detected on intermediate Frames or timeout errors where an ABTS is not requested. However, such errors are reported to an upper level and to the Sequence Initiator by transmitting a P_RJT with a reason code. The Recipient will request that the Sequence be aborted (by setting Abort Sequence Condition to "Abort Sequence, perform ABTS" (b'01')) and will follow other "Discard a single Sequence" rules if the first or last Frame of the Sequence is missing. A missing last Frame is detected by the Sequence timeout mechanism, as usual.

Class 3 Sequence Error Handling

In Class 3 Sequence transmission, since there are no Link Control Frames, errors can only be detected by the Sequence Recipient. The Exchange Originator can still set the Exchange Error policy. In the discard policies, the Sequence Recipient will discard Sequences using the same rules as in Class 1 and 2, including discarding Frames received following an error. Individual FC-4s can decide whether to recover data for the full Sequence or only the data discarded.

Class 3 Sequence errors are detected based on Sequence timeout or internal malfunction. In either case, the Recipient will abnormally terminate the Sequence, update the Sequence and Exchange Status Block, and notify the FC-4 level. In either of the "Discard multiple Sequences" Error Policies, later Sequences in the Exchange won't be delivered. Notification of the error condition to the Sequence Initiator is up to the FC-4 level. As described in the "Class 3 Sequence Timeout" section, on page 223, the Initiator of a Class 3 Sequence must use some external mechanism, such as ABTS or RES transmission or some ULP mechanism to determine whether any errors have occurred.

Sequence Recovery

Fibre Channel provides several mechanisms for implementing Sequence recovery. Actual usage of them depends on the FC-4 or upper level and on the specific implementation. The facilities provided at the Fibre Channel level include procedures to terminate or abort a Sequence, to recover end-to-end Credit, to detect missing or invalid Frames, and to allow determination of another Port's Sequence or Exchange status to remedy any disparities.

Sequence recovery can be implemented by the FC-4 or upper level when a Sequence completes abnormally. There is a mechanism whereby a Sequence can be completed while it has only been partially received. This mechanism is termed the "Stop Sequence Protocol," and is described in the "Stop Sequence Protocol Overview" section, on page 239. Otherwise, a Sequence is terminated abnormally if it is incomplete or open when any of the following events occurs:

- the Sequence Recipient receives ABTS from an Initiator invoking the Abort Sequence protocol,
- a Class 1 Dedicated Connection is broken by transmission or reception of a Primitive Sequence,
- an ABTX Extended Link Service request is received for the Exchange of

the Sequence, or

- either the Sequence Initiator or Recipient initiates a Logout.

Error Recovery Strategy

The most difficult part of Sequence Recovery in any communication protocol is never the actual retransmission of the data. The most difficult part is making sure that the transmitter and the receiver agree on what data was actually delivered successfully and what was not.

The fundamental problem is that a lost Data Frame and a lost ACK initially appear exactly the same to the Sequence Initiator (a Frame was sent out, but nothing came back). At the Recipient, however, one situation yields a corrupted Sequence and the other yields an intact Sequence. This ambiguity between source and destination views of Sequence delivery status, combined with the complex interdependencies between Data Frames and Link Control Frames in the same and in different Sequences which may or may not be retransmitted, makes determination of a complete status extremely difficult. It is clear that only very simple and high-level error recovery processes have a likelihood of being implemented with reasonable hope of success.

Fibre Channel takes a strategy of using two basic forms of error recovery, which only operate at the Sequence and at the Exchange level (no recovery of individual Frames) and which either allow very restricted retransmission under Fibre Channel control, or allow retransmission under FC-4 or ULP control, with help from the Fibre Channel levels in determining communication status.

All error recovery is initiated by the Sequence Initiator, either under its own initiative or following a Request by the Recipient in the Abort Sequence Conditions of an ACK or other Link Control Frame. Other than setting these bits, the only action by the Recipient of a corrupted Sequence is to mark the Sequence as being abnormally terminated by the Recipient in a Sequence Status Block, to release any link facilities associated with the Sequence, and to discard any Frames in the Sequence except an ABTS Frame.

Abort Sequence Protocol for Sequence Recovery

The basic form of Sequence recovery is the use of the Abort Sequence protocol, wherein an ABTS Basic Link Service Request is sent by the Sequence Initiator to the Recipient, and the Recipient returns a BA_ACC Reply to the Initiator, indicating the status and further processing.

The name of the ABTS Basic Link Service Frame is a bit misleading because it implies that the Initiator of the ABTS Frame is notifying the Recipient to Abort the Sequence, when this is really not the case. The ABTS Initiator actually cannot tell the Recipient to abort anything because it does not know what has been received intact. All the ABTS Initiator knows is that it has not received a proper reply, such as an ACK Frame. It does not know whether the Data Frame arrived intact and was received and delivered to the FC-4 level and ULP on the Recipient N_Port or not. The ABTS Frame therefore functions more as a request for information to clarify a situation that looks like a possible error than as a command to abort a Sequence.

The Recipient of an ABTS Frame will return an ACK and/or R_RDY for the Frame and will determine the current status. The most important information in the ABTS Frame is actually the pair of X_ID fields, since it is only within an Exchange that the SEQ_CNT fields are continuous. The ABTS Recipient examines the state of the Sequences within the Exchange indicated and returns the completion information to the ABTS Initiator.

A basic understanding of the operation of the ABTS Sequence can be gained by considering three examples. None of these examples will be described in complete detail, but they give the flavor of what is required in the actual processing.

In each of the three cases, the Sequence Initiator has not received an ACK for a Frame of one of several outstanding Sequences of an Exchange. Streaming of Sequences is allowed over a Fabric which doesn't guarantee in-order delivery, so the Sequence Initiator does not know the receipt status of any of the Sequences. The actual situation, at the Recipient, is different in the three cases.

The reply to the ABTS Frame is a BA_ACC Basic Link Service Reply, as shown in Figure 11.2. The fields in the BA_ACC indicate the situation at the Recipient and determine the procedures which will be carried out for the rest of the Exchange. The SEQ_ID field of the BA_ACC contains the SEQ_ID of last deliverable Sequence, the low SEQ_CNT field contains the SEQ_CNT of the last known valid Frame, and the high SEQ_CNT field contains the SEQ_CNT of the first known valid Frame. (Yes, this is complicated, but it will make more sense soon.)

Example 1: Determination of Intact Delivery. In the first example, all Sequences have actually been received complete and are deliverable, but an ACK has been corrupted. The Sequence Initiator sends an ABTS Frame with OX_ID and RX_ID fields for the Exchange in question. Since Sequences are being streamed, the SEQ_CNT for the ABTS is 1 higher than the last SEQ_CNT value on the last Sequence sent.

The ABTS Recipient examines the status of all Sequences, determines that they are all complete and deliverable, and returns a BA_ACC Frame.

Since all Sequences have been delivered intact, the SEQ_ID field in the BA_ACC contains the SEQ_ID of the last Sequence transmitted, and the "Low SEQ_CNT" and "High SEQ_CNT" fields both contain the SEQ_CNT of the ABTS Frame. When the ABTS Initiator receives this BA_ACC back, it sees that the last SEQ_ID sent was delivered intact, and that there is no window of invalid SEQ_CNT values in between Low SEQ_CNT and High SEQ_CNT values. Having assured intact transmission, the Sequence Initiator knows that it can close the Sequences normally. Transmission of the ABTS has determined that no error recovery is required.

This method allows an Initiator to determine the status of transmitted Class 3 Sequences, as described in the "Class 3 Sequence Timeout" section, on page 223. In addition to the ABTS Basic Link Service Frame, which allows a Sequence Initiator to determine the Recipient status and initiate recovery for incomplete Sequences, an N_Port can determine Recipient status without initiating any recovery by using the Read Exchange Status (RES) and Read Sequence Status (RSS) Extended Link Service Requests, as described in the "Read Exchange Status Block (RES)" section, on page 207 and the "Read Sequence Status Block (RSS)" section, on page 209.

Example 2: Aborting Incomplete Sequences. In the second example considered, the Sequence Initiator has sent three streamed Sequences, with SEQ_IDs of 5, 6, and 7. In the example, these are all single-Frame Sequences, so the SEQ_CNT values are 5, 6, and 7, too. Sequence 5 was delivered intact, but something happened to Sequence 6, and Sequence 7 was transmitted before the Initiator detected a problem. Now the Initiator has detected a missing ACK for Frame 6, so it sends out an ABTS with the OX_ID and RX_ID values for the Exchange, with SEQ_CNT = 8. It may either be sent as part of Sequence 7 or as part of a new Sequence with SEQ_ID = 8.

The Recipient looks at the completion status of its received Sequences, determines that Sequence 5 was delivered intact, but Sequence 6 was not. Sequence 7 may or may not have been received depending on whether it had an error and depending whether the Exchange Error Policy was "Discard a single Sequence" or "Discard multiple Sequences." The Recipient returns a BA_ACC with SEQ_ID = 5, Low SEQ_CNT = 5, and High SEQ_CNT = 8, which is the SEQ_CNT of the ABTS Frame.

The Initiator, on looking at this BA_ACC, knows that Sequence 5 was delivered but that Sequence 6 was not, so it needs to clear up the status of Sequence 6, and perhaps Sequence 7, again depending on the Exchange Error policy.

The Initiator and Recipient then each build a "Recovery_Qualifier," with Low SEQ_CNT = 5 and High SEQ_CNT = 8 and with OX_ID and RX_ID set for the Exchange. The Recovery_Qualifier gives the N_Ports an indication of a range of invalid Frames. From then on, until after an R_A_TOV timeout period when all invalid Frames must be either delivered or discarded, any Data or Link Control Frames received in the Exchange with a SEQ_CNT value in the invalid range of the Recovery_Qualifer will be discarded.

After the R_A_TOV timeout period, the Initiator will know that no invalid Frames can be delivered, and it can, under FC-4 direction, go about retransmitting whatever data requires retransmission. The Recovery_Qualifer can be canceled at this point, since it is no longer necessary, using the Reinstate Recovery Qualifier (RRQ) Extended Link Service Command, described in the "Reinstate Recovery Qualifier (RRQ)" section, on page 213.

But wait, it's not quite as easy as that! Remember, the Initiator still may not know what happened to Sequence 7, since the processing of Sequence 7 depends on complicated things like Exchange Error Policy and out-of-order ACK or Data Frame delivery, etc. What we need is a really simple Error recovery policy that lets the two N_Ports know unambiguously what happened and what to do next. This is the reason for the mechanism described the third example.

Example 3: Aborting an Entire Exchange. In the third example, the Sequence of operations is the same as in the second case up to where the Initiator detected a problem with SEQ_ID 6 and transmitted the ABTS Frame. The Recipient, however, instead of trying to give notification of individual Sequences within the Exchange, just aborts the whole Exchange. This may invalidate Sequences that were delivered correctly, but the error rate is low enough that it's worth it to discard a bit of good data every so often to simplify recovery processing.

The Recipient in this case returns a BA_ACC with the Last_Sequence F_CTL flag set to 1 to indicate that it is the last Sequence in the Exchange. It also sets the End_Connection bit to remove a Class 1 Connection, if there is one established. It sets the SEQ_ID to the last valid SEQ_ID delivered for the Exchange, which is 5 in this case. It sets Low SEQ_CNT to x'0000' and High SEQ_CNT to x'FFFF,' to invalidate every Frame in the Exchange.

Following the R_A_TOV timeout period, where both N_Ports were discarding every outstanding Frame in the Exchange, they can both clean up any record-keeping for the Exchange, and know that any Sequences beyond SEQ_ID 5 are invalid, and must be recovered. This basic procedure, with more details included, is termed the "ABTS-LS" error recovery procedure, for "Abort Sequence - Last_Sequence," and is what will be used for many Fibre Channel installations.

Sequence Retransmission for Class 1 Recovery

Fibre Channel does provide a way to retransmit Sequences, but, for the reasons described above, it is a very limited retransmission facility.

Retransmission can only occur for Class 1 Sequences which are initiated in an Exchange that indicates the "Discard multiple Sequences with retransmission" Exchange Error policy, which only uses Class 1 Sequences. Both the Sequence Initiator and Recipient must be able to support it, and there are cases where particular error scenarios can prevent the Sequence Initiator from unambiguously retransmitting Sequences.

A Recipient N_Port is not required to support retransmission and will revert to "Discard multiple Sequences" behavior within an Exchange for which "Discard multiple Sequences with retransmission" was requested by the Originator.

If a missing Frame error occurs, and the Recipient can determine the SEQ_ID of the first non-deliverable Sequence (which is, for example, in general impossible if a single-Frame Sequence is lost), it will transmit all further ACKs for that SEQ_ID with the Abort Sequence Condition bits set to "Immediate retransmission requested" (b'11'). ACKs for later Sequences will be transmitted with Abort Sequence Condition set to "Abort Sequence, perform ABTS."

On receipt of an ACK marked with "Immediate retransmission requested" (b'11'), the Sequence Initiator can retransmit the Sequence, after determining that the SEQ_ID of the ACK matches the SEQ_ID of the first non-deliverable Sequence. This determination may require looking at its own set of Sequence Status Blocks and also possibly determining the status of the Recipient Sequence Status Blocks using the ABTS, RES, or RSS Link Service Requests.

The retransmitted Sequence Frames must have the "Retransmission" flag (F_CTL bit 9) set to 1 and must use their original SEQ_CNT values from the original transmission. The Retransmission flag will be set to 1 in all Frames of the retransmitted Sequence and all following streamed Sequences until the first retransmitted Sequence is determined to be deliverable. Later Sequences are not marked with the Retransmission flag, whether or not they have been retransmitted.

There are several other more complex rules concerning setting of F_CTL bits, interaction between Frames, use of Frame delimiters, capability for handling ABTS while retransmission is going on, reverting to "Discard multiple Sequences" behavior when further errors occur, and what to do when the Initiator didn't restart where the Recipient expected it to start, but it's not worth describing them here. Please look it up if you're interested; it's an endlessly fascinating subject. The simpler Abort Sequence protocol described in the "Abort Sequence Protocol for Sequence Recovery" section, on page 234 will be used a good deal more commonly, though.

Stop Sequence Protocol Overview

The final possible response code in the Abort Sequence Condition F_CTL bits for an ACK Frame is a Request for the "Stop Sequence" protocol. The Stop Sequence protocol can be used by a Sequence Recipient to ask the Initiator to stop sending Frames for a Sequence, without invoking one of the recovery protocols. In this protocol, the Recipient sets the Abort Sequence Condition F_CTL bits to "Stop Sequence" (b'10') in an ACK, and when the Initiator receives and interprets this Frame, it will set the next Data Frame transmitted to be the last of the Sequence, regardless of how many Frames were originally going to be transmitted.

This protocol has a number of uses and is particularly useful since the length of a Sequence is effectively unlimited. An example of its use could be as a form of "stop and restart" flow control mechanism in a video data transmission application, where there is huge amount of data to be sent to a Recipient with limited buffering space at the FC-4 or ULP level. The Initiator could start sending the data in one huge Sequence, and when the ULP buffering space started filling up, the FC-4 could tell the Fibre Channel levels to stop the Sequence. The Sequence Initiator would terminate the Sequence normally as soon as reasonable, and data transmission would stop, with no error recovery needed. A separate Sequence from the original Sequence's Recipient to its Initiator could tell the Initiator where and when to restart sending the data, in a new Sequence. (This example is totally invented and is not meant to bear any resemblance to any real implementation, but it demonstrates the intended usage of the protocol.)

After the Recipient returns an ACK with the Stop Sequence indication, it must set the same indication in any other ACK Frames transmitted. When the Sequence Initiator receives the first ACK with Stop Sequence indicated, it will terminate the Sequence normally using End_Sequence (F_CTL bit 19) and the **EOFt** or **EOFdt** delimiter, if it hasn't already been terminated.

Any Sequence errors detected for Frames with SEQ_CNT higher than the SEQ_CNT of the Stop Sequence ACK will not be reported and the Frames may be discarded, but the Recipient will handle end-to-end and buffer-to-buffer Credit normally. The Stop Sequence protocol does not specify what should happen to the Sequence Initiative, determining which N_Port will send the Sequence following the stopped Sequence for the Exchange. This is entirely under FC-4 control. It is also up to the FC-4 and ULP to determine the reason for stopping the Sequence.

Link Error Status Block Rules

An N_Port should provide some facility for tracking what errors have occurred. This lets the N_Port monitor the overall integrity of the link for detecting when human-initiated recovery processes (such as replacing connectors, etc.) should be started. The actual errors accumulated and stored are dependent on the implementation, but some minimal set of errors should be counted, for compatibility with other N_Ports on the network. An N_Port or F_Port may choose to log other errors as well, for internal diagnostics and simpler debugging.

At a minimum, an N_Port must keep the 32-bit error counts returned in the ACC to the Read Link Error Status Block (RLS) ELS Request. This is described in the "Read Link Error Status Block (RLS)" section, on page 212, and includes counts of instances of failure shown in Figure 12.1.

Figure 12.1
Reply format to Read Link Error Status Block Request.

Formal of the ACC to the Read Link Error Status Block (RLS) ELS Request

Word \ Bit	33222222 10987654	22221111 32109876	11111100 54321098	00000000 76543210
1	Link Failure Count			
2	Loss of Synchronization Count			
3	Loss of Signal Count			
4	Primitive Sequence Protocol Error Count			
5	Invalid Transmission Word Count			
6	Invalid CRC Count			

Conditions under which the first four values are incremented are shown in Figure 6.3. The Invalid Transmission Word counter increments once for every invalid transmission word received, unless either the receiver is in the Loss of Synchronization state (described in the "Transmitter and Receiver States" section, on page 85) or the Port is in the OLS Receive (OL2) or Wait for OLS (OL3) state (described in the "Offline States (OL1, OL2, and OL3)" section, on page 103). The Invalid CRC Count increments once for every received Frame which arrives at a Port in the Active state with an erroneous CRC field that isn't terminated with either the **EOFni**, **EOFdti**, or **EOFa** delimiter.

Chapter 13

Optional Headers and Special Functions

Introduction

This chapter describes the use of optional headers in Data Frames. It also describes some advanced functions defined in Fibre Channel for use in special circumstances.

Optional Headers

There are currently three possible optional headers in Fibre Channel Data Frames :

Network_Header: Used for including source and destination Name_Identifiers in a Sequence which can have a more general scope than the 3-byte Fibre Channel Address Identifiers, for addressing networks with different name spaces;

Association_Header: Used for addressing an Image, which is one of a process or group of processes which share a single N_Port and also used for associating Frames from different Exchanges together and for locating Exchange information when an X_ID field is invalidated; and

Device_Header: Used for including specific ULP header information.

A fourth header, termed the "Expiration Security Header" was defined, but since it wasn't being used, the space was reclaimed. If optional headers are present, they must be in the order listed as shown in Figure 13.1. A Frame may contain any or all of these headers, subject to the "where present" guidelines shown in Figure 13.2.

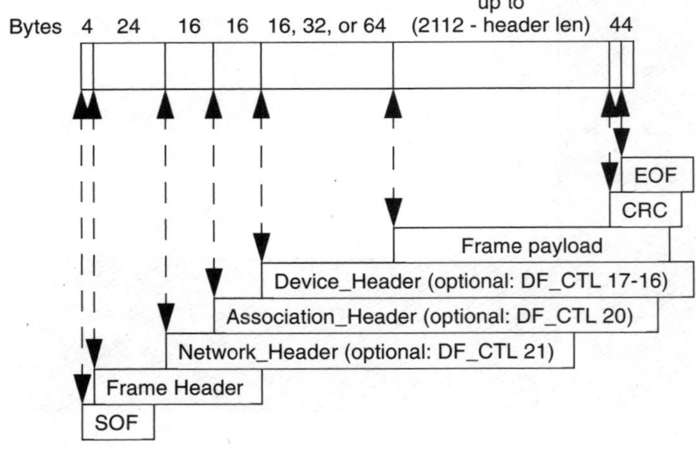

Figure 13.1
Placement of optional headers in Frame.

Figure 13.2
Summary of optional header usage.

Optional Header	Where present	Applicability	Receiving N_Port action
Network_Header DF_CTL[21]	First Data Frame of a Sequence	All Sequences except Basic and Extended Link Services	Skips if not supported or required
Association_Header DF_CTL[20]	First Data Frame of a Sequence	Only in Sequences in Association_Heade r management protocol	Described in text
Device_Header DF_CTL[17-16]	First Data Frame of a Sequence	All Sequences except Basic and Extended Link Service	Skips if not needed. FC-4 may reject if not supported by the ULP.

The presence or absence of each of the four headers is indicated in bits 22 to 20 and 17 and 16 of the DF_CTL field in the Frame Header, as shown in Figure 13.2. If any particular optional header is present, the corresponding DF_CTL bit is set to 1, and space is reserved between the Frame Header and Payload. If no optional header is present, the payload immediately follows the Frame header. The maximum length of a Frame payload plus any optional headers is 2112 bytes, so if the maximum optional header space was used, the maximum Payload size would be (2112 – 3×16 – 64) = 2000 bytes.

In actual practice, many systems can be built without using optional headers at all. The Device_Header and Network_Header, particularly, are FC-4 and ULP constructs, which could as easily be incorporated into the ULP payloads transparently to the Fibre Channel level. Exposing these upper level constructs to the Fibre Channel level blurs the separation between ULP data structures and Fibre Channel data transport mechanism which is such a significant feature of the Fibre Channel architecture, and requires ULP designers to partially understand Fibre Channel rules and operations. For these reasons, it is quite possible for many Fibre Channel installations to operate quite well without using any optional header fields. Descriptions of these fields is not, therefore, covered in detail here.

The Association Header field is discussed in more detail because it provides greater functionality than the base Fibre Channel constructs alone provide, in terms of addressing flexibility and support for more complex operations. The Association Header is therefore discussed in more detail in this chapter. Usage of the Association Header in linking pairs of Images behind N_Ports was also discussed in the "Procedure to Estimate End-to-End Credit" section, on page 176.

Network_Header

The Network_Header, shown in Figure 13.3, may be used by a bridge or a gateway Node which interfaces to an external network, such as an Ethernet or token ring network, or to a Fibre Channel network with a different Fabric address space. This field holds network addressing information which is valid for the other network or for both external and Fibre Channel networks. The Network_Header contains Name Identifiers for Network_Destination_Address and Network_Source_Address using the formats shown in Figure 9.7. Usage of the 60-bit Network_Destination/Source_Address field depends on which Network_Address_Authority format is used for the Port_Name.

Figure 13.3 Format of Network_Header.

3322 1098	2222 7654	22221111 32109876	11111100 54321098	00000000 76543210
D_NAA	Network_Dest_Address (high-order bits)			
Network_Destination_Address (low-order bits)				
S_NAA	Network_Source_Address (high-order bits)			
Network_Source_Address (low-order bits)				

For IEEE Registered Extended, each Network Address is 8 bytes longer than shown here.

Notes:
D/S_NAA: Destination/Source Network Address Authority
0000 - ignored
0001 - IEEE 48-bit
0010 - IEEE extended
0011 - locally assigned
0100 - IP (32 bit)
0101 - IEEE Registered
0110 - IEEE Registered Extended

Association_Header

The Association_Header can be used for either or both of two functions. The first is for associating Sequences with "Images," which are separate processes or groups of processes behind a Fibre Channel Port. Login with an Image is through the Process Login, and usage of the Originator and Responder Process Associators for addressing Sequences between Image pairs is described in detail in the "Overview of Process Login/Logout" section, on page 174. Since Process Login and the usage of the Process Associators were described in detail there, this section only describes usage of the Originator and Responder Operation Associators. The Association Header format is shown in Figure 13.4.

Operation Associators are defined for use on systems, such as mainframes, where the mapping of operations onto Fibre Channel constructs requires the association of multiple Exchanges together into higher level constructs.

Figure 13.4 Format of Association_Header.

3322 1098	2222 7654	22221111 32109876 11111100 54321098 00000000 76543210
vvvv	xxxv	Originator Process_Associator (high-order bytes)
Originator Process_Associator (low-order bytes)		
reserved		Responder Process_Associator (high-order bytes)
Responder Process_Associator(low-order bytes)		
Originator Operation_Associator (high–order bytes)		
Originator Operation_Associator (low–order bytes)		
Responder Operation_Associator (high–order bytes)		
Responder Operation_Associator (low–order bytes)		

Word 0, bits 31–28 indicate validity of Associator fields
Word 0, bit 24 indicates that the Processor Associator is to be used as a multicast Process_associator, across all images within the Multicast group specified by the Process_Associator

The general approach with these types of systems is for a small number of Exchanges (roughly dozens) to be used for transmitting data from a large number (roughly thousands) of I/O operations. The I/O operations have a longer life than the individual Exchanges do, so a number of Exchanges may be necessary to implement the operation. The Exchange facilities, however, are multiplexed, with a particular Exchange Identifier and other Exchange facilities being reused repeatedly for multiple operations.

The usage of the Association Header Operation_Associator fields is generally as follows. In the first Data Frame of the Exchange, the Exchange Originator would include an Association Header with a valid Originator Operation_Associator field, identifying the Operation identifier and the Exchange identifiers currently being used for it. In the first Data Frame of the first Sequence returned, the Exchange Responder would include an Association Header with the Originator Operation_Associator field echoed back to the Originator and the Responder Operation_Associator field assigned.

At this time, the operation is uniquely identified by a combination of 64-bit Originator and Responder Operation Associator fields. Inclusion of these fields in the Association Header in the first Data Frame of a Sequence makes the Exchange Identifiers (OX_ID and RX_ID) unnecessary for operation identification. They are still necessary for associating Sequences together, transferring Sequence Initiative, etc., but the Exchange Identifiers can be changed to new values within an ongoing Operation. This action is called "X_ID reassignment" and is carried out through the use of the "Invalidate X_ID" and "X_ID reassigned" flags (F_CTL bits 15 and 14).

Device_Header

The Device_Header, if present, is used only in the first Data Frame of a Sequence. A Device_Header may be used by a ULP type. For that ULP type, the Device_Header must be supported, although it may be ignored and skipped, if not needed. If a Device_Header is present for a ULP which does not require it, the related FC-4 may reject the Frame with the reason code of "TYPE not supported" (b'0000 0111'). The format of the Device_Header is entirely ULP-specific, and is not specified by the Fibre Channel documentation.

Special Functions

Data Compression

One of the primary goals for Fibre Channel is that of providing high-bandwidth data transmission. One method of achieving high effective bandwidth over fixed-bandwidth lines is to compress the data before it is transmitted and decompress it after it is transmitted. Fibre Channel specifies the use of a particular data compression algorithm, and maps it into the Frame format.

The data compression algorithm used in Fibre Channel is termed the "Adaptive Lossless Data Compression Lempel Ziv-1" (ALDC LZ-1). This particular algorithm has the advantage of providing quite good compresion on a wide variety of types of data, while being possible to implement in quite fast and efficient hardware. For example, ALDC hardware speeds are approximately an order of magnitude faster than other contemporary LZ-1 algorithms and requires under 10 mm^2 of silicon chip area. Compression is carried out on a per Information Category basis within a Sequence, immediately preceding segmentation to Frame Payloads. F_CTL bit 11 in the Frame Header indicates whether the Payload is compressed or not.

Encryption

The actual encryption is beyond the Fibre Channel scope, but the documentation indicates a recommendation of RSA's RC4 algorithm with 90-bit distribution keys and 75-bit secret keys, unless faster operations or export restrictions require shorter keys. These recommendations are subject, of course, to the possible development of new and better encryption algorithms.

Chapter 14

Class 1 Connection Management

Introduction

The establishment of Class 1 Dedicated Connections for providing guaranteed access and interference-free bandwidth over a multi-N_Port Fabric environment is one of the most distinctive features of the Fibre Channel architecture. It can also be one of the more difficult features for a vendor to support, since it can involve some delicate design problems, it involves close interaction between switch and N_Port elements of the network, and it is an unusual facility for communications networks to provide. It is for these reasons that a good portion of the standards work has been based on assuring that Class 1 Dedicated Connection service can be provided without unduly hurting performance. Class 1 Dedicated Connection management will be covered in somewhat more detail in this chapter than some other parts of the protocol have been.

Utility of Class 1 Dedicated Connections is strongly dependent on the topology used for interconnecting the N_Ports. A Class 1 Dedicated Connection provides a guarantee of service to a pair of communicating N_Ports, ensuring that the full link bandwidth is available to both. In a Point-to-point topology, this is largely guaranteed anyway, since there is no contention for transmission resources with other N_Ports. Similarly, on an Arbitrated Loop topology, provision of full transmission resources is straightforward since the architecture of the Loop arbitration, described in Chapter 16, assures that communication between two N_Ports is essentially identical to point-to-point communications, once Loop access is won. However, on a Fabric topology, establishing a Dedicated Connection requires coordination between switch elements and both N_Ports in the Connection. It also requires coordination of facilities within a switch element between Class 1 and Class 2 or 3 communication resources, which may be handled quite differently. This chapter describes the procedures required for establishing, maintaining, and removing Class 1 Dedicated Connections over Fabric, point-to-point, and Arbitrated Loop topologies.

During a Connection, the Fabric is assumed to operate essentially as a "wire," with no Fabric-level buffering or flow control visible to the attached N_Ports. It is possible for a vendor to actually implement a Class 1 Dedicated Connection using the equivalent of a wire, with a direct electrical connection between Connection Initiator and Connection Recipient. If this is the case, then the clocking of the received data is synchronized with the transmission clock of the connected N_Port. If implementation is done this way, then switching between connected N_Ports or Class 1 and Class 2 Frames in Intermix mode implies a discontinuity in the phase or frequency of the receive clock. This may, depending on the implementation of clock recovery circuitry and the amount of phase or frequency difference, cause a temporary performance degradation as the N_Port's receiver synchronizes to the new input data stream.

Overview: Normal Procedures

The following sections describe the overall operations for establishing, maintaining, and removing a Class 1 Dedicated Connection. More details on Connection operation are discussed in the "Class 1 Dedicated Connection: Detailed Operation" section, on page 254.

Establishing a Connection

When the FC-2 level receives a request from an ULP to initiate a Class 1 Sequence when a Dedicated Connection does not exist, the N_Port will also establish a Class 1 Connection with the destination N_Port as part of the Sequence initiation.

A source N_Port indicates its desire to establish a Dedicated Connection with a particular destination N_Port by transmitting a "connect-request," which is a Data Frame with a **SOFc1** delimiter. The Frame must also be the first in a new Sequence but may or may not be the first in a new Exchange. The source N_Port cannot transmit any more Frames for the pending Connection until a proper ACK Frame for the connect-request is received.

The source N_Port, termed the "Connection Initiator," transmits this Frame over the link. The destination N_Port of the Frame is the potential "Connection Recipient." If the N_Port is on a Point-to-point or Arbitrated Loop topology, the connect-request will be received by the destination N_Port directly. Otherwise, the connect-request will be received directly by the Fabric.

In the connect-request, the Connection Initiator can specify whether it is requesting a unidirectional or bidirectional Connection, by setting the unidirectional Transmit flag (F_CTL bit 8) in the connect-request Frame. In a Unidirectional Dedicated Connection, only the Connection Initiator can transmit Data Frames, with the Connection Recipient transmitting only Link Control Frames.

On receipt of a connect-request, the Fabric will begin establishing a unidirectional or bidirectional Dedicated Connection, making the Connection "pending," and changing the availability of the Connection Initiator and Connection Recipient N_Ports to connect-requests from third-party N_Ports.

The connect-request is delivered to the Connection Recipient either directly over the point-to-point or Arbitrated Loop topology or by the Fabric, on a Fabric topology. If the Connection Recipient is available for the Connection, it will do the internal processing required to set it up and then return an ACK Frame with a **SOFn1** delimiter.

On receipt of the ACK with **SOFn1** delimiter, the Fabric will finish establishing the Dedicated Connection and will route the Frame to the Connection Initiator N_Port. When the Initiator receives the ACK Frame, the Connection is considered to be established by the Connection Initiator, Fabric, and Connection Recipient, and further Class 1 Frames can be transmitted.

During a Connection

Immediately following the establishment of the Connection, Sequence transmission can resume. There are several possibilities, depending on options in the connect-request Frame. It may have been that the connect-request Frame had the **EOFn** EOF delimiter, indicating that more Frames were available for the Sequence. In this case, the Initiator will continue transmitting the Sequence. If the connect-request used the **EOFt** delimiter, then the connect-request was a single-Frame Sequence, and the next Frame the Initiator transmits will initiate a new Sequence.

Either the Connection Initiator or the Connection Recipient can initiate new Sequences, subject to the Sequence Initiative limitation within bidirectional Exchanges over bidirectional Connections. Other than this, Exchange origination and Sequence Initiation follow normal rules described in the "Sequence Management" section, on page 144 and the "Exchange Management" section, on page 149. Sequence Initiation is independent of Connection Initiation.

Removing a Connection

There are several different ways of removing a Dedicated Connection by either the Connection Initiator or Connection Recipient. The normal method of removing a Connection is for the two N_Ports to negotiate removal of a Connection, using End_Connection (F_CTL bit 18) to request transmission of a Frame terminated with **EOFdt**. Removal of a Connection can be a very disruptive operation if it occurs during Sequence transmission, so it is worthwhile for both N_Ports to ensure that there are no Frames in flight when the Fabric receives the **EOFt** delimiter. Under normal operations, both N_Ports know when a Dedicated Connection should be removed. The Sequence Initiator of the last Sequence of the last Exchange of the Connection will set the F_CTL bits to indicate that it is done with the Connection, and the Recipient will return an ACK terminated by **EOFdt**.

Several other, more disruptive, methods are available for more urgent situations. An N_Port may request the normal termination of a Dedicated Connection by transmitting the Remove Connection (RMC) Extended Link Service Request. If the Request is accepted, the ACC Frame will be terminated with the **EOFdt** delimiter. More disruptively, either N_Port may terminate any Data or Link Control Frame with the **EOFdt** delimiter. The Fabric, if present, and the other N_Port should terminate the Connection, whether or not the Frame's Sequence or Exchange or any other active Sequences or Exchanges have been transmitted. Most disruptively, either N_Port can begin the Link Reset protocol, continuously transmitting the LR Primitive Sequence. This will abnormally terminate any existing Connections, Sequences, or Exchanges.

Dedicated Connection Recovery

In all the stages of Dedicated Connection establishment, management, and removal, link errors can cause situations where the state of a Dedicated Connection cannot be unambiguously determined. An example of this would be if N_Port A was engaged in a Dedicated Connection with N_Port B and received a Class 1 Data Frame or connect-request from N_Port C. N_Port A would not know whether the N_Port B understands the Dedicated Connection with A to be established or not. The number of possible cases which could case an indeterminate Connection status is impossible to enumerate here.

Rather than attempt to do status determination on an established Connection by transmitting a Read Connection Status (RCS) ELS Request to the Fabric or to a specific N_Port, the general procedure for solving these kinds of problems is to unambiguously terminate all possibly indeterminate Connections by performing the Link Reset protocol, as described in the "Link Recovery Protocols" section, on page 104. The Port begins transmitting the LR Ordered Sequence continuously. This causes the attached Port to be reset, removing any pending or established Connections. Errors within any active Sequences are recovered according the normal rules for handling Sequence errors described in the "Sequence Recovery" section, on page 233.

Dedicated Connections over Various Topologies

Connection Initiator and Recipient N_Ports may either be directly attached to one another through a point-to-point link, around the circumference of an Arbitrated Loop topology, or through a Fabric. Handling of Dedicated Connections is different in these cases, since the influences of third-party N_Ports and the duration of dedicated communication are different.

Point-to-Point and Arbitrated Loop Topologies

Two N_Ports attached in a Point-to-point topology may choose to either establish one Dedicated Connection for the duration of an operating period or establish and remove Dedicated Connections dynamically, as the need to communicate arises. This choice will really be implementation-dependent and will depend on the relative overhead of establishing and removing Class 1 Connections versus the difficulty of developing procedures which are optimized for the particular topology. Since there is no competition for link resources, and no way for Frames to be delivered out of order, there is much less difference between Class 1 and Class 2 behavior on a Point-to-point topology than for a Fabric topology and less requirement for Class 1 transmission.

The considerations for Arbitrated Loop topology are similar. As described in Chapter 16, the arbitration mechanism for Loop access assures that once a pair of L_Ports gain access to the Loop, no other L_Ports can compete for transmission resources without performing very disruptive initialization procedures. Also, Frames transmitted in Class 2 around a Loop will always be delivered in order. Also, L_Ports are generally expected to open and close communications with multiple other L_Ports on the Loop, so transmitting in Class 1 would add an additional level of overhead to the arbitration overhead required every time a communication partner was changed. Clearly, in an Arbitrated Loop topology, there are not many advantages and several disadvantages to Class 1 transmission. Again, however, for some implementations, the advantages in building one set of procedures for assigning Class of service that is used over all topologies may outweigh the advantages of optimizing performance for the individual topologies.

Fabric Topology

The different operational characteristics of Class 1 versus Class 2 and 3 are most different on Fabric topologies, so it can provide significant performance advantages on Fabric topologies to tune the choice of Class of service to the required transmission characteristics. The operation of Class 1 Dedicated Connections versus transmission in other Classes of service on a Fabric topology is based on the model that the Fabric provides certain behaviors for which the N_Ports can determine their own best procedures.

The relevant parts of the topology for Class 1 Dedicated Connection operation are shown in Figure 14.1, with N_Ports A and B attached to the Fabric and operating as Connection Initiator and Connection Recipient, and N_Port X operating as a third-party N_Port unrelated to the Connection. The "internal side" and "link side" terminology, indicating where an F_Port receives a signal from, is used for this discussion only. An F_Port may receive requests for establishing Dedicated Connections on both its internal side and its link side, and must arbitrate between both. On this topology, the N_Ports expect the Fabric to implement the following behavior.

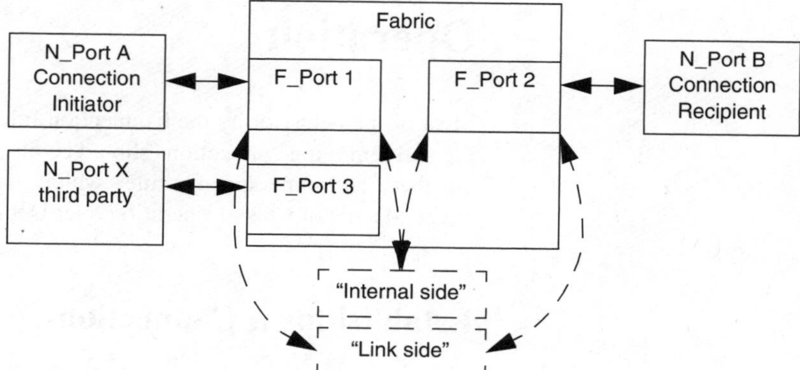

Figure 14.1 Example topology for Class 1 Dedicated Connection discussion examples.

When a particular F_Port is not engaged in a Connection, it can accept a connect-request from the attached N_Port and begin processing it, to reserve Fabric resources required for the requested Dedicated Connection. Once it has accepted a connect-request, it will be busy regarding requests from the internal side, and will generate "Fabric Busy" F_BSY Frames if necessary. It will try to establish a Connection through the internal side of the F_Port attached to the requested destination N_Port to the destination N_Port. If it encounters a rejectable condition either at the remote F_Port or Connection Recipient N_Port, it can return an F_RJT with the **EOFdt** delimiter to the requesting N_Port to prematurely terminate the Connection. If it encounters a busy condition, it will either return an F_BSY with the **EOFdt** delimiter immediately or may delay retrying F_BSY for a retry period if the Fabric supports "Stacked connect-requests," as described in the "Stacked Connect-Requests" section, on page 257.

Once an F_Port has accepted a connect-request from its internal side, it passes the request on to the destination N_Port and waits for an ACK with **SOFi1** delimiter to be returned. At this point, it will discard any connect-requests from the link side, unless it supports Stacked Connect-requests. If "simultaneous" connect-requests arrive at the link and internal sides (whatever simultaneous happens to mean for the particular F_Port implementation), it will accept the connect-request on the internal side as the more established Connection.

If some failure requires a Dedicated Connection to be removed, the identifying entity will notify the Fabric, which notifies each of the involved F_Ports, which then notifies the attached N_Port. Generally this will be carried out using the Link Recovery Primitive Sequence Protocols, as described in the "Link Recovery Protocols" section, on page 104.

Class 1 Dedicated Connection: Detailed Operation

Most of the behavior by the Connection Initiator and Recipient in establishing and removing connections should be clear from the previous discussions, but there are some specific rules which may not be obvious. This section covers details of Class 1 operation over Dedicated Connections.

Establishing a Connection

First, the connect-request Frame, although having a "Class 1" **SOFc1** delimiter, travels through the network as if it were a Class 2 or Class 3 Frame. This means that it will generate R_RDY responses and that its Payload size is limited by the same maximum buffer-to-buffer receive data field size specified in the PLOGI or FLOGI Common Service parameters for Class 2 and Class 3 Data Frames. Following transmission of the connect-request Frame, no other Frames can be transmitted for this Sequence. However, R_RDY will be returned, so if the Fabric and N_Port support Intermix, the N_Port can transmit Class 2 and Class 3 Frames if there is buffer-to-buffer Credit available.

After N_Port A sends a connect-request to N_Port B with the **SOFc1** delimiter, it can receive one of a number of responses, which are shown in Figure 14.2.

A Link Response to the connect-request Frame (F_BSY, P_BSY, F_RJT, or P_RJT) will have the **EOFdt** delimiter to indicate that the Connection was disconnected (before it was connected).

On receipt of a connect-request Frame, an N_Port that returns a P_BSY or P_RJT will terminate the Link Response Frame with **EOFdt**. If it returns an ACK, the ACK will have the **SOFi1** and **EOFn** or **EOFt** delimiters, and the N_Port will consider the Connection established.

In the case of connect-request collisions over a Fabric, the connect-request that is more established is given priority. This means that, for the topology of Figure 14.1, if N_Port A transmits a connect-request to N_Port B, then receives connect-request from N_Port X before the ACK from N_Port B returns, it knows that the Fabric has given N_Port X's connect-request priority and will re-queue its own request for transmission at a later time, unless the Fabric supports stacked connect-requests.

On a connect-request collision over a Point-to-point topology, where an N_Port receives a connect-request after sending its own and before receiving an ACK, the N_Port with the higher-valued N_Port identifier gets priority, and the other N_Port re-queues its request for transmission at a later time.

Unidirectional versus Bidirectional Connections

A Connection is established as unidirectional when the Connection Initiator asserts Unidirectional Transmit (F_CTL bit 8) in the connect-request. If the Connection Initiator ever sets Unidirectional Transmit to 0 in the first or last Frame of a subsequent Sequence, the Connection becomes bidirectional and cannot be made unidirectional again. The Connection Recipient can request that the Connection Initiator make the Connection bidirectional by setting Unidirectional Transmit to 0 in any Frame transmitted to the Initiator — the Initiator may comply with the request and make the Connection bidirectional by setting Unidirectional Transmit bit to 0 on a first or last Frame of a Sequence.

The unidirectional transmit facility has three primary uses: (1) for compatibility with lock-down stacked connect-requests, as described in the "Stacked Connect-Requests" section, on page 257, (2) to prevent overflow during temporary buffer allocation problems, such as when the Connection Initiator knows that for some reason it can't provide the buffer-to-buffer Credit specified at Login, or (3) for usage by an N_Port that may only receive or transmit Data Frames at any one time.

Figure 14.2
Possible responses to a **SOFc1** connect-request Frame.

Frame Type	SOF	S_ID	D_ID	EOF	Cause: N_Port action
					Connect-request Frame
Data Frame	**SOFc1**	A	B	**EOFn**	Dedicated Connection Requested: Transmit connect-request, wait for response
					Responses
ACK_1 or ACK_N	**SOFn1**	B	A	**EOFdt**	Dedicated Connection established: continue transmitting Sequence
ACK_1 or ACK_N	**SOFn1**	B	A	**EOFt**	Dedicated Connection established: Sequence ended, do next Sequence
F_BSY	**SOFn1**	B	A	**EOFdt**	Connection failed - busy in Fabric: try again later
P_BSY	**SOFn1**	B	A	**EOFdt**	Connection failed - busy in N_Port: try again later
F_RJT	**SOFn1**	B	A	**EOFdt**	Connection failed - Fabric reject: Check reason and action codes, follow indications
P_RJT	**SOFn1**	B	A	**EOFdt**	Connection failed - N_Port reject: Check reason and action codes, follow indications

Data Frame	**SOFc1**	B	A	**EOFn**	Collision of connect-requests in point-to-point -- Response depends on relative Port IDs:
					if A > B in value:
					Discard Frame from B and wait for ACK
					Dedicated Connection established with A as initiator
					if A < B in value:
					Respond with **SOFn1** on ACK
					Dedicated Connection established with B as Initiator
					Retransmit original connect-request Frame as a Class 1 Frame (**SOFi1**)

Data Frame	**SOFc1**	X	A	**EOFn**	Collision of connect-requests in Fabric: Requeue request associated with the connect-request (unless stacked connect-requests are supported) Respond with **SOFn1** on ACK_1 or ACK_N to X Dedicated Connection established with X as Initiator
-	-	-	-	-	Timeout, no response Frame: Perform Link Reset Protocol

Stacked Connect-Requests

Stacked connect-requests is a feature which may be provided by a Fabric. Support for stacked connect-requests is determined by an N_Port and by the Fabric during Fabric Login. Intermix must also be functional in order to use this feature. If stacked connect-requests are functional (i.e., supported at both Fabric and N_Port), an N_Port may transmit one or more connect-requests without regard to whether a Connection is pending, established, or complete.

Stacked connect-requests allow the Fabric to work on establishing a Connection while an N_Port is busy servicing another Connection and may enhance system performance. The Fabric may process multiple connect-requests in any order, as facilities become available. It will be able to provide the status of a Stacked connect-request via the Read Connection Status (RCS) ELS Command. An N_Port uses the E_D_TOV timeout period after a connect-request has been transmitted (part of the Sequence timeout) whether the connect-request is a normal or a stacked request. That is, an ACK response will be received or an F_BSY will be returned from the Fabric to the N_Port within an E_D_TOV timeout period for a connect-request, even if it is stacked. If neither condition is met within E_D_TOV, the N_Port detects a Sequence timeout, and Connection recovery is performed (described in the "Dedicated Connection Recovery" section, on page 251). A Fabric which supports Stacked connect-requests must be able to deal with any associated race conditions which occur during the establishment and removal of Connections.

Due to limited timing relationships involved, there are two methods of implementing stacked connect-requests by a Fabric:

transparent mode: When the SOFc1 Data Frame is delivered to the destination N_Port, the return path of the bidirectional circuit is established in the same manner as exclusive Dedicated Connections. This means that the destination N_Port of the SOFc1 is able to transmit Data Frames immediately following transmission of the ACK Frame in response to the SOFc1 Frame, if the connection is not unidirectional.

lock-down mode: When the SOFc1 Data Frame is delivered to the destination N_Port, the return path of the bidirectional circuit is not necessarily established to the source N_Port of the **SOFc1**. The **SOFc1** Data Frame must have Unidirectional Transmit (F_CTL bit 8) set to 1 in order to prevent the Connection Recipient from sending any Data Frames back on the connection until after the Connection Initiator has made it bidirectional.

In either case, the destination N_Port will be unaware that the connect-request had been stacked.

The determination of what mode of connect-request is functional is based on the FLOGI request and ACC reply from the Fabric Login and is contained in bits 29 and 28 of the first word of the Class 1 Service Parameters, as shown in Figure 9.8. Bit 29 is a request by the N_Port for transparent Stacked connect-requests. Bit 28 is a request by the N_Port for lock-down Stacked connect-requests. An N_Port may request either or both modes. However, a Fabric cannot support both modes and will support either transparent or lock-down or neither.

This function can also be more controlled at the N_Port level than at Fabric level, with the N_Ports requesting that particular connect-requests be stackable, rather than having the Fabric allow stacking of every connect-request. Similarly, the N_Port can request that the Camp-on capability be used, as described in the "Camp-On" section, on page 259. These capabilities are requested using the CS_CTL Class-specific control field, as shown in the "Class Specific Control: CS_CTL" section, on page 119. This capability helps clarify processing for a Connection Initiator, which can have better information on the status of pending connections, and also may improve fairness of connection establishment.

Removing a Connection

For requesting removal of a Connection, a Sequence initiator will set the End_Connection F_CTL bit to 1 on the last Data Frame of a Sequence to indicate that it won't send any more Sequences and is requesting the other N_Port to complete any active sequences and remove the Connection by transmitting an ACK with **EOFdt** delimiter. This Sequence Initiator can still receive Frames in the Connection and can begin reception of new Sequences newly initiated by the other N_Port of the Connection.

If either N_Port receives a Data Frame with End_Connection set to 1, it will complete any active Sequences before returning the ACK with **EOFdt**, but it won't Initiate any more Sequences, even if it believes it has data to send. Any further Sequences can be sent following establishment of another Connection.

In the case of a "Connection removal collision," where one or the other N_Ports in a Connection transmits a last data Frame with End_Connection set and then receives a last Data Frame with End_Connection set before receiving the ACK with **EOFdt** or **EOFdti**, the Connection Initiator will take priority on transmitting the **EOFdt** or **EOFdti**. The Connection Recipient will transmit a normal **EOFt** to prevent attempted removal of already-removed Connection.

There are no specified rules on when a Connection should be removed. It will be generally up to the ULP to decide when to remove a Connection. For maximum system utilization, in general it is best to remove a Connection if it will not be used soon, in order for the Port to be available for Connection to another N_Port. Normally, an N_Port chooses to remove a Dedicated

Connection when it has no Sequences to transmit to the connected N_Port and it has a request to transmit Sequences to a different N_Port. The Continue Sequence Condition F_CTL bits can help indicate how soon the next Sequence of an Exchange is expected to be transmitted. Since the timing estimation is based on the time to remove and reestablish a Connection, this indication can help in the decision.

Camp-On

Camp on is proposed as a optimization for the following fairness problem. N_Port A may send a connect-request to the Fabric for connection with N_Port B, which happens at the time to be busy with a connection to N_Port C. The connect-request will be busied, and it may be that N_Port B will break off the connection with N_Port C and establish another connection with N_Port D before N_Port A retries the connect-request. Under bad timing conditions, N_Port A could conceivably never achieve a connection to N_Port C. The Camp-On mechanism lets the Fabric queue connect-requests to each destination N_Port, so that the connect-requests can be granted in the order that they were requested. A connect-request is specified as Camp-On connect-request by setting the COR bit in the CS_CTL field, as shown in the "Class Specific Control: CS_CTL" section, on page 119.

Camp-On is very similar to Stacked Connect Requests, in that they both allow a Connect Request to be sent to the Fabric regardless of the status of existing connections. The main difference is that Stacked Connect Requests can be Accepted in any order (i.e., when each Recipient becomes ready to accept a connection), where Camp-On requests are always accepted in the order they were issued by the Initiator. A connect-request Frame can request one or the other, but requesting both on the same connect-request is a protocol error.

Preemption

Preemption is the capability to remove an existing dedicated connection to establish a new dedicated connection. This is primarily for real-time systems, which require interconnection networks that provide guaranteed bandwidth, guaranteed in-order delivery, and guaranteed latency. To meet the requirements placed on real-time systems, both priority (as described in the "Priority and Preemption" section, on page 129) and preemption at the network level are extremely useful, if it's managed and used carefully.

Chapter 15

Flow Control

Introduction

Flow control is a tremendously important factor influencing the performance of any data communications architecture. This is particularly true in a multiplexed networking environment, where the aggregate bandwidth of multiple sources sending simultaneously would overwhelm the reception capacity of any practical destination receiver implementation, if no flow control were imposed.

This chapter covers the flow control mechanisms used in the Fibre Channel architecture for assuring flow control at the link, Frame, N_Port, and Fabric levels. Several of the various flow control mechanisms which have been developed for network data communications are described in the "General Strategies" section, on page 262. The strategies used in the Fibre Channel architecture are enumerated in the "Fibre Channel Flow Control Strategies" section, on page 263. In particular, the end-to-end flow control strategies used between N_Ports is described in the "End-to-End Flow Control" section, on page 265, and the link-level buffer-to-buffer flow control mechanism is described in the "Buffer-to-Buffer Flow Control" section, on page 270. Finally, the "Integrated Flow Control for Class 2 Service" section, on page 272 describes the interaction of buffer-to-buffer and end-to-end flow control mechanisms for Class 2 traffic.

General Strategies

In any communication system, a great deal of attention must be paid to flow control, to assure that no Port is flooded by more data than it can handle. This is particularly important on a switched network communication Fabric, where multiple sources may be attempting to send data to a single destination simultaneously, generating more transmission data than can possibly be received with any practical receiver design.

There are several strategies for flow control implemented in high-speed computer communications.While there are numerous variations, they can be broadly grouped into two categories.

In the first, a source can make a request for a certain level of transmission capacity. This request must be handled by both the Fabric, if present, and by the destination Node. If the request is granted, the Fabric and the destination Node will have guaranteed that no traffic originating at other Nodes will interfere with the transmission. This is essentially a source-based, or request-based method, since the source requests a certain amount of transmission capacity, based on the requirements of the data to be transmitted.

The second method, which is essentially a destination-based method, can be referred to as a token-based or credit-based flow control mechanism. In this case, the destination has a fixed number of reception resources available, each marked by a token. When one frees up, the destination sends the token back to the source. The token lets the source send a single "unit" of information, whatever is capable of being handled by the reception resource. The source can send no more units of information to that reception resource until the destination sends the token back.

In general, request-based flow control is more flexible, since the source can tell the destination Node exactly what type of information is being transmitted, while token-based flow control is faster, since the handshaking, which assures that the data can be handled, is done before the actual transmission. Thus, token-based flow control is used for the basic units of transmission, and request-based flow control is used for higher-level constructs.

In some networks which are more oriented towards telecommunications, such as ATM networks, there is a distinction made between credit-based and rate-based flow control strategies. In a rate-based strategy, a source can transmit a fixed amount of data per time period, rather than transmitting when it has been assured that the previously transmitted data has been received. This mechanism simplifies flow control management, since it requires less handshaking, but it can be extremely inefficient when the rate of data transmission is not constant and well-known in advance. Rate-based flow control is well-suited to telephone-type communications, where the data rate is fixed and relatively constant over the lifetime of a single communications operation. It is less well suited for data communications networks, where the traffic is more bursty. In keeping with it's targeted function in data communications, Fibre Channel uses no rate-based flow control, although the Class 4 capability describe in the "Class 4 — Fractional" section, on page 192, provides somewhat equivalent capabilities.

Fibre Channel Flow Control Strategies

Fibre Channel uses both types of flow control in different places. A token-based type of flow control is used at the Frame level, to ensure data buffering space between link-connected N_Ports or F_Ports (buffer-to-buffer Credit) and between source and destination N_Ports (end-to-end Credit). A request-based type of flow control is used in Class 1 service, to ensure that resources exist both at the Fabric and the destination N_Port to handle the high-speed continuous data flow that occurs under a Class 1 Connection.

Between every two Ports, whether they are F_Ports or N_Ports, there is defined a buffer-to-buffer Credit value (BB_Credit). This value describes the number of available buffers there are for transmission Frames. The token passed back after a receive buffer is emptied is the "R_RDY" Ordered Set. One R_RDY is passed back over the link when resources are available to receive another Frame.

Between every two N_Ports which have logged in with each other, there is defined an end-to-end Credit value (EE_Credit). This value describes the number of available buffers at the destination N_Port for Frames from the source N_Port. The ACK_1, ACK_N, and ACK_0 Link Control Frames are used for acknowledging cleared receive buffers in this case — the different ACK types allow implementation-dependent variations in flow control granularity.

Before transmission of a Class 1 Sequence can begin, the source Node sends an initial Frame with a **SOFc1** SOF delimiter. The source may not send any more Frames of the Sequence until this first Frame is acknowledged, with an ACK_1 returned Frame. If the destination Port is unable to handle the Class 1 transmission, for example, if it already has a Class 1 Connection set up with a different N_Port, the destination Node will respond with a "P_RJT" Frame. The source must then retry the Connection at a later time.

Class 1 Frames use end-to-end flow control only — they do not use buffer-to-buffer flow control, since it is assumed that there is no Fabric-level buffering necessary for Class 1 Connections and traffic. Class 2 Frames and Class 1/**SOFc1** Frames use both end-to-end and buffer-to-buffer flow control. Class 3 Frames use only buffer-to-buffer flow control. Applicability of the flow control mechanisms to the different service Classes of service is shown in Figure 15.1.

Figure 15.1
Applicability of the flow control mechanisms to different Classes of service.

Applicability Mechanism	Class 1	Class 1/SOFc1 connect-request	Class 2	Class 3
End-to-end Credit	Yes	Yes	Yes	-
Buffer-to-buffer Credit	-	Yes	Yes	Yes
ACK_1 or ACK_N	Yes	Yes	Yes	-
ACK_0	1 per Seq.	Yes	1 per Seq.	-
R_RDY	-	Yes	Yes	Yes
F_BSY to a Data Frame	-	Yes	Yes	-
F_BSY to a Link Control Frame (except to a P_BSY)	-	-	Yes	-
F_RJT	-	Yes	Yes	-
P_BSY	-	Yes	Yes	-
P_RJT	Yes	Yes	Yes	-

End-to-End Flow Control

End-to-end flow control is used to pace the flow of Class 1 and Class 2 Frames between N_Ports. Management of end-to-end flow control is carried out through the transmission of Link Control Frames (ACKS, RJT, and BSY Frames) in response to Data Frames. The Fabric, if present, is not involved in end-to-end flow control, except to carry the Data and Link Response Frames. Any Fabric-level buffering that may be involved in end-to-end Credit (versus the link level credit handled using the buffer-to-buffer mechanism) is handled transparently to the N_Ports.

End-to-end Credit is initialized using a field in the N_Port Class Service Parameters of the ACC to the PLOGI Sequence, described in the "N_Port and F_Port Class Service Parameters" section, on page 171. Before N_Port Login, the minimum and default value of 1 applies. A unique value is assigned at each source N_Port acting as Sequence Initiator for transmission in each Class of service to each destination N_Port. The maximum end-to-end Credit value, termed "EE_Credit," indicates the maximum number of Frames that can have outstanding ACKs to the corresponding destination in the corresponding Class of service.

The optimal EE_Credit value for a particular topology and implementation is the value which is just large enough to allow continuous streaming of Frames. Any larger value of EE_Credit would be wasted and might waste valuable receive buffering resources at the destination N_Port. The optimal EE_Credit value can be determined using the Estimate Credit Procedure, described in the "Procedure to Estimate End-to-End Credit" section, on page 176. The Advise Credit (ADVC) ELS Request can also be used as a stand alone procedure to request revision of the EE_Credit.

During normal operation, each N_Port must keep track of a count of the number of Frames with outstanding ACKs. This count is termed the "EE_Credit_CNT." The EE_Credit_CNT is incremented with each Frame transmitted and decremented with each ACK received. If the EE_Credit_CNT value ever matches the EE_Credit value, no more Frames may be sent in that Class to that Recipient N_Port until an ACK Frame returns.

The Recipient N_Port does not keep track of EE_Credit_CNT and only has the responsibility of returning ACK Frames as soon as is reasonable, to ensure that the EE_Credit_CNT value is kept low.

The allocation by a source N_Port of end-to-end Credit between the multiple Sequences or Exchanges that may be active at one time is unspecified. For example, a source N_Port with several Class 2 Sequences active in different Exchanges to a destination N_Port might choose to allocate equal credit value to each active Sequence, allowing them equal transmission priority, or it could choose to assign the full EE_Credit count in turn to each single Sequence, allowing each access to the maximum available credit to allow faster completion. The allocation strategy gets particularly complicated when different Sequences may be completed at different times or may be aborted. Allocation of end-to-end Credit between Sequences or Exchanges for maximum efficiency is left up to implementation ingenuity.

The EE_Credit_CNT value is reset to 0 at the end of N_Port Login or re-Login or following transmission of the Link Credit Reset (LCR) Frame, as described in the "Link Credit Reset (LCR) Frame" section, on page 134.

An Initiator and Recipient N_Port pair implementing the ACK_0 acknowledgment model are essentially assumed to have infinite receive buffering capacity or to have flow control handled at the buffer-to-buffer level, so end-to-end flow control and EE_Credit_CNT management are not applicable under ACK_0 usage. This is independent of the value of the History bit.

Receive Buffer Allocation for EE_Credit Assignment

If an implementation of the receiver Portion of an N_Port has a specific number of receive buffers to receive Frames that are governed by end-to-end flow control, the count of total allowable EE_Credit for the N_Port acting as a Recipient has to be allocated among the Initiator(s) that may be sending Frames to it. The strategy for allocating buffers to assign EE_Credit values depends on the particular implementation, and depends on the Class of service as well.

In Class 1, only a single Initiator can have a Dedicated Connection at a time, so the end-to-end Credit can be shared between Initiators. For example, if a particular Recipient has eight buffers reserved for receiving Class 1 Frames, it may log in with an end-to-end Class 1 Credit value of eight to every source N_Port which may initiate Frames to it. As long as the Recipient assures that all 8 buffers are free before returning the ACK to a connect-request Frame, the buffers cannot be overrun.

In Class 2, on the other hand, allocation of end-to-end Credit for a destination N_Port which has a specific number of buffers is more delicate. Since Frames may be multiplexed from multiple Initiators simultaneously, there are a range of possibilities for end-to-end Credit allocation. One possibility is for the fixed number of receive buffers to be distributed among all N_Ports which might need to transmit Frames, which could diminish performance drastically and would be impossible on a large network with more logged-in source N_Ports than receive buffers. Another possibility is for the total of end-to-end Credit values on all source N_Ports to be greater than the number of receive buffers, which naturally eliminates much of the function of end-to-end flow control. If the actual traffic patterns are such that all the sources aren't simultaneously sending Frames to the destination, this behavior may be fine. Actual implementations must use other mechanisms, particularly buffer-to-buffer flow control, to solve these types of problems without allowing buffer overrun.

Events Affecting EE_Credit_CNT

The events that affect the end-to-end Credit count (EE_Credit_CNT) value are shown in Figure 15.2. Some events that are not applicable to end-to-end Credit, but which might be expected to be, are included as well, to help clarify any confusion that might arise.

Figure 15.2 End-to-end flow control management rules.

Event	Effect on EE_Credit_CNT
N_Port transmits a Class or Class 2 Data Frame	+1
N_Port receives F_BSY for a Data Frame, F_RJT, P_BSY, or P_RJT	−1
N_Port receives ACK_1 (History bit = 1)	−1
N_Port receives ACK_1 (History bit = 0)	−1 for ACK_1, also −1 for each unacknowledged Frame with lower SEQ_CNT
N_Port receives ACK_N (History bit = 1)	−N
N_Port receives ACK_N (History bit = 0)	−N for ACK_N, also −1 for each unacknowledged Frame with lower SEQ_CNT
N_Port receives ACK with **EOFt** or **EOFdt** delimiter	−1 for each unacknowledged Frame in the Sequence
N_Port transmits a Link Control Reset	Reset to 0
N_Port receives an LCR as Sequence Recipient	no change — reset any receive buffers

No effect — reason

N_Port receives ACK_0 or transmits a Frame with ACK_0 in effect —
Reason: ACK_0 doesn't participate in end-to-end Credit

N_Port receives a Data Frame (Class 1, Class 2, or Class 3) —
N_Port transmits ACK, P_BSY, or P_RJT —
N_Port receives F_BSY for a Link Control Frame —
Reason: No Recipient EE_Credit_CNT

N_Port transmits a Class 3 Data Frame —
Reason: All Class 3 end-to-end flow control at ULP level
Reason: No Recipient EE_Credit_CNT

As shown in the figure, an ACK_0 does not affect the EE_Credit_CNT value but does indicate that the Sequence has been received successfully (or unsuccessfully, if the History bit = 1, indicating that a previous ACK has not been transmitted).

When an ACK is received with the History bit set to 0, or when an ACK with the **EOFt** or **EOFdt** is received, then the Sequence Initiator can decrement the EE_Credit_CNT value for all previously transmitted Frames with outstanding ACKs. This mechanism is useful for clearing Frame acknowledgment status under conditions of out-of-order Frame or ACK delivery in Class 2 or in the case of lost ACK Frames in Class 1. If this is implemented, the N_Port must be able to recognize by SEQ_CNT and ignore an ACK for an already-acknowledged Frame that arrives later.

When a Sequence is terminated, all credit for Frames with outstanding ACKs in that Sequence can be recovered. This applies even if the Sequence is terminated abnormally. Class 1 EE_Credit is also recovered for Frames with outstanding ACKs when a Dedicated Connection is removed by either **EOFdt** or by the Link Reset Protocol.

Sequence Recipient Responsibilities

In order for end-to-end flow control management to operate correctly at the Sequence Initiator N_Port, the Sequence Recipient must follow some specific rules in handling ACK Frames. The rules followed depend on whether ACK_1, ACK_N, or ACK_0 Link Control Frames are being used, on the Class of service being used, and on whether Sequences are streamed. These responsibilities are clarifications of the ACK rules described in the "ACK Frames" section, on page 135, as applied to simplifying end-to-end Credit management.

ACK_1 Usage. For each Data Frame received which requires acknowledgment, the Sequence Recipient will return an ACK_1 Frame, with the ACK_CNT in the Parameter field set to 1. The History bit follows the normal usage, indicating whether there is anything worrisome in the Sequence history or not.

The transmission of the last ACK_1 of the Sequence depends on the Class of service used. In Class 1, the last ACK_1 — which must have End_Sequence (F_CTL bit 19) set to 1 and History set to 0 and must use the **EOFt** or the **EOFdt** delimiter — must be withheld until all ACKs to previous Frames of the Sequence have been transmitted. In Class 2, the last Data Frame of the Sequence might arrive before one of the previous Frames. The Sequence Recipient can either withhold the last data Frame's ACK, essentially following Class 1 procedure, or it can transmit an ACK_1 for the Sequence's last Data Frame, using End_Sequence = 0, History = 1, and the **EOFn** delimiter. Only when all Frames have been accounted for will it finally send its final ACK_1 with End_Sequence = 1 and the **EOFt** delimiter to terminate the Sequence.

If the Sequence Recipient is caught withholding the last ACK for a Sequence when a Sequence error is detected or the E_D_TOV expires before receiving one of the previous Frames, a Sequence Recipient supporting one of the three discard error policies will transmit the last ACK with the Abort Sequence F_CTL bits set to either "Abort Sequence, perform ABTS" (b'01') or "Immediate Sequence retransmission requested" (b'11'), depending on the Exchange Error Policy used for the Exchange. The Sequence Recipient supporting the "Process with infinite buffering" Policy will transmit the last ACK without setting the Abort Sequence bits.

ACK_N Usage. Under ACK_N usage, each ACK Frame indicates the number of Data Frames being acknowledged in the ACK_CNT field (Parameter bits 15 to 0). The History bit is used for indicating status of previous ACK transmission. The actual value N is implementation dependent and may be 1.

Whenever a missing Frame is detected, an ACK_N must be sent to acknowledge Data Frames received with SEQ_CNT values lower than that of the missing Frame. Handling of the End_Sequence and History bits under Class 1 and under Class 2 with possible Frame reordering is the same as for handling of the ACK_1 Link Control Frame.

ACK_0 Usage. When ACK_0 is used, end-to-end flow control is not active. A single ACK_0 per Sequence indicates successful or unsuccessful Sequence delivery except under the conditions specified in the "ACK Frames" section, on page 135. The ACK_0 Frames should have History = 0 (ignored) and ACK_CNT = 0 and should use either the **EOFt** delimiter (Class 1 or Class 2) or the **EOFdt** delimiter (Class 1).

ACK Usage with Streamed Sequences. All of the normal flow control rules are followed when streamed Sequences are used. In addition, in the case of the "Abort, discard multiple Sequences" Exchange error policy, the last ACK for a succeeding Sequence will be withheld until all the previous Sequences are complete and deliverable. This additional withholding, to wait for previous Sequences to complete and be deliverable, is not applicable to the case of "Abort, discard a single Sequence" error policy, since the Sequences are independently deliverable.

Buffer-to-Buffer Flow Control

Buffer-to-buffer flow control governs the transmission of Frames over individual links in a Point-to-point or Fabric topology and between source and destination Ports in an Arbitrated Loop topology. The basic operation of buffer-to-buffer flow control is for an R_RDY Primitive Signal to be sent to the transmitting Port of a link whenever there are sufficient resources to receive a single Class 2, Class 3, or Class 1 connect-request Frame on the receiving side of the link. Buffer-to-buffer flow control only applies to connectionless service.

To manage these operations, each transmitter manages a BB_Credit value, which denotes the total number of receive buffers at the attached Port, and BB_Credit_CNT value, which denotes the number of Frames outstanding to the attached Port without R_RDY acknowledgments returned. The term "receive buffer" indicates the capability of receiving Frames and may describe both storage space and processing resources. The values of BB_Credit are conveyed in the buffer-to-buffer Credit field of the common Service Parameters, shown in Figure 9.5, of either the FLOGI, in a Fabric topology, or the PLOGI, in a Point-to-point topology. As long as the BB_Credit_CNT value is below the BB_Credit value, Frames can be transmitted.

Buffer-to-buffer Credit governs flow of both Data Frames, and of ACK, BSY, RJT, and LCR, Link Control Frames. It operates identically on N_Port-to-N_Port and N_Port-to-F_Port links, in both directions. The BB_Credit_CNT value can be reset to the login value by performing the Link Reset protocol, described in the "Link Recovery Protocols" section, on page 104, or by performing a Fabric re-Login.

The events that affect the buffer-to-buffer Credit count value are shown in Figure 15.3.

Figure 15.3 Buffer-to-buffer flow control management rules.

Event	Effect on BB_Credit_CNT
N_Port or F_Port transmits a Class 2, Class 3, or Class 1 connect-request Data Frame or Link Control Frame	+1
N_Port or F_Port receives R_RDY	–1

No effect

N_Port or F_Port receives a Class 2, Class 3, or Class 1 connect-request Frame
N_Port or F_Port transmits R_RDY

Alternate Buffer-to-Buffer Flow Control

On a Fabric topology, with a switch between N_Ports, BB_Credit is established with the Buffer-to-Buffer credit field in the Fabric Login Link Service Command (FLOGI, shown in Figure 9.5). On point-to-point topology, without a switch, BB_Credit is established with the corresponding field in the N_Port Login Link Servicecommand (PLOGI, also in Figure 9.5). In both of these cases, BB_Credit indicates the total number of link-level receive buffers available, which are all assumed to be available when the link is brought up for immediate reception of Class 2 or Class 3 Data Frames, or Class 1 connect-request Frames.

On a Loop topology, neither of these values is directly applicable, since an Arbitrated Loop acts (as far as link-level flow control is concerned) as a point-to-point link, where the L_Ports of the link keep changing as they're granted access to the loop. There is no Fabric, and the Buffer-to-buffer credit that an L_Port uses at any time depends on which other L_Port on the loop it has established a circuit with.

To handle this situation, the alternate Buffer-to-buffer credit management can be used. In this case, the PLOGI Buffer-to-Buffer credit field contains, not the total number of receive buffers, but only the total number of receive buffers guaranteed to be available when the circuit is opened. The L_Port may have more receivers buffers available, as signalled with R_RDY Primitive Sequences, but the sender can only the number of Frames up to the PLOGI Buffer-to-buffer credit limit, and has to wait until it has received more R_RDY Primitive Sequences before it can transmit more Data Frames. Use of this capability in an Arbitrated Loop environment is described in the "Use of Alternate Buffer-to-Buffer Flow Control Management" section, on page 293.

It is always possible for an L_Port or N_Port to advertise Buffer-to-Buffer Credit of 0, in which case it doesn't have to guarantee to have any free buffers available when the circuit is opened. In this case, the opposite end of a circuit cannot send any Data Frames until it has received R_RDY Primitive Signals, which causes an extra link round-trip delay on a Loop topology.

As discussed in the "Performance and Timing" section, on page 294, an N_Port Login Buffer-to-Buffer credit value of more than 1 doesn't usually make much sense. While the first Data Frame is being transmitted, R_RDY Signals for any other available receive buffers can be returned, allowing immediate full-bandwidth data transmission as circuits are opened between L_Ports or N_Ports.

Integrated Flow Control for Class 2 Service

A summary of the factors affecting end-to-end Credit and buffer-to-buffer Credit for Class 2 service is shown in Figure 15.4. Since Class 2 incorporates both types of flow control, factors affecting Class 1 (which only uses end-to-end Credit) or Class 3 (which only uses buffer-to-buffer Credit) can be inferred from the figure, as well as from Figure 15.2 and Figure 15.3.

Figure 15.4
Summary of combined flow control management for Class 2 transmission.

Event	N_Port		F_Port
	EE_Credit_CNT	**BB_ Credit_CNT**	**BB_ Credit_CNT**
Port transmits a Class 2 Data Frame	+1	+1	+1
Port transmits F_BSY, F_RJT, P_BSY, P_RJT, or ACK		+1	+1
Port receives R_RDY		–1	–1
Port receives F_BSY for Data Frame, F_RJT, P_BSY, or P_RJT	–1		
Port receives ACK_1 (History bit = 1)	–1		
Port receives ACK_1 (History bit = 0)	–1 for ACK_1, also –1 for each unacknowledged Frame with lower SEQ_CNT		
Port receives ACK_N (History bit = 1)	–N		
Port receives ACK_N (History bit = N)	–N for ACK_N, also –1 for each unacknowledged Frame with lower SEQ_CNT		
Port transmits an LCR Frame	Reinitializes to Login Value	+1	+1
Port receives an LCR Frame	no change—reset receive buffer		

No effect / Not applicable
- Port receives any Class 2 Data Frame
- Port transmits R_RDY
- Port receives F_BSY for a Link Control Frame
- Port receives ACK_0

Chapter 16

Arbitrated Loop

Introduction

The original Fibre Channel architecture had only the Point-to-Point and switched network (i.e., Fabric) topologies defined. However, both of these topologies have significant disadvantages for a number of communications applications. The Point-to-point topology guarantees maximum bandwidth availability and freedom from collision or congestion, but it is usually a relatively inefficient usage of hardware, since very few systems or applications can effectively use a 100 MBps link for any reasonable fraction of the time, and because the number of required links grows very quickly (N**2) versus the number of attached ports. The Fabric topology provides the connectivity and link bandwidth utilization flexibility, but can be a quite expensive topology, since the switches are complex components that will be sold in relatively low volumes. The Arbitrated Loop provides a compromise between these two architectures, with many advantages of both.

Overview

In the Arbitrated Loop topology, the outgoing and incoming fibers to a Port are split off to attach to different remote Ports, such that the aggregation of fibers forms a unidirectional Loop which passes through every Port exactly once. The general picture of traffic generation on this topology is that (1) a single Port arbitrates for usage of the entire Loop, (2) once it obtains Loop access, it opens up communications with another Port on the Loop, and then (3) transmits normal Fibre Channel traffic at the full guaranteed link bandwidth as if they were attached point-to-point until it is done, when it (4) releases the Loop for usage by another Port to begin another round of full bandwidth communication. During communications between two Ports on the Loop, the non-participating Ports forward the data around the loop, monitoring the traffic to watch for the next chance to communicate but not interfering in the transmission. There is minor increased latency (but no loss of bandwidth) caused by the fact that the paths between communicating Ports may pass through other Ports on the Loop.

This topology and network architecture has several significant advantages. The primary one is that it provides multi-Port connectivity between up to 126 NL_Ports without the extra expense of a switch. The aggregate bandwidth for the whole network is 100 MBps, which can be several factors less than the aggregate bandwidth of a switched Fabric, but the arbitration for access to the entire Loop means that communication between any pair of communicating Ports during the time that they're communicating occurs at the full Fibre Channel bandwidth of 100 MB/s. This type of architecture is not particularly good for applications such as video transmission, which require fixed, guaranteed bandwidth between multiple communicating pairs. However, it is ideal for traffic patterns which are very bursty, requiring very high bandwidth transmission for relatively short periods. These kinds of traffic patterns match the typical traffic patterns on LANs pretty well, and the traffic patterns on SANs very well. In particular, this interface is particularly well-sited for providing access to disk arrays, with a small footprint and relatively low aggregate hardware costs.

For these reasons, the use of the Arbitrated Loop topology in Fibre Channel devices is very common. A number of early Fibre Channel devices that were built before switches were widely available even left Fabric-attach capabilities altogether, and would only work on loops, but this type of device is rarely built now.

To provide even more flexibility, the Arbitrated Loop architecture also recognizes the concept of "Private Loop" and "Public Loop" devices, as well as "Private" and "Public" networks. Public Loop topologies are Arbitrated Loops where one of the participating L_Ports is on a switch. Such a switch Port that implements both L_Port Arbitrated Loop functionality and F_Port switch functions is termed an "FL_Port." By sending Frames through such an FL_Port, Public Loop devices would be able to communicate around the Loop but would also have connectivity over a Fabric, using the same protocols and transmission hardware. Private Loop devices are L_Ports that are not able to do a Fabric Login or send data through an FL_Port.

A major factor affecting the performance of such a Loop topology is the arbitration mechanism used for accessing the shared medium. On true shared medium transmission topologies, such as Ethernet, the bandwidth utilization can be quite low, well below 50%, since each Port has limited knowledge of operations happening over the link. With the Arbitrated Loop topology, the signal passes through every Port on the Loop, with each Port acting as a repeater. This means that each Port can monitor every word as it passes by, and can also modify the words before passing them on. This capability for very closely monitoring and modifying Loop activity allows the Ports to use very efficient arbitration and utilization protocols, yielding better than 80%–90% bandwidth utilization with optimal traffic patterns.

A major focus of the design effort for the Arbitrated Loop topology protocol was to prevent, as much as possible, modifying the other parts of the standard, in order to simplify interoperability between different types of Ports. This particularly meant not modifying the structure of Sequences, Frames, Transmission words, etc. This has largely been achieved. The only additions to the Fibre Channel topology required for implementing Arbitrated Loop functions are (1) addition of several new Primitive Sequences and Signals for loop initialization and arbitration, (2) addition of some new link service Sequences, and (3) addition of some new functions, termed FC-AL functions, which are described below. These extra capabilities can be implemented between the FC-1 and FC-2 levels with minimal disturbance to their normal operation. The only real change to Fibre Channel operations done on an Arbitrated Loop topology compared to the Point-to-point or Fabric topologies is that the buffer-to-buffer Credit mechanism is modified.

Since Arbitrated Loop is mostly a superset of other Fibre Channel operations, particular devices can usually operate as either Arbitrated Loop or non-Arbitrated Loop devices, depending on the wiring topology and the set of other devices they're attached to. The NL_Ports and switch FL_Ports can discover at network initialization what types of Ports they are attached to, in what topoloperform a set of self-discovery operations at initialization time to determine what topology they are attached to and to interoperate accordingly, without external intervention.

Extra Functions Required for FC-AL Operation

As stated earlier, the Arbitrated Loop functionality was added without affecting the format of Frames or the mechanisms for Frame, Sequence, and Exchange management, with the exception of incorporating an alternative buffer-to-buffer Credit mechanism. Some additions were required to extend the architecture to the Arbitrated Loop topology.

N_Ports which implement the additional functions required for Arbitrated Loop operation are termed "NL_Ports," and switch Ports or F_Ports which implement the required functions are termed "FL_Port." NL_Ports and FL_Ports are perfectly capable of operating as normal N_Ports and F_Ports when they are connected in Point-to-point and Fabric topologies.

The biggest change is the additional concept of a "Physical Address," or AL_PA. An AL_PA address is a single-byte value used to address Ports on a Loop. It therefore provides a more restricted and simpler address space than the 3-byte range of S_ID and D_ID identifiers. There are a total of 127 valid AL_PA values, of which one is reserved for an FL_Port on the Loop, so there can be a maximum of 126 NL_Ports participating at once. The AL_PA addresses are selected out of the data codes that have neutral disparity 8B/10B codes, as shown in Figure 5.4 through Figure 5.6.

There are nine new Primitive Signals, used for arbitration, opening and closing Ports for communication, and synchronization. There are also eleven new Primitive Sequences, used for Loop initialization and reset and for enabling and disabling bypass of Ports that must be logically separated from Loop operations.

There are a set of new Link Service Frames Sequences defined, for initializing the Loop and assigning AL_PA addresses, that are transmitted in Class 3. There are several new concepts that are required for understanding link initialization and arbitration, including Fill words, and Access Fairness.

There is a new set of state machines and other functionality implemented between the FC-1 and FC-2 levels, which governs the Port's behavior in handling traffic arriving at the Port over the Loop. This set of functions is termed the "Loop Port State Machine," or "LPSM," and includes recognition of the AL Ordered Sets and link service Sequences, plus procedures for monitoring and modifying traffic that may be forwarded around the Loop. A Port must incorporate LPSM to operate in an Arbitrated Loop environment.

All these constructs fit naturally into the standard Fibre Channel mechanisms, constructs, operations, and procedures described in the preceding sections of the book.

Arbitrated Loop Hubs

The Arbitrated Loop does have some fairly serious disadvantages, primarily in the area of (1) wiring and (2) reliability. Hubs are built to address the wiring problem, and to help address the reliability problem.

The wiring problem with the loop topology without hubs has to do with the fact that the connectors, as shown in Chapter 4, are bidirectional (both transmitter and receiver signals go through the same connector), but the cable signals in the ring are unidirectional (receiver receives from the loop neighbor in one direction, transmitted signals go the other way). This would mean that each cable must fan-out, and since the other ends of the cables are also connected to bidirectional connectors, a wiring harness must be created with exactly as many connectors as there are L_Ports on the loop. Adding or subtracting an L_Port would require replacing all the cabling. In an environment where extra storage must be added every few days, weeks, or months, this is obviously ridiculous.

There are several solutions to this problem. FDDI, for example, used the solution of dual-counter-propagating rings, with bidirectional cables going both directions around the ring. This simplifies the wiring, and also adds a (limited) amount of extra redundancy, but doubles the transceiver, connector, and cable costs. A hub is a less expensive solution, and simplifies the topology.

A hub looks, from the outside, like a switch: a box with a set of cable plugs that the cables from the L_Ports plug into. The internal wiring, however, as shown in Figure 16.1, is a simple loop, with port bypass circuitry allowing each L_port to be connected onto the loop (enabled) or bypassed. Hubs can also be cascaded: attaching a port of one hub to a port of another hub creates a loop twice as large. Typically, one or two LEDs at each hub port tell whether a device is attached and/or generating valid FC-AL signals.

There are a variety of types of hubs available, targeted at different market requirements. The most robust hubs are "managed hubs," which have more sophisticated status and protocol monitoring capabilities, and which allow

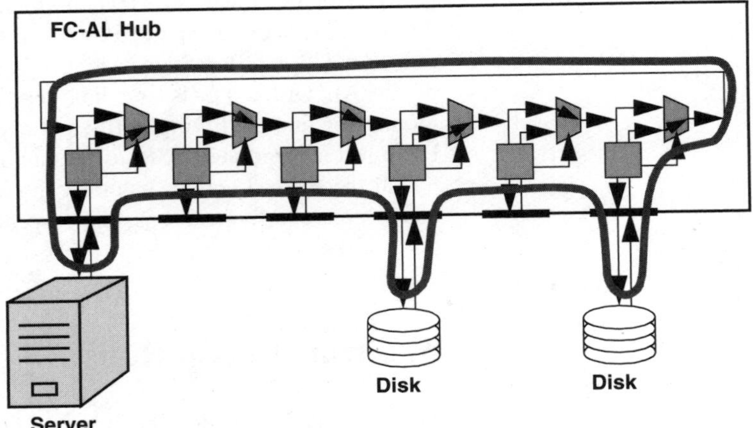

Figure 16.1
FC-AL Hub, showing internal port bypass circuitry and loop topology.

remote management through separate external Ethernet connections to LANs, using Web browser-based or SNMP-based management software.

The reliability issue with the hub-less Arbitrated Loop has to do with the fact that every component on the loop must operate correctly, forwarding signals around the loop, or the entire loop operation fails. In other words, every component on the loop is a single point of failure for the entire loop, which is the worst possible design, from a reliability standpoint. With a hub in the loop, however, the reliability can be much better - if an L_Port fails, the port bypass circuitry in the hub can be used to simply remove it from the loop, either manually, by pulling out the cable (in a low-end unmanaged hub) or electronically (in a more sophisticated managed hub). High-availability installations typically use devices with 2 or more L_Ports per device, attached to separate redundant loop hubs, to prevent the hub or it's internal port bypass circuitry from being single points of failure in the overall system.

AL_PA Physical Addresses

On the Fabric topology, which might provide connectivity to thousands of Ports through a multi-stage multi-hop switch Fabric, there is a need for a large 3-byte Port address space, along with quite complex functions for routing Frames through the switching Fabric in all Classes of service.

In a Loop topology, there is a need for a much smaller range of addresses, with a much simpler addressing mechanism and routing scheme. These considerations limit the size of the AL_PA field to a single byte.

A further limitation in determining the valid range of AL_PA values relates to assuring neutral Running Disparity in the Primitive Sequences and Primitive Signals. An examination of the Ordered Sets for Arbitrated Loop in Figure 6.2 shows that the Ordered Sets may have zero, one, or two AL_PA addresses (as AL_PS and AL_PD) contained in them. The rules for valid Running Disparity generation between Frames specify that the disparity following an EOF delimiter and the disparity preceding the next SOF delimiter must both be negative. Since one or more Ordered Sets may be inserted between Frames, and may be removed or altered during circulation around the Loop, the total disparity of all Ordered Sequences must be neutral.

To assure this neutral disparity of all Ordered Sequences, the disparity of all AL_PA values must be neutral as well. An examination of the encoded values in Figure 5.4 through Figure 5.6 shows that 134 data bytes are coded with neutral disparity. Of these 134, the value x'00' is reserved for use by an FL_Port, the value x'F0' is reserved as a lowest-priority address for the access priority mechanism, the x'F7' and x'F8' values are reserved for special use by the loop initialization primitives (LIP), and x'FF' is reserved for use in broadcast functions. This leaves 126 possible AL_PA values for NL_Ports, which are ranked in order from highest priority at x'01' to lowest NL_Port priority at x'EF.'

The Arbitrated Loop "Physical Address" actually has nothing to do with any physical locations, and it is perfectly valid to have a random ordering of AL_PA values at sequential locations around a Loop. As described in the "Loop Initialization" section, on page 284, AL_PA values can be assigned in several different ways. using either hardware or software mechanisms, so the actual ordering of AL_PA values may or may not match the physical ordering of Ports along the Loop.

AL_PA addresses for NL_Ports are assigned in one of four ways, with a hierarchy of preference for assignment method. First, a Fabric may have assigned a specific AL_PA address to an NL_Port. If a Fabric has not assigned an address to the NL_Port, the Port can use the address it used previously. If it never had a previous address, or does not remember it, it can use a preferred address, or hard-wired address, which is set by a user through some hardware means such as switches or jumper cables. If there is no mechanism for hard-wiring a preferred address, or if there is a collision in the preferred address, and an NL_Port still does not have an AL_PA address, it will pick one of the 126 unused AL_PA addresses on the Loop. FL_Ports always use AL_PA x'00,' which is the highest-priority address.

In the Arbitrated Loop topology, no routing of Frames is done based on either Port identifier or AL_PA addresses. This is in contrast to the Fabric topology, where routing of Frames is based on destination Port D_ID value. Instead, the source and destination L_Ports on the loop arbitrate for access to the full loop. When arbitration is completed, other Ports are left in a monitoring state, where they can only forward Frames on to the two open Ports, adjusting for clock frequency differences. The open Ports operate essentially as if they were connected in a Point-to-point topology, simplifying FC-2 and higher-level operations.

Primitive Signals and Sequences for AL

Inclusion of Arbitrated Loop functionality requires that the Port recognize and be able to generate nine new Primitive Signals and eleven new Primitive Sequences. In Chapter 6, it was stated that if a Port does not recognize a received Ordered Set, it will treat the Ordered Set as if it were an Idle. A Port which incorporates the Arbitrated Loop functionality can tell therefore whether an attached Port implements AL functionality or not by sending it one of these new Ordered Sets (presumably one of the Loop Initialization Primitives) and monitoring the response. If the Ordered Set is recognized and treated properly, rather than treated as an Idle, then the Port knows that the attached Port also incorporates AL functionality.

Primitive Signals

The nine new Primitive Signals for an Arbitrated Loop are used for arbitrating for access to the Loop, for opening communications with other specific Ports on the Loop once access is gained, for opening multicast or broadcast communications with other Ports, for closing communications, and for providing a synchronization signal. Discussions on some of these signals make reference to the LPSM state diagram shown in Figure 16.4. Full discussion of this state machine is reserved for the "Loop Port State Machine Operation: An Example" section, on page 288.

Arbitrate (ARBx) and ARB(F0). Arbitration on the Loop is handled using the ARBx and ARB(val) Primitive Signals, with val=F0. The formats, as shown in Figure 6.2, are "K28.5 D20.4 AL_PA AL_PA" or "K28.5 D20.4 D16.7 D16.7" since D16.7 corresponds to x'F0.' A Port will insert the ARB Primitive Signal containing its own AL_PA value into free spaces in the forwarded bit stream, to notify the other Ports on the Loop that it is requesting Loop access.

When a Port does not have the Loop and wishes to acquire it, it will monitor the incoming data stream to look for Idles and ARB Primitive signals for the Port with lower-priority AL_PA values. If an Idle or a lower-priority

ARB transmission word is detected, the LPSM will replace it with an ARBx containing its own AL_PA in characters 3 and 4.

A Port which has the Loop will constantly be replacing all ARBx signals with ARBx containing its own AL_PA. Thus, the ARBx arriving back at the requesting Port will contain the AL_PA of the owning Port. When the owning Port is done with the Loop, it will close communications and stop replacing ARBx or Idle signals with its own ARBx. Then the ARBx for the requesting Port will circulate the Loop and arrive back at the originating Port (provided that no higher-priority Port is also requesting the Loop by replacing ARBx signals as well). When the requesting Port receives an ARBx containing its own AL_PA value, it knows that it owns the Loop and goes into the OPEN state (shown in Figure 16.4), replacing all ARBx signals with its own ARBx.

The ARB(F0) is used by a Port which owns the Loop to test to see if any other Ports desire access. It transmits ARB(F0) as a fill word, between Frames. If another Port is requesting access, the ARB(F0) will be replaced with ARBx for the requesting Port before returning, since x'F0' is the lowest-priority ARB Primitive signal. If the owning Port receives an ARB(F0) back, it knows that no other Port is requesting the Loop, and it can set ACCESS to 1, which indicates that all Ports requesting the Loop in the current arbitration window have had the chance for a turn.

Open Full-Duplex (OPNyx) and Open Half-Duplex (OPNyy). OPNyx and OPNyy are used by a Port which owns the Loop to open communications with a specific other Port on the Loop. The choice of which signal to use indicates to the other Port whether the communication will be in full- or half-duplex mode. Functionally, this choice does not affect the LPSM, except to indicate whether it will be sending Data Frames or will not be in the opened communication.

Open Broadcast Replicate (OPNfr) and Open Selective Replicate (OPNyr). OPNfr and OPNyr are used by a Port which owns the Loop to open communications with multiple other Ports on the Loop. When a Port in the MONITORING state receives a OPNr signal, it will go into the OPENED state, with REPLICATE = 1. This means that it will accept Frames with a D_ID it can respond to and will also replicate the Frame and pass it on. If a Port owning the Loop wants to communicate with a specific subset of Ports on the Loop, it can transmit OPNyr signals to each one, so that they will be OPENED and ready to receive Frames addressed to a recognized D_ID and will also pass the Frames on around the Loop.

Close (CLS). The CLS signal is used to begin closing communications with whatever other Port on the Loop is open. It is only sent by either an OPEN or OPENED Port (one of the two Ports in an open communication). When it is received, the Port goes into a RECEIVED CLOSE state, which allows it to complete any pending Frame transmission and close any open Class 1 Dedicated Connections before returning a CLS Signal and going to the MONITORING state.

Dynamic Half Duplex (DHD). The DHD signal transmitted by the L_Port in the OPEN state to indicate to the port in the OPENED state that it has no more data to send, and can start receiving half duplex data. Dynamic Half Duplex capability must be extablished at Login, as described in the "N_Port and F_Port Service Parameters" section, on page 166, for both L_Ports, for this to be used. When the OPENED L_Port is finished transmitting data frames, it will typically send the **CLS** signal that finishes communication.

Mark (MRKtx). The MRKtx Primitive signal is used as a synchronization primitive. It can be transmitted by any Port and is forwarded by Ports in the MONITORING, ARBITRATING, XMITTED CLOSE, and TRANSFER states. As long as the transmitting Port can assure that all other Ports are forwarding the signal (by being the Loop Port in the OPEN state and assuring that no other Ports are OPENed), it can provide a synchronization signal to all Ports on the Loop. Usage of this facility is somewhat vendor dependent and yet to be determined, but proposals have been made for usage as a clock synchronization signal (x'00') and for a disk spindle synchronization signal (x'01'). The "t" value in MRKtx is a neutral Disparity Data Character indicating the type of synchronization, so the facility is available for 132 more different types of synchronization primitives.

Primitive Sequences

The new Primitive Sequences incorporated for AL are used for initializing the Loop and for enabling and bypassing Ports on the Loop. On a Public Loop, with a switch FL_Port attached, the Primitive Sequences only go around the loop - they don't affect the rest of the topology attached through the switch.

Loop Initialization (LIP). The Loop Initialization Primitive Sequences (LIP) are used to begin initialization of the Loop. The initialization procedure, described in the "Loop Initialization" section, on page 284, is used to assign AL_PA values around the Loop and to stop any previously-ongoing communication. There are seven types of Loop Initialization Primitive Sequences.

LIP (F7, F7) (no valid AL_PA) indicates that the transmitting L_Port is attempting to acquire an AL_PA.

LIP(F8, F8) (Loop failure, no valid AL_PA) indicates that the transmitting L_Port has detected a Loop failure and does not have a valid AL_PA.

LIP(F7, x) (valid AL_PA) indicates that the L_Port is attempting to reinitialize the Loop, to restore the Loop to a known state, and that it has the valid AL_PA value x.

LIP(F8, x) indicates that the L_Port has detected a Loop failure and does have a valid AL_PA value x.

LIPyx indicates that the originating L_Port, x, is attempting to reset the NL_Port identified by the AL_PD y. The actual meaning of "reset" will depend on the particular device and vendor for each device on the loop.

LIPfx indicates that the originating L_Port, x, is attempting to reset all the NL_Ports on the loop except itself.

LIPab is reserved for future use, with the provisions that a and b are any of the neutral disparity values in Figure 16.3, and the values are not already have been used by another LIP Primitive Sequences.

Loop Port Enable (LPEyx, LPEfx) and Bypass (LPByx, LPBfx). The LPE and LPB Primitive Sequences are used to handle situations where an L_Port must be logically removed from the Loop. This may be required when an L_Port is physically removed, is powered-off, or is failing. Full bypass implementation requires both a physical bypass circuit, which is a two-way multiplexer either transmitting data from the L_Port or forwarding data from the previous L_Port to the next L_Port, and a logical bypass function, which prevents the L_Port from transmitting data. L_Ports that are bypassed are different from non-participating L_Ports in that they may retain an AL_PA address.

When an L_Port receives the LPBfx Sequence or the LPByx Sequence with its own AL_PA for y, it will go into a bypass mode, where it transmits no words and only monitors the Loop for the LPEyx, LPEfx, or LIP Primitive Sequences. If it receives either the LPEfx Sequence or the LPEyx with y set as its own AL_PA, it will go into enabled mode in the MONITORING state and can participate in the Loop. If it receives any of the LIP Primitive

Sequences, it will give up its AL_PA value and can only be reenabled with the LPEfx Primitive Sequence.

Typically, the component sending LPE or LPB Primitive Sequences will either be the server in a server+storage loop, or the management entity, in a managed hub.

Loop Initialization

Initialization of an Arbitrated Loop requires a set of procedures to allow NL_Ports to acquire AL_PA addresses, using the hierarchy of assignment mechanisms described in the "AL_PA Physical Addresses" section, on page 278. Successful completion of the AL_PA address assignment means that all L_Ports on the loop are operating correctly.

The mechanism for doing this requires selecting a temporary initialization Loop master, which will either be the FL_Port with the highest-priority (lowest number) worldwide name or, if there are no FL_Ports on the Loop, the NL_Port with the highest-priority worldwide name. The initialization Loop master then circulates four different kinds of Sequences around the Loop, collecting AL_PA assignments by the four different assignment mechanisms, to determine which AL_PA addresses are used on the Loop. It then circulates a Sequence to all the Ports on the Loop, collecting the sequential locations of all the AL_PA addresses around the Loop, and then circulates another Sequence around to notify all the Ports about the locations of all the AL_PA addresses.

The initialization procedure is begun following transmission of one of the four LIP initialization Primitive Sequences. On receipt of the LIP Primitive Sequence, every Port goes into an INITIALIZATION state and begins transmitting the LISM (Loop Initialization Select Master) Sequence. The procedure described here uses normal Frames and Class 3 behavior with no flow control. All Ports either continuously initiate Sequences or receive, modify, and forward Sequences, until the initialization procedure is done. The single-Frame Sequences used for initialization, shown in Figure 16.2, have a fixed set of Frame Header fields and can be divided into three groups. The LISM is used for selecting an initialization Loop master; the LIFA, LIPA, LIHA, and LISA Sequences are used for assigning or selecting AL_PA identifiers; and the LIRP and LILP Sequences are used for collecting and reporting the AL_PA positions on the Loop.

Figure 16.2
Initialization sequences for Arbitrated Loops.

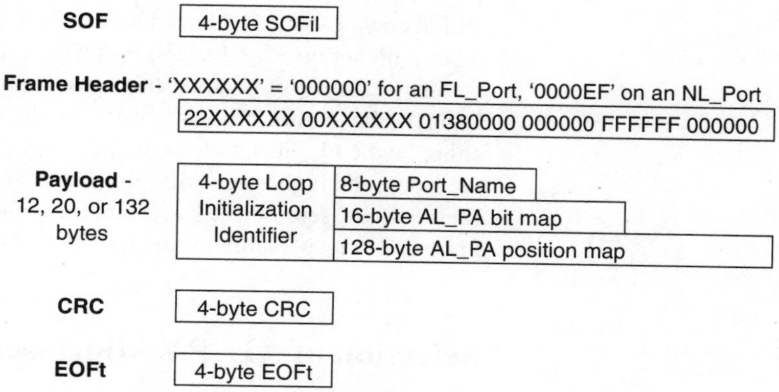

Identifier	Name	Usage	Length
x'11010000'	LISM	Select Master based on 8-byte Port_Name - to select initialization master	12 bytes
x'11020000'	LIFA	Fabric assigned bit map - to gather all Fabric-assigned AL_PAs	20 bytes
x'11030000'	LIPA	Previously acquired bit map - to gather all previously used AL_PAs	20 bytes
x'11040000'	LIHA	Hard assigned bit map - to gather all hardware-requested AL_PAs	20 bytes
x'11050000'	LISA	Soft Assigned bit map - to gather all soft-assigned (remaining) AL_PAs	20 bytes
x'11060000	LIRP	Report AL_PA position map - to gather relative positions of AL_PAs on Loop	132 bytes
x'11070000'	LILP	Loop AL_PA position map - to inform all L_Ports of AL_PA position map	132 bytes

Selecting a Temporary Loop Master

After transitioning into the INITIALIZATION state, all Ports continuously send out LISM Frames and simultaneously monitor incoming Frames to be either forwarded or discarded. The Frames initially sent out have the D_ID and S_ID fields set to x'00 00XX,' where x'XX' is the initial AL_PA value and x'00' is for FL_Ports and something else, defaulting to x'EF' for NL_Ports. The Payloads of the LISM Frames contain a Loop initialization identifier and an 8-byte worldwide name. On receiving an LISM Frame, each Port looks at the D_ID field and at the worldwide N_Port_Name in the Payload, with the intention of either forwarding the Frame or replacing it with its own LISM. The Port discards the Frame and sends its own instead if it has a higher priority than the Port which sent the Frame.

Since every Port is doing this, eventually the only Frame circulating the Loop is the Frame originally sent by the highest priority Port. This will either be the FL_Port with the highest priority or, if there are no FL_Ports on the Loop, the highest-priority NL_Port. When an L_Port receives the same Frame that it sent out, it knows that it is the highest-priority L_Port on the Loop and is responsible for the rest of the initialization.

Selection of AL_PA Addresses

The initialization master L_Port then must determine AL_PA Port assignments. To do this, it sends the LIFA, LIPA, LIHA, and LISA Frames around the Loop, after setting the bit corresponding to its AL_PA and resetting all the other AL_PA bits. The mapping of the 126 AL_PA values to the 128 bits in the 16 byte AL_PA bit map in the Payload for these Sequences is shown in Figure 16.3.

As each non-master Port receives the LIFA Frame, it reads in the entire Frame, and if it had been previously assigned an AL_PA by a Fabric, it sets the bit corresponding to that AL_PA in the Frame Payload if it is not already set, and then forwards the Frame. After passing through all the Ports on the Loop, the initialization Loop master receives the LIFA Frame back, with all the bits for AL_PA addresses assigned previously by Fabrics asserted. It then uses that Payload as the initial Payload for the LIPA Frame. Similarly, the non-master Ports receive the LIPA Frame, and if they have a previously assigned AL_PA address and the bit is not set by an earlier assignment in the LIFA Frame or an earlier L_Port in the Loop, then set the bit. When the Port master receives the LIPA Frame, it uses that as the initial Frame for the LIHA Frame and circulates that around the Loop for Ports to set their preferred hardware addresses, if they can. On the LISA Frame, any Ports which have not already been able to acquire AL_PA values through the other three methods can set the first un-set bit of the LISA Frame to acquire one of the free AL_PA addresses.

Use of this four-message hierarchy for assigning AL_PA addresses assures that every Port gets its preferential AL_PA address, if possible. Also, it is impossible for there to be collisions. Even if two Ports have a hard-wired preference for address x'01,' for example, only one will get it, and the other will get one of the soft-assigned addresses available when the hard-assigned ones have been taken. The only way that any Ports will end up without assignments is if there are more than one FL_Port or more than 126 NL_Ports on the Loop. Any Ports without AL_PA address assignments are

Figure 16.3
Mapping of the 127 neutral parity (F/N)L_Port AL_PA addresses onto Initialization Sequence bit positions.

AL_PA hex code	Bit Map Wd Bit	AL_PA hex code	Bit Map Wd Bit	AL_PA hex code	Bit Map Wd Bit	AL_PA hex code	Bit Map Wd Bit
Login Req'd	0 31	3C D28.1	1 31	73 D19.3	2 31	B3 D19.5	3 31
00 D00.0	0 30	43 D03.1	1 30	74 D20.3	2 30	B4 D20.5	3 30
01 D01.0	0 29	45 D05.2	1 29	75 D21.3	2 29	B5 D21.5	3 29
02 D02.0	0 28	46 D06.2	1 28	76 D22.3	2 28	B6 D22.5	3 28
04 D04.0	0 27	47 D07.2	1 27	79 D25.3	2 27	B9 D25.5	3 27
08 D08.0	0 26	49 D09.2	1 26	7A D26.3	2 26	BA D26.5	3 26
0F D15.0	0 25	4A D10.2	1 25	7C D28.3	2 25	BC D28.5	3 25
10 D16.0	0 24	4B D11.2	1 24	80 D00.4	2 24	C3 D03.6	3 24
17 D23.0	0 23	4C D12.2	1 23	81 D01.4	2 23	C5 D05.6	3 23
18 D24.0	0 22	4D D13.2	1 22	82 D02.4	2 22	C6 D06.6	3 22
1B D27.0	0 21	4E D14.2	1 21	84 D04.4	2 21	C7 D07.6	3 21
1D D29.0	0 20	51 D17.2	1 20	88 D08.4	2 20	C9 D09.6	3 20
1E D30.0	0 19	52 D18.2	1 19	8F D15.4	2 19	CA D10.6	3 19
1F D31.0	0 18	53 D19.2	1 18	90 D16.4	2 18	CB D11.6	3 18
23 D03.1	0 17	54 D20.2	1 17	97 D23.4	2 17	CC D12.6	3 17
25 D05.1	0 16	55 D21.2	1 16	98 D24.4	2 16	CD D13.6	3 16
26 D06.1	0 15	56 D22.2	1 15	9B D27.4	2 15	CE D14.6	3 15
27 D07.1	0 14	59 D25.2	1 14	9D D29.4	2 14	D1 D17.6	3 14
29 D09.1	0 13	5A D26.2	1 13	9E D30.4	2 13	D2 D18.6	3 13
2A D10.1	0 12	5C D28.2	1 12	9F D31.4	2 12	D3 D19.6	3 12
2B D11.1	0 11	63 D05.3	1 11	A3 D03.5	2 11	D4 D20.6	3 11
2C D12.1	0 10	65 D05.3	1 10	A5 D05.5	2 10	D5 D21.6	3 10
2D D13.1	0 9	66 D06.3	1 9	A6 D06.5	2 9	D6 D22.6	3 9
2E D14.1	0 8	67 D07.3	1 8	A7 D07.5	2 8	D9 D25.6	3 8
31 D17.1	0 7	69 D09.3	1 7	A9 D09.5	2 7	DA D26.6	3 7
32 D18.1	0 6	6A D10.3	1 6	AA D10.5	2 6	DC D28.6	3 6
33 D19.1	0 5	6B D11.3	1 5	AB D11.5	2 5	E0 D00.7	3 5
34 D20.1	0 4	6C D12.3	1 4	AC D12.5	2 4	E1 D00.7	3 4
35 D21.1	0 3	6D D13.3	1 3	AD D13.5	2 3	E2 D02.7	3 3
36 D22.1	0 2	6E D14.3	1 2	AE D14.5	2 2	E4 D04.7	3 2
39 D25.1	0 1	71 D17.3	1 1	B1 D17.5	2 1	E8 D08.7	3 1
3A D26.1	0 0	72 D18.3	1 0	B2 D18.5	2 0	EF D15.7	3 0

said to be in "non-participating mode" and can't send or receive any words until another Initialization procedure lets them acquire an address.

Building an AL_PA Address Map

The LIRP and LILP Sequences are used to build a mapping between AL_PA values and participating Port positions on the Loop. This lets the Ports know where each of the AL_PA addresses is physically located on the Loop.

To build this address map, the Loop master initiates a Frame with a 132-byte Payload field with each byte set to x'FF.' The first four bytes identify the LIRP Sequence, the next byte is an offset counter, the next 126 bytes will contain the AL_PA addresses at the possible 126 hops to participating L_Ports on the Loop, and the last byte is reserved.

The Loop master sets the offset counter byte to x'01,' sets the byte at offset x'01' to the value of its AL_PA, and forwards it to the next Port on the Loop. The second Port receives and checks the Frame, increments the offset counter to x'02,' sets the byte at offset x'02' to its own AL_PA value, and forwards it on around the Loop. When the Frame arrives back at the Loop master, the Payload contains a mapping of AL_PA values indexed by hop count. It also shows the size of the Loop, since if there are only five L_Ports on the Loop, only the first 5 bytes of the address map field will be set.

The Loop master then does the initialization step by transmitting an LILP Frame containing the address map built up in the LIRP Frame around to all the L_Ports on the Loop.

At this point, all L_Ports have been assigned AL_PA addresses, and all L_Ports know what other addresses are participating on the Loop and have a mapping of where the AL_PA addresses are on the Loop. The Loop master closes all the Ports and places them all in the MONITORING state by transmitting the CLS Primitive Sequence, and when the CLS circulates back, it goes into the MONITORING state itself. At this point, all L_Ports are in the MONITORING state, transmitting Idles and waiting for their LPSM levels to receive commands to transfer data from the FC-2 levels to start transmitting data.

Loop Port State Machine Operation: An Example

Operation of the Ports on an Arbitrated Loop architecture requires that Ports be able to determine when to only monitor and forward the data received, when to initiate data transmission, when to accept data directed towards it, etc. These states of operation are controlled through a Loop Port State Machine, or LPSM. The LPSM is implemented between the FC-1 level and the FC-2 level and has as inputs signals derived from the incoming data and from operations requested by the FC-2 level.

The actual LPSM is a fairly complex design that incorporates 11 states, 23 state transitions, and over 45 different input signals, but the basic idea of its operation can be given here. A simplified picture of the LPSM state diagram, showing the major states and most common transitions, is shown in Figure 16.4.

Figure 16.4
Loop Port State Machine state diagram, simplified.

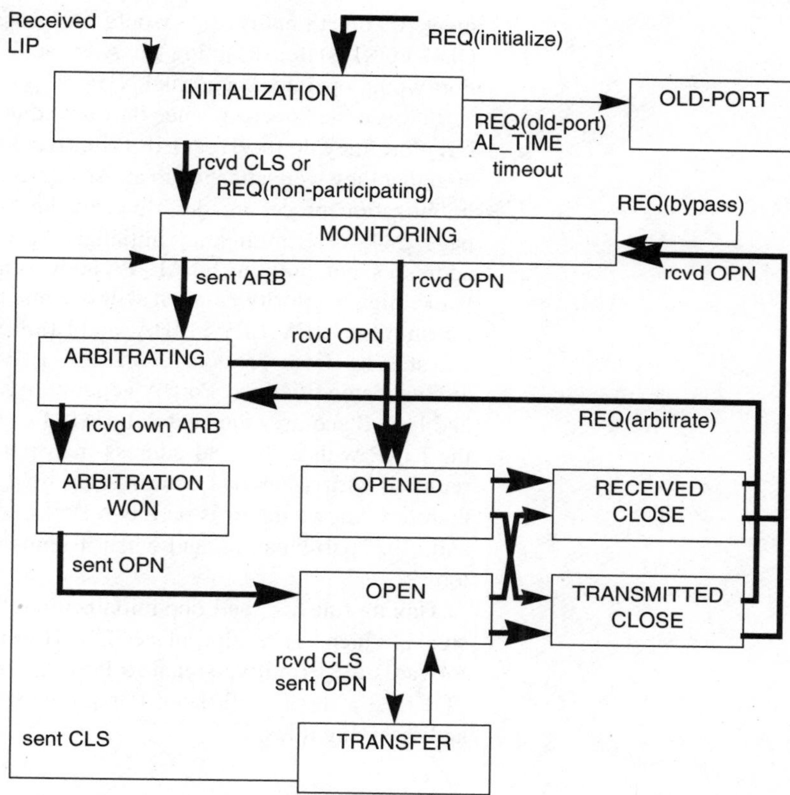

In this figure, the most common transitions are shown in thicker lines, with less commonly used transitions shown in thinner lines, and some rarely used transitions are left out altogether. The conditions triggering the conditions are shown on the state diagram, with a key at the base describing the notation.

The following discussion describes the operation of the Loop Port State Machine for normal operations up through a set of communications. For the purpose of discussion, assume there are two NL_Ports, labeled A and B, that are attached into a two-Port Loop. They start out having just been powered on, with no knowledge of their environment.

Initialization of the Two-Port Loop

When they are powered on, both Ports go into the INITIALIZATION state, where they transmit a LIP continuously and monitor for an LIP on the incoming fiber. If they had been attached to a Point-to-point topology with an N_Port that didn't recognize LIP Primitive Sequences and didn't implement AL functionality, they would time out after some time and go into the OLD-PORT state, disabling all Arbitrated Loop functionality, so that the Port would operate as a normal N_Port.

In this case, however, since they are connected in a topology with another NL_Port, they do receive a LIP Primitive Sequence, and having self-discovered that they were attached to an Arbitrated Loop, both go through the Loop initialization procedure described in the "Loop Initialization" section, on page 284, determining an initialization Loop master, acquiring AL_PA addresses, and building an AL_PA address map. Since Port A has the lower-value, higher-priority name, it will become the initialization master and will transmit the LIFA, LIPA, LIHA, and LISA Sequences. Since both NL_Ports are starting from scratch, the LIFA, LIPA, and LIHA Frames circulate around unmodified, so Port A acquires AL_PA x'01' on the LISA Frame, and Port B acquires the next valid AL_PA, which is x'02.' Port A sends out the LIRP with a Payload address map field of x'0101FFFF..FF,' and it is returned with a Payload address map field of x'020102FFFF....FF.' Port A therefore knows that it is on a two-Port Loop, with AL_PA addresses x'01' and x'02' participating, and when it sends out the LILP, Port B knows that too.

Having finished the Loop initialization, Port A sends out a CLS Primitive signal, which, on receipt, places Port B in the MONITORING state. Port B forwards the Primitive signal to Port A, which place Port A in the MONITORING state as well. Both Ports, having no work to do, are transmitting and receiving Idles.

Arbitration and Initiation of Communication

Now suppose the LPSMs at both Ports got signals from their upper levels to begin arbitrating for Loop access to send data at exactly the same time. This is of course highly unlikely, but iut makes a more instructive example to see what happens when there are simultaneous transmission requests for both Ports to send Sequences to the other.

At the moment, the Loop is filled with Idle Primitive Sequences, which the Ports have been receiving and forwarding. When they get the REQ(arbitrate) signal from FC-2 level, they transition into the ARBITRATING state and begin sending ARB Primitive signals, with Port A sending an ARB(1) and Port B sending an ARB(2). Given the size of the Loop, they each get time to send about one Primitive Sequence before receiving the ARBx from the other. Port A examines the ARB(2) received and discards it, since the source AL_PA has lower priority than it does. Port A may send another ARB(1). Port B examines the ARB(1) it received, finds that the source has a higher priority than it does, and forwards the ARB(1) on, staying in the ARBITRATING state.

Port A receives the forwarded ARB(1) and knows that it has won the arbitration, so it transitions to the ARBITRATION WON state. It needs to open communications with Port B, participating as AL_PA x'02,' so it will send an OPN Primitive signal. Let's assume that the operation to B is a bidirectional Exchange, so it will need full-duplex communications and will need to send an OPN(2,1) Primitive Signal on the Loop. Once it sends the OPN(2,1), it transitions to the OPEN state and begins sending Frames.

Port B, being still in the ARBITRATING state with the REQ(arbitrate) input from FC-2 raised, is waiting for a chance to get on the Loop. However, while it is doing this, it is monitoring the words passing by and sees an OPN directed to it. This causes it to take the transition shown with a thin line in Figure 16.4 around to the OPENED state, indicating that it was opened for communication by another L_Port.

At this point, Port A is in the OPEN state, Port B is in the OPENED state, and communication can almost go on exactly as if A and B were N_Ports connected in a Point-to-point topology. The fact that they actually are physically connected point-to-point is beside the point, since they are NL_Ports; if there were other L_Ports on the Loop, the others would be in MONITORING state at the time, merely examining and forwarding the transmission words as they flowed by, increasing the point-to-point latency between Ports A and B marginally without affecting the bandwidth at all.

Ports A and B can do all the normal Fibre Channel operations, such as transmitting Frames and Sequences in Classes 1, 2, or 3, with normal flow control and acknowledgment rules. There is no explicit timeout on how long Ports A and B could hold the Loop, even if there were other L_Ports on the Loop that might have to do communications as well. It is up to the OPEN and OPENED Ports to monitor the ARBx signals to know that other L_Ports may want Loop access and up their "good citizenship" to be fair about allowing access to other Ports on the Loop.

Arbitration and Access Fairness

The only difference between Arbitrated Loop communication from this point and communications over other topologies is in the Primitive signals transmitted between Frames. On the Arbitrated Loop topology, instead of Idles, the OPEN L_Port will transmit the ARB(F0) Primitive Sequence and monitor for it returning between Frames. Since x'F0' is the lowest-priority address (lower than any address used by Ports on the Loop), any L_Port which is ARBITRATING will replace ARB(F0) with ARBx with the requesting address. In the current situation, Port A will send out ARB(F0), and since Port B is OPENED, it won't do the replacement. However, if there were a third L_Port on the Loop which was ARBITRATING, Port A would see the ARBx for that third L_Port's AL_PA returned. If Port A transmits ARB(F0) in between Frames, and receives ARB(F0) back, it knows that no other Port is ARBITRATING for the Loop and will start transmitting Idles to begin a new round of fair arbitration for the Loop.

An L_Port which is guaranteeing fair access to the Loop will wait, after it has had a chance to be OPEN or OPENED for Loop access, until it sees Idle Primitive Signals, to know that every other L_Port on the Loop has had a turn at Loop access, before ARBITRATING again. However, allowing fair access is an optional feature, which L_Ports with critical data to send might override. There is nothing in the LPSM to prevent the L_Port with AL_PA x'01' from always having priority over other NL_Ports. This is probably a feature, given that there is usually a real priority distinction between Ports on a Loop, and manufacturers of Loop topology components can use the AL_PA assignment to allocate priority where it makes sense.

In the case where an L_Port is not granting access to other L_Ports on the Loop, it is possible for L_Ports to force an OPEN Port to give up the Loop by transmitting a LIP Primitive Sequence. This is quite a disruptive operation, since it causes a complete Loop initialization procedure, with reassignment of AL_PA addresses and regeneration of an AL_PA address map. However, it is available as an option for interrupting communication that is in progress, and a particular LIP, the Loop initialization with valid AL_PA (K28.5 D21.0 D23.7 AL_PS), carries with the connotation of transmission because of a perceived performance degradation.

Finishing Communications

When the operation is over with, either the OPEN or OPENED Port can close the communication down. In Figure 16.4, the thick lines show that the OPEN L_Port, Port A, has determined that the communication is over and has sent the CLS signal. When it sends the CLS, it goes into the XMITTED CLOSE state, to wait for another CLS to be returned.

On receiving the CLS from Port A, Port B transitions to the RECEIVED CLOSE state and gets ready to return the CLS back to Port A. However, there is a slight problem.

Use of Alternate Buffer-to-Buffer Flow Control Management

The slight problem is that Port B may not be finished with all the processing for Frames received from Port A. If it is still working on Frames such that there are no Frame buffers available when it returns the CLS to A and returns to the MONITORING, and then receives an OPN on the next cycle, it may violate buffer-to-buffer Credit rules, if the OPENer on the next open communication is expecting a non-zero buffer-to-buffer Credit value.

There are two solutions to this problem. The first, and less interesting, is to begin with a beginning buffer-to-buffer Credit value of 0 and to transmit an R_RDY just before going into the OPEN or OPENED state. This is less interesting because it requires a round-trip time over the entire circuit before any activity can begin on an open communication.

The second solution, termed the "alternate buffer-to-buffer Credit" mechanism, is for the Port to delay closing an open communication until it can guarantee buffer-to-buffer receive resources availability for the next communication. This capability is also described in the "Alternate Buffer-to-Buffer Flow Control" section, on page 271.

In this case, we'll assume the latter. If Port B is advertising a buffer-to-buffer Credit of 1 (the default), then it will delay returning the CLS Primitive Sequence to Port A until it has a buffer, with associated resources, free. If the alternate buffer-to-buffer Credit mechanism is being used, Port A would have done the same thing, delaying the transmission of the CLS after communication was done until it had ensured that the BB_Credit value worth of buffers was available.

Continuing Communications

Once Port B has cleared a buffer and transmitted the CLS Primitive Sequence, both L_Ports will transition back to the MONITORING state. However, remember that Port B had a communication operation pending with Port A. While Port A is finished and goes into the MONITORING state, transmitting Idles, Port B still has the REQ(arbitrate) input set and transitions to the ARBITRATING state. Port B begins replacing any received Idles with ARB(2) and sending out ARB(2) in any available free space on its own. Port A is MONITORING, so it forwards the ARB(2) back to Port B, which then transitions through the ARBITRATION WON state and sends an OPN to Port A. This OPEN could be any of the four, but we can assume it's an open full-duplex, OPN(1,2). Port A receives and recognizes the OPN(1,2) and transitions from the MONITORING state directory to the OPENED state.

The Open communication from here proceeds identically to the previous one, except that now Port B is the one in the OPEN state, and Port A is the OPENED one. Again, either Port can transmit the CLS Primitive signal when it has enough buffers free to match the advertised buffer-to-buffer Credit and is done communicating or wants to get off the Loop to allow any other Ports to acquire the Loop.

Transfer to Another Port

Another set of transitions are shown in Figure 16.4 at the transition between the OPEN and the TRANSFER states. The TRANSFER state is provided as an optimization for an OPEN Port which has traffic to send to multiple L_Ports on the Loop. This would be particularly advantageous, for example, for a SCSI initiator or for an FL_Port, which might have multiple operations occurring simultaneously with different L_Ports. If a Port is OPEN, and it is done communicating with one L_Port but needs to open communications with another L_Port, it can send the CLS, but rather than transition to the XMITTED CLOSE state, it can transition to the TRANSFER state. Then, when it receives the CLS from the first L_Port, it can immediately send another OPN and transition directly back into the OPEN state without incurring the overhead required for doing an arbitration cycle.

Clearly again, an unfair L_Port could use this mechanism to keep permanent access to the Loop. This could be problematic, especially in combined network and storage Loops, where different L_Ports might have very different data transmission requirements. Manufacturers of Fibre Channel Arbitrated Loop solutions must take care to provide Loop access priority that is compatible with the transmission needs of all L_Ports on the Loop.

Performance and Timing

In gaining an intuitive understanding of the operation of Arbitrated Loop topologies, it is very useful to consider some elements of performance, particularly those related to how long various types of operations take. The results are often quite surprising to those used to working with communications systems of different types.

It's useful to keep some basic numbers in mind.

- **Port Delay:** First, the Arbitrated Loop specification states that the maximum latency for a Transmission Word to be forwarded through an L_Port must be six Transmission Words or less. For a full-speed Loop, this is a latency of 240 ns per hop for logic processing. The actual time must be at least 1 transmission word, for clock frequency difference compensation, and is usually at least 2 words, or 80 nsec. at full speed.

- **Fibre Delay:** The signal propagation speed through cables is roughly 4 ns/m in electrical cable and 5 ns/m in optical fibers. Signals travel slower in optical fiber than in electrical cable since the dielectric constant (related to index of refraction) is higher in glass than in electrical cables. For 2-m or shorter links, the transmission latency is less than 10 ns per hop, so essentially all the latency in traversing a Loop within an enclosure is in the monitoring and forwarding logic of the intermediate Ports.
- **Frame Length:** It's useful to compare this transmission latency with the transmission time for a typical Frame. For a Frame with a 2,048-byte Payload, the number of characters between the beginning of one Frame and the next is 2,108 characters, which is the equivalent of 21,080 ns over full-speed links.

Frames, therefore, are very long relative to transmission latency. For example, on a 64-Port Loop, which is a fairly large Loop, a transmission from one L_Port all the way around the Loop to the previous L_Port would require 240 ns per hop × 63 hops = 15,120 ns. Since the Frame is over 21,000 ns long, when the SOF character arrives at the destination L_Port, at least one-quarter of the Frame is still in the source L_Port's buffers (more, if the port delay is lower), and the Frame is spread out over every intermediate Port in the Loop.

In this kind of environment, it makes little sense to use buffer-to-buffer Credit values larger than 2, since the end-to-end transmission latency is significantly less than the length of a Frame. Any higher value of buffer-to-buffer Credit, therefore, would be used, for example, for compensating for the latency incurred at the receive Port in processing a Frame after it has arrived.

This receive processing latency can be fairly significant, due to the usage of the a single CRC field covering both the Frame header and payload. Proper processing of received Frames depends on being assured that all the Frame Header fields are correct. This is not assured, however, until the CRC covering the Frame header is received and checked. A receiver cannot commit any processing operations, therefore, until the CRC at the end of the Frame has been received, without incurring the danger of, for example, sending invalid data up to a ULP.

An example of implications can be seen in examining the Loop initialization process described in the "Loop Initialization" section, on page 284. It would be possible to build hardware that would do the entire modification of LISM, LI(F/P/H/S)A, and LIRP Frames on the fly as they are forwarded through the Port, but it would be a significant investment in hardware design. On the other hand, allowing on-the-fly modification of initialization Frames improves initialization times significantly.

The Loop Initialization protocol requires complete circulation of six Frames around the entire Loop. On a 64-Port Loop with on-the-fly modification giving a latency per hop of 240 ns, the entire latency for the Loop is approximately 15 μs. Transmission of the six initialization Frames, plus initialization processing and other round trips for Primitive Sequences and Signals, would yield an initialization time of less than 200 μs.

On the other hand, if processing of the initialization Frames is carried out in software, and the software algorithm depends on the entire Frame being received for CRC verification before Frame modification and retransmission, then the latency per hop becomes equal to the hardware latency (240 ns), plus the transmission time for the Frame (roughly 21 μs per Frame), plus whatever software processing latency can't be overlapped with data reception. This increases the latency per hop by a factor of roughly 100, bringing Loop initialization time to well over 20 ms. In an initialization procedure, this factor of 100 may not be critical, but in other types of processing, where latency is more important, the usefulness of flow-through processing with back-end recovery in case of error may be absolutely vital toward giving reasonable performance, even with the added complexity.

Arbitrated Loop Advantages and Disadvantages

The primary advantages of the Arbitrated Loop topology make it a good match for combined connection of peripheral storage devices and local area network interfaces. The Fibre Channel Arbitrated Loop architecture provides a number of attractive features relative to other communications and I/O architectures and relative to the different topologies.

- First, it promises high-bandwidth communication over an arbitrated shared medium, with 100 MBps × 2 duplex bandwidth, with physically and architecturally feasible extensions to 200 MBps × 2 and 400 MBps × 2 available as economics permit. The arbitration and traffic monitoring facilities allow a significant fraction of this bandwidth to be used.
- It allows distributed control, with every Port implementing capabilities for acting as Loop master for the few situations where a Loop master function is needed.
- It allows interconnection of up to 126 NL_Ports per Loop, plus connection of a single switch Port, for connectivity to externally switched Fabric networks. More Ports than this can be physically connected to the Loop in non-participating mode, to operate, for example, as backup devices.
- It has a very small physical footprint, requiring only one incoming or out-

going optical or coaxial electronic cable. This feature will become increasingly important as disk density and read/write bandwidths increase, and disk drive size decreases below the size of parallel electronic connectors.

- Since each link can be extended with single-mode optical fiber to a length of 10 km, very large Loops can be built with very simple wiring requirements, no central control mechanisms or hardware, and no change in network operations relative to Loops which are routed within a single enclosure. The difference in signal propagation speeds over long distances available with optical fiber (roughly 0.05 ms/km) versus the times required for protocol processing and drive access (5–10 ms) means that, for example, the latency in accessing data separated by several 10 km links is negligibly higher than the latency in accessing data in the same enclosure.

There are, however, a number of disadvantages with Arbitrated loop. These include the following.

- Reliability. If a single port fails, the whole loop fails. This problem is largely solved by using Hubs, but of course Hubs negate some of the supposed cost advantages of the Arbitrated Loop topology.
- ACK Buffering: When an NL_Port issues a CLS to close the connection with another NL_Port on the loop, it cannot send any more Frames. This includes ACK Frames. Therefore, if one NL_Port sends a CLS, and then receives a Data Frame, it has to hold the ACK until it can re-arbitrate for the loop, and OPEN again with the sender. Since there is no immediate guarantee of winning arbitration, and no assurance of not being OPENed by a different NL_Port on the loop, each NL_Port may have to buffer ACKs for a very long time, in a way that does not prevent FIFO head-of-line blocking. The issue is even worse on the other side, where the sending NL_Port sent a Data Frame, then received a CLS, and never got an ACK back -- it has to hold the state for the Data Frame's in limbo, waiting for the ACK to come back, with no guarantee about when it will ever return. In essence, this implies that Class 2 traffic not possible on an Arbitrated Loop. Class 1 also provides no benefit, Class 4 is meaningless, and Class 6 is meaningless - so Class 3 is the only Class of Service which can be implemented on Arbitrated Loops. For SCSI and FCP, this is not a disadvantage - FCP works better without the overhead of Fibre Channel ACKs anyway - but it limits the use of Arbitrated Loop with other FC-4s.
- Address Assignment: Operating systems in some situations will use a SCSI target number, which is set physically (e.g., by setting jumpers), to uniquely identify disks. Since Arbitrated Loop allows 4 different ways to assign addresses (LIFA, LIPA, LIHA, and LISA) with every loop initialization, it is possible that two drives on a loop with conflicting Hard Addresses will be assigned non-conflicting Soft Addresses on one loop initialization, and opposite non-conflicting Soft addresses on a later loop initialization -- causing the operating system to corrupt the data. This can be

prevented by just assuring that a drive always gets assigned the Hard Address that it's physically configured for, but it requires understanding of the protocols and careful management of the loop, since collision in Hard Addresses may not be immediately obvious.

- Multiple round trips for arbitration and closing of the loop on every operation mean that bandwidth of the loop cannot be used efficiently, even including the inherent inefficiency of the FCP duplex protocols.

These disadvantages, and the many additional advantages arising from a switched fabric, mean that over time, Arbitrated Loop will probably increasingly be used for "inside-the-box" applications, with external links using switched Fabric mechanisms.

Chapter 17

Switch Fabrics

Introduction

In the previous chapters of this book, we have been primarily concerned with the N_Ports, and with their interaction with other N_Ports, with an Arbitrated Loop, or with an F_Port providing an interface to a switching Fabric. The internal operation of the switch Fabric has been largely hidden — as far as the N_Ports are concerned, a Fibre Channel Fabric operates only as a resource for transporting data to other N_Ports, and for performing services, such as, for example, those described in the "Fabric Login" section, on page 161. In this chapter, however, we'll examine the internal operation of a switch Fabric, both in terms of the interaction between the various elements of the Fabric, and in terms of the services offered by the Fabric as a whole.

In the simplest cases, a Fabric consists of a single switch. This is, for example, the picture shown in Figure 2.3, on page 17. In this environment, there is no issue of Inter-Switch Links (ISLs), and no issue of inter-switch interoperability. On small-scale networks, where a single switch from a single vendor is used, this configuration is entirely adequate.

Increasingly, however, Fibre Channel installations appear as in Figure 17.1, with multiple switches connected together in a multi-stage Fabric. which may even contain switches from several different vendors.

This more complex Fabric structure has a number of advantages relative to the single-switch Fabric. These advantages of multi-stage fabrics include (1) the capability to support more N_Ports, (2) better fail-over and reliability capabilities, and (3) easier upgradability, since adding new switch capabilities doesn't require replacing the old switches.

Building a multi-stage switch Fabric is, however, significantly more complex than building a single-stage switch Fabric, since (a) switch ports must be configured to attach both to nodes and to other switches, and (b) the switches in the Fabric must cooperate to manage the whole fabric. This requires both more complex management, and interaction between switches that isn't required in a single-switch network. Fibre Channel networks, in their fullest implementation, may contain support for transmitting all classes of traffic, and for all the Generic services described in the "Well-Known Generic Services" section, on page 309. However, this is a complex set of functions. In practice, switches can be successfully built and deployed which support a subset of the full capabilities.

In this chapter, the characteristics of Fibre Channel switches that are specific to multi-switch Fabrics are discussed. A large part of the discussion is on how Fibre Channel switches differ from, for example, Ethernet and InfiniBand switches, showing how Fibre Channel switches provide more and better functionality, by some measures, than, for example, Ethernet and InfiniBand switches, which makes them more powerful, but also more complex. The discussion in the "Class F Service" section, on page 305, shows an

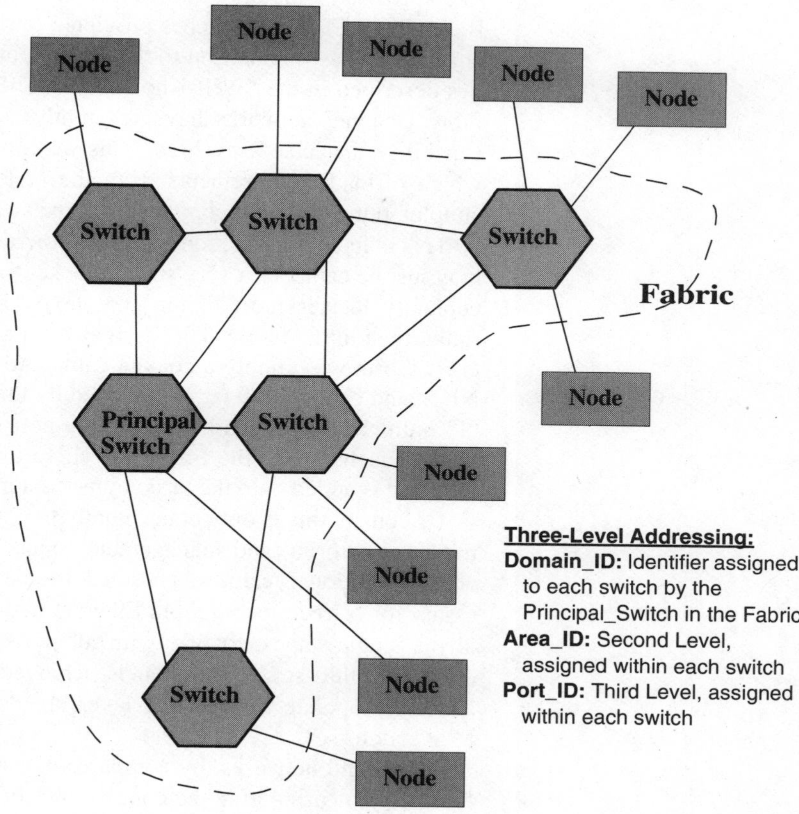

Figure 17.1
Example multi-switch Fabric

outline for how some forms of inter-switch communication work. In the "Special Switching Functions" section, on page 308, and the "Well-Known Generic Services" section, on page 309, a more detailed view of the services that a Fibre Channel switch Fabric may provide to the attached N_Ports is given.

Network Characteristics

This section describes some of the overall characteristics of Fibre Channel switch Fabrics, particularly in comparison with other types of networks.

Comparisons with other Networks

In comparison with other network technologies, such as Ethernet and InfiniBand, Fibre Channel switches provide a fairly rich set of capabilities to the attached Nodes. In addition to the set of centralized network-level capabilities described in the "Well-Known Generic Services" section, on page 309, Fibre Channel networks have inherently a set of assumptions on switch capability that are fairly high. This makes the fabric more sophisticated, which offloads requirements from the N_Ports and Nodes, making them simpler, but adds to the complexity of the switches.

This is in contrast to, for example, Ethernet switches, where the "Fabric" may just be cable, or a very simple packet forwarding engine. Ethernet conceptually locates most of the complexity in the network interface cards, rather than in the fabric. This derives from a history where the original Ethernet fabric was simply a coaxial cable. Addresses are hard-coded into the NICs, and cannot be flexibly assigned by the Fabric. Services such as flexible multicast, or network security are not possible or are implemented by nodes attached onto the Ethernet switches. Flow control is extremely limited, and is added onto the basic transmission capabilities.

Of course, this is only conceptually true. Modern Ethernet switches have traffic monitoring and management capabilities, and can be improved by adding additional features. These additional features are not, however, necessary for correct operation. In Ethernet networks, every switch is conceptually a single-stage network, with all ports connected to the same switch element. Multi-stage operation is achieved at the IP level, where the IP addresses operate as a second layer of addressing on top of the Ethernet MAC addresses.

InfiniBand networks, by comparison, implement support for multi-stage fabrics by incorporating the concept of a Fabric Manager, which is not part of the switch network, but is attached to the network of switches, and which can explore the network to discover the topology and set up routing tables.

Fibre Channel, on the other hard, moves all of the capabilities for network management conceptually into the switches themselves. This means that the switches must be capable of operating as both managers of their own single-switch fabric, and also as part of a cooperative set of switches, forming a larger fabric.

A further problem is the capacity for supporting both Connection-oriented and packet-oriented data transmission, on the same Fabric. Practically, the most interoperable networks only have support for Class 2 and Class 3 delivery.

It's not clear at this point where the best place to put the functionality is: on the Nodes or in the switches. This decision is partially dependent on the overall network design, and partially dependent on the type of system into which the network is installed.

Types of Switch Ports

Each switch port may be attached to N_Ports, or to other switch ports over inter-switch link, or to Arbitrated Loops. Each of these attachments requires a different set of capabilities, and a particular port implementation may not support all these capabilities. There are, therefore, a number of different types of switch ports, with different names. These are listed below.

F_Port: attaches to N_Ports

E_Port: attaches from one Fabric Element to another Fabric Element, or "Inter-element Expansion Port"

G_Port: can operate as either an F_Port or as an E_Port. It can't attach to an F_Port, since F_Ports only attach to N_Ports.

FL_Port: can act as an F_Port, and can also sit on an Arbitrated Loop

GL_Port: can act as a G_Port, and can also sit on an Arbitrated Loop

There is also what is sometimes termed an "**S_Port,**" which stands for Service Port. An S_Port exists in name only, has no physical implementation, and is used for generic services.

For the switch ports which can operate in a variety of ways, the decision of which way they actually operate in a particular system will depend on what they are attached to. A GL_Port, for example, may act as an E_Port, if it's attached to another switch across an interswitch link, or as an F_Port, if it's attached to an N_Port, or as an FL_Port, if it's attached to an Arbitrated Loop. In general, the Ports will try to act in the various modes, and see what the attached Port is, in order to decide in which mode it should operate.

Switch and Addressing Hierarchy

To simplify the problem of determining precedence between switches, without adding the concept of a separate Fabric Manager, Fibre Channel has the concept of a "Principal Switch." This switch assumes the role of Domain Address Manager, assigning Domain_ID values to the other switches in the Fabric.

The Principal Switch is denoted as the switch with the lowest numerical value combination of "Switch_Priority" concatenated with Switch_Name. Since each Switch_Name is unique worldwide, and the Switch_Priority is configurable by the network administrator, this mechanism can always result in a deterministic, yet configurable, definition of which switch acts as the Principal Switch.

During network initialization, the Principal Switch assigns an 8-bit Domain_ID to each other switch in the Fabric. Each other switch then uses the Domain_ID as the top 8 bits in the address identifiers it assigns to the N_Ports attached to it, as shown in Figure 7.5, on page 117.

Class Dependency

One major factor in the development of multi-stage switch fabrics has been the difficulty of supporting all of the traffic classes on them. Primarily this is due to the fact that on a multi-stage fabric the Nodes are not fully-connected — that is, performance or functionality of transmission between two Nodes may depend on the status of transmission between two or more other Nodes in the system, as well as on the internal topology of the Fabric.

An example of this can be seen in Figure 17.2, which shows two switches, each supporting 3 external N_Ports, connected together over enough inter-switch links to support the maximum possible connectivity between Nodes A, B, and C, and Nodes D, E, and F. There are a number of issues in this topology which don't arise in single-stage networks.

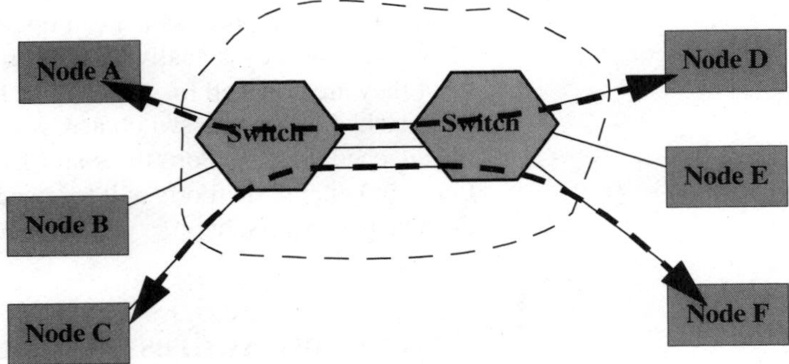

Figure 17.2 Traffic interaction in a multi-stage switch Fabric

- Configuration of a large multi-stage network with many ports may or may not be less expensive than a single-stage network with the same number of ports, depending on the port count used in the multi-stage network, and the topology used. The topology used depends on the traffic locality, i.e., how much bandwidth must be supported between the various ports. For example, the topology shown in Figure 17.2, with a 6-port Fabric built using two 6-port switches, would probably be more expensive than supporting the same 6 N_Ports on just 1 of the switches. But using 4-port switches instead might be cheaper, since smaller switches and only 1 ISL would be used, and the performance would be acceptable if the traffic were primarily localized to the (A, B, C) and (D, E, F) Port groups.
- If the top and bottom ISLs are highly loaded with A-to-D and C-to-F traf-

fic, then B-to-E traffic would have to traverse the middle ISL to achieve good performance. This routing restriction would, in general, not be present on a single-stage network.

- Class 1 Dedicated Connections, in particular, are limited, since they may contend for shared resources with other existing connections, or with Class 2 or 3 traffic.
- Behavior is quite different for the different Classes of Service. For Class 2 and Class 3, the switches will buffer Frames, and transmit them when the inter-switch links become free, such that the traffic can progress. Class 1 traffic, on the other hand, must wait until a Connection is made, which may depend on when other connections between separate Nodes are broken.
- Intermix of Class 1 and Class 2 or 3 data becomes much harder, since successful transmission becomes dependent on interdependent resource allocation between multiple switches.

All of these issues are solvable, but they are more complex than in single-stage switch networks.

Class F Service

Class F is only used between switches, for internal communications within a multi-switch Fabric. The N_Ports don't see Class F traffic, so discussion of it has been reserved to this chapter. Class F is distinguished from the Class 1, 2,... Services, which are collectively referred to as "Class N."

Class F operates like somewhat restricted Class 2. Frames may be multiplexed on Frame boundaries, and may be intermixed within Class 1 traffic if doing so does not cause loss of any Class 1 Frames. Each Frame is Acknowledged, using only ACK_1 Frames. The switches don't have to deliver Class F frames in order — each Frame in a Class F Sequence is numbered, with the first Frame being numbered zero. A separate End-to-End Credit between switches is established for Class F traffic.

Class F Frames can be inserted between Class 2 or Class 3 Frames, using Buffer-to-Buffer flow control, if necessary, to ensure adequate bandwidth for Class F Frame transmission. Alternatively, and in the case of Class 1 Dedicated Connection, the switches can insert a limited amount of Class F traffic between Class N Frames, using the fact that the source N_Port must insert at least 6 Primitive Signals between Frames, and the destination N_Port can receive as few as 2 Primitive Signals between Frames.

Switch Fabric Internal Link Services

There is a set of Login procedures for Class F. These let an E_Port exchange parameters with the E_Port that it's connected to. The format and algorithms used for the Exchange Link Parameters operation are very similar to the Login and Logout procedures described in the "N_Port Login" section, on page 163, and in the "N_Port and F_Port Service Parameters" section, on page 166.

The Switch Fabric Internal Link Service operates with Request and Response (Accept or Reject) Frames transmitted between an E_Port and the attached E_Port in the Fabric. Each Request is transmitted with R_CTL = x'02', and with TYPE=x'22', as shown in Figure 7.3, on page 114, and the responses use R_CTL = x'03.' Each Switch Fabric Internal Link Service Request sent across an Inter-Switch Link to the connected E_Port is responded to with either a corresponding format of the ILS Accept (ACC) Frame or an ILS Reject Frame, giving a Reason code.

One surprising thing about the following list is the identification of various modes of flow control in the ELP (Exchange Link Parameters) service request. The Fibre Channel standards specify flow control regarding N_Ports in detail, but the mechanisms for flow control between switches of a fabric are left largely up to the vendors.

The types of requests include the following. Clearly, a number of other types of services would be useful or necessary to provide full functionality — some of these are still under standardization.

Exchange Link Parameters. The exchange of link parameters between two E_Ports establishing the capabilities of the switches that are connected to those E_Ports. This includes specifying parameters such as Resource_Allocation and Error_Detect timeout values, E_Port_Name, Switch_Name, Service Parameters for Classes F, 1, 2, and 3, specifying an Interswitch Link Flow Control Mode, and parameters for the flow control.

Exchange Fabric Parameters. This is used for letting the switches notify each other about which switch will act as the Principal Switch, assigning Domain_ID values to the other switches. The payload contains the name and priority level of the switch which the sending switch believes to be the current Principal Switch. The Responder compares the priority in the Request with the priority of the switch that it believes is acting as the Principal Switch, such that the highest-priority switch can become the Principal Switch across the combined switch Fabric. Along with the request and the Accept is included a set of Domain_ID_List records, that list the

Switch_Name and the associated Domain_ID for the lists managed by the Principal Switch.

Announce Address Identifier. This request contains the originating Switch_Name, as well as the originating Address Identifier. The ACC contains corresponding information for the responder. This exchange allows each switch to address the other by its address identifier, rather than using the generic x'FFFFD' Fabric Controller address. If a switch transmits an AAI request, it indicates that it can satisfy requests for address identifiers from the responding E_Port.

Request Domain_ID. If a switch has received an AAI request from the connected E_Port, it can request a Domain_ID from the Domain Address Manager. The requestor can either set the field value to a requested Domain_ID (which is confirmed or re-assigned in the ACC), or to zero, to let the Domain Manager specify the Domain_ID.

Build Fabric. This service is used to request a non-disruptive reconfiguration of the entire Fabric. This will not cause loss of any Class N Frames.

Reconfigure Fabric. This service is used to request a complete reconfiguration of the entire Fabric, including possible disruption and abnormal termination of active sequences and Dedicated Connections.

Disconnect Class 1 Connection. This service is used by a switch acting as F_Port for one end of a Class 1 Connection to request the switch at the other end of the connection to perform a Link Reset to abort the connection with its attached N_Port. It is only used if Link Failure or Link Reset is detected on the Connection path.

Detect Queued Class 1 Connection Request Deadlock. This service is used to detect loops of queued Class 1 dedicated connections (Camp-On).

Special Switching Functions

This section describes some capabilities that become possible if a switch Fabric is employed between N_Ports. Some of these functions are not widely used, but can be extremely useful in particularly circumstances or configurations.

Aliases

Aliases are a mechanism for allowing a particular N_Port to be known to the Fabric and to other N_Ports by a variety of names. Doing this allows a single computer to both operate as a target for standard traffic and as, for example, the Time Server for the Fibre Channel network. Implementing this function requires both that the N_Ports be able to accept data sent to them using a variety of different D_IDs and that the Fabric record a variety of Address Identifiers for a particular physical N_Port. Allowing this function enables several interesting capabilities, as shown below.

Multicast

If a Fabric includes capability to handle aliasing, it may not require much extra complexity to enable multicast, which is a restricted broadcast across a subset of the N_Ports on the network. In multicast, a number of N_Ports may register together under a single address identifier, termed an "Alias_ID." Frames sent to the multicast group's Alias_ID can be routed simultaneously to all N_Ports in the multicast group. Only Class 3 Frames can be multicast, to avoid the significant problems in handling the return of multiple (possibly different) Link Control Frames for a single Data Frame, and the Sequence Initiative cannot be transferred, since it would cause all the N_Ports in the multicast group to generate the responding Sequence.

Fibre Channel includes 2 kinds of multicast: unreliable multicast, where the Fabric only has to duplicate frames across the multicast group, and reliable multicast, where the Fabric has to aggregate responses to provide an acknowledged service. Unreliable multicast re-uses Class 3 service and Frame delimeters — the main difference from Class 3 are that the Frames use the Alias_ID as the destination ID. For reliable multicast, a new class of service, Class 6, is defined. The operation of Class 6 is described in the "Class 6 — Uni-Directional Dedicated Connection" section, on page 194.

Multicast does not work for other service classes. Class 2 Frames addressed to an Alias_ID are rejected by the Fabric, and Class 1 Frames are indistinguishable from Class 6 Frames.

Hunt Groups

Under the hunt groups mechanism, a number of N_Ports working together under a Common Control Entity register together under a single Alias_ID. When a subsequent Frame is sent to or requested from that Alias_ID, the Fabric can route it to whichever N_Port of the group happens to be free at the time. That N_Port will handle the operation, allowing other members of the group to handle other requests to the hunt group's Alias_ID. This functionality would be useful, for example, when several N_Ports had redundant copies of an important piece of data, for reliability reasons. Then, any outside requests for copies of the data could be fielded by any one of the N_Ports, allowing simultaneous service and more efficient utilization of the fiber links.

Rotary Groups

A Rotary Group is similar to a Hunt Group, in that any of a number of N_Ports commonly controlled can respond to other N_Ports. Implementing rotary groups requires that the N_Ports all be coordinated by a Common Controlling Entity, and that the Switch have Rotary group support, to route traffic to one of the Rotary Group N_Ports from Ports outside the group.

The difference between Rotary Groups and Hunt Groups is that in Hunt Groups, the Ports outside the group are aware that they are communicating with a Hunt Group (there is a Hunt Group ID as the destination ID in the Frames), whereas in a Rotary Group, the existence of multiple N_Ports is hidden from outside the group (a normal N_Port D_ID is used in the Frames). As with Hunt Groups, the objective is to create an object which, at the Fibre Channel FC-2 level, appears as a single N_Port, but with greater capability (higher BW, supporting multiple simultaneous connections, etc.).

Well-Known Generic Services

The Generic Services, includes the Name Service, Time Service, Alias Service, and Security Key Distribution Services, are services provided by the Fabric to the attached N_Ports. In some contexts, these services are known as "S_Ports," for service ports, since they are addressed using well-known Port Identifiers, as shown in Figure 7.5. This does not imply that they are implemented as physical Ports in the Fabric. There is no explicit implication on how or where these services are to be implemented — they might be handled by a switch element, by a distributed set of switch elements, by a special N_Port known to the switch elements, or in some other way.

The important point is that an attached N_Port can send a Request to, for example, the Name Server D_ID (x'FF FF FC', as shown in Figure 7.5), and sometime later, it will get a Response back, in the format corresponding to the Request that it sent, without concern for how the service is implemented. These generic services at well-known addresses are summarized in the "Address Identifiers: S_ID and D_ID" section, on page 115. The only required services are the F_Port, and Fabric Controller services.

The mechanism for generating these responses, which in some cases (e.g., Broadcast) require cooperation between all the switches in the Fabric, is not completely specified, which is an important issue regarding interoperability of switches from different vendors.

Broadcast — x'FF FFFF'

A Frame addressed to this D_ID will be replicated by the Fabric, and distributed to every N_Port and NL_Port attached to the Fabric, including the source N_Port. The service is only applicable to Class 3 Frames, since Class 1 Dedicated Connections and Class 2 Responses aren't meaningful for broadcast traffic. This service is especially useful for management of the Nodes attached to the Fabric.

F_Port — x'FF FFFE'

This entity assigns, confirms, or reassigns N_Port address identifiers and notifies N_Ports of the operating characteristics of the Fabric, as described in the "Fabric Login" section, on page 161.

Fabric Controller — x'FF FFFD'

A Fabric Controller is a logical entity within the Fabric that controls the general operation of the Fabric. This includes Fabric initialization, Frame routing, generation of F_BSY and F_RJT link responses, and setup and tear down of Dedicated Connections. Link Service Requests used to, for example, read the status of a connection, or find out that a Camp-On exists, or that the Login state of another Node on the Fabric has changed are directed to or from the Fabric Controller Address.

Name Service — x'FF FFFC'

The objective of the Name server is to provide to the attached N_Ports information on any or all of the other N_Ports attached to the Fabric. This information includes the following parameters:

- Port Type — N_Port, NL_Port, F/NL_Port, E_Port, etc.
- Port Identifier — 3-byte S_ID/D_ID values
- Worldwide Names, of up to 8 bytes in size, for each Port and Node, as in Figure 9.7
- Symbolic names for the Port and the Node, of any length up to 255 bytes.
- An Initial Process Associator, if the Port requires one.
- IP Addresses for the Node and the Port, of either 4 (IPv4) or 16 (IPv6) bytes.
- Class of Service and TYPE bit maps, describing which Classes of service and which TYPEs of traffic (as shown in Figure 7.3) are supported by the port.

The Name Server takes requests packets directed to the Name Server ID, and returns the specified response. There are full set of defined requests for registering a set of parameters, and a full set for getting information mapping between each of these parameters. The registration requests, for example, take the form of "Register IP Address x as belonging to Port Identifier y." Similarly, the get requests to the Name Service take the form of "Respond with the Worldwide name corresponding to Port Identifier y," or "Respond with the Node IP Address correspond to Node Name x."

The most useful Get request is probably one called "Get all next," which takes the form of "Respond with all the parameters for the Port with the next higher registered Port ID than Port ID x." By repeatedly issuing this request to the Name Service, starting with Port ID and substituting the response Port ID in the following request, an N_Port can build an itemized ordered list of all Ports in the Fabric that have correctly registered with the Name Server.

Time Services — x'FF FFFB'

The Time Service allows a Port to request that a response be returned specifying the current time. The format includes a 4-byte number specifying the number of seconds relative to January 1, 1990, universal time, and a second 4-byte number specifying the fractional part of the time value. The specification indicates that multiple Ports should have clocks synchronized to an accuracy of 2 seconds, although tighter synchronization could be allowed and would be useful in some situations.

The mechanism for a much more accurately synchronized time value is described in the "Clock Synchronization Server — x'FF FFF6'" section, on page 314.

Management Server — x'FFFFFA'

The Management server address is used for denoting the entity which would be used to report information on how the network is being utilized and how it is behaving overall.

Quality of Service Facilitator — x'FFFF9'

When a Fabric implements Class 4 Virtual Circuit service, to allow specifying a Dedicated Connection with a reserved fraction of the available link bandwidth, this address is used to address the server that manages this function. Class 4 Service is described in the "Class 4 — Fractional" section, on page 192.

Alias Services — x'FF FFF8'

The Alias Server manages the registration and deregistration of Alias IDs for both Hunt Groups and Multicast Groups. As described in the "ELS Requests for Alias IDs" section, on page 213, it is a generally quite useful function to be able to have a single N_Port known by several different IDs, since this allows packets with different D_ID values to be routed to the same Port, and handled differently once they get there. This address is used as, for example, the D_ID for Extended Link Service requests dealing with multicast (GAID, FACT, FDACT, NDACT).

Security Key Distribution Services — x'FF FFF7'

Secure (encrypted) communications are very important in many application areas. Secure communications require encryption keys to be distributed among the communication nodes such that the sender can encrypt the data in a way that the receiver can decrypt it.

The Security Key Distribution Server operates as a facility for securely distributed authenticated private keys for Port pairs so that they may securely exchange encrypted data. The centralization of key distribution allows keys to be authenticated, and allows the management of keys to be centralized in one facility, rather than to be separately managed by every node requiring secure communications services. The security key distribution service operates as follows.

Initially, each client has a unique private key, called a "distribution key," and the Security Key Distribution Server also has a copy of the distribution key for each client. The mechanism for distributing these unique distribution keys is beyond the scope of Fibre Channel, and is very important to the overall security of the system. A further feature that is important to the overall security is that distribution keys are long (75 to 90 bits), and are used (along with a good, slow encryption algorithm) to encode a separate shorter (40 to 75 bits) key, called a "secret key," that is used for data encryption between an originator and a client node.

The Originator client sends a KEY_REQUEST.request to the Security-Key Server, specifying the Originator and Respond address identifiers, and a unique request id. The Security-Key Server validates the request, and generates a secret key to that originator/responder pair and that request. The Security-Key Server then encrypts the secret key, the unique request id, and the responder id with the originator's larger distribution key. It also encrypts the secret key and the originator_id using the responder's distribution key, and transmits this all to the originator as a KEY_REQUEST.reply.

The originator receives the reply, and decrypts the secret key, the responder id, and the unique request id using its distribution key. It then sends the remainder of the reply (the secret key and the originator id, encrypted with the responder's distribution key) to the responder, along with an encryption_id. The responder decrypts this message, to retrieve the originator id and the secret key used for data encryption.

To authenticate that the secret key came from the originator client, the responder generates a different unique identifier, encrypts it using the secret key, and transmits it to the originator. The originator decrypts it, using the secret key, subtracts 1 from the value, and re-encrypts it using the secret key before transmitting it back to the responder. When the responder decrypts the received value and generates 1 less than the identifier it transmitted, it is assured that the secret key came from the originator client, rather than some other client, and can then begin secure data transfer in both directions, using the secret key for encryption and decryption.

Clock Synchronization Server — x'FF FFF6'

The Clock Synchronization server uses the natural synchronization that occurs between transmitters and receivers to achieve very accurate network-wide clock synchronization. This mechanism has much better accuracy than can be achieved with software-based clock distribution protocols. Typically networks use clock distribution protocols based on estimated path delays between endnodes to synchronize time across the network. With software-based protocols, this accuracy is limited by the variability between when data arrives from the network and when it is made available to the software, with variations due to, for example, interrupt service latencies and memory hierarchy variations.

In the Fibre Channel Clock Synchronization service, the Clock Synchronization Server embedded in the Fabric has a clock reference that operates at the speed of the N_Ports being served. Since the 8b/10b code allows clock recovery from the transmitted data stream, the server can operate a clock at the exact frequency of the client N_Port. To initiate the service, the server and client cooperate to measure the propagation time on the link. The server sends a request to the client, addressed to the CSS D_ID (x'FF FFF6'). The client receives the request and returns an accept Frame, with a payload indicating the time between the reception of the request and the transmission of the ACC. When the server receives the ACC, it measures the total roundtrip time and subtracts the amount of time that the client inserted in the payload. The remaining value is equal to the roundtrip propagation time over the link, of which half is assumed to be consumed in each direction.

Having measured the propagation delay over the link, the Clock Synchronization Server can then add this propagation delay to the absolute time in its clock before transmitting time values to the client. With implementation in hardware, and with links that provide an equal propagation time in each direction (so that the unidirectional delay is very close to half the roundtrip delay), it is reasonable to get the clocks synchronized to better than a microsecond, and to better than 100 nanoseconds in well-controlled environments.

Multicast Server — x'FF FFF5'

Service used in Class 6 Uni-Directional Dedicated Connection which acts as the Destination Port for responses and aggregates them properly to provide a single response to the multicast originator. This service acts to provide a reliable, acknowledged unidirectional multicast, while preventing the originator from having to aggregate responses from all the responders in the multicast group. More details on this operation can be found in the "Class 6 — Uni-Directional Dedicated Connection" section, on page 194.

Chapter 18

FC-4: SCSI and IP over Fibre Channel

Introduction

A large part of the work done in developing the Fibre Channel architecture has been concentrated in assuring that the architecture could efficiently and naturally operate as a transport or delivery mechanism for a wide variety of well-established Upper Level Protocols, or ULPs. Since much of the investment in current operating systems is at the device-driver level, the incremental cost in transferring systems over to a Fibre Channel data communication level decreases if the interfaces can be made as similar to previously-existing interfaces as possible. This allows new capabilities to be added with minimal changes to currently-available interfaces.

The general picture of interaction of Fibre Channel with ULPs is for the Fibre Channel levels to act as a data transport mechanism for ULP logical constructs. The logical constructs are termed "Information Units" and include unidirectional, continuous blocks of bytes, such as are used for commands, for data streams, and for status. These Information Units, acting as commands or data streams or for status, are generally mapped onto the Fibre Channel level as Sequences. Related Information Units, such as are required in an I/O operation having command, data, and status information units, are mapped as a single Exchange at the Fibre Channel level.

The Sequence and Exchange structures are general enough and have enough tunable options concerning flow control, error recovery policy, and interaction between Sequences that FC-4 mappings can be made to many already-existing ULPs. For example, there are currently stable FC-4 mappings defined for Fibre Channel delivery under IPI-3, which is a disk and channel I/O interface used largely in the mainframe market, for SBCCS, which is the command code set for operation over ESCON systems, for HIPPI, which is a channel and network attachment interface used largely in the supercomputer market, and for several others.

Here, we will concentrate on describing the interfaces to two of the major ULPs: IP and SCSI. IP is the network layer for the TCP/IP protocol, which is a widely used protocol for communications and networking. Most of the Internet uses the TCP/IP protocol stack, which is flexible enough to be implemented over almost any reliable or unreliable data delivery mechanism. SCSI is a widely used general-purpose I/O interface to a bus-type hardware layer which has seen its largest usage for connecting storage peripherals such as disk drives, CD-ROMS, and tape drives to workstations and high-end desktop machines. The SCSI-3 architecture is separable into the hardware description of bus, connectors, and signal levels and the software description of command, response, and data logical constructs, which are transported over the hardware. Mapping to the Fibre Channel level is carried out by replacing the SCSI physical and data transport layers with Fibre Channel levels, which carry the SCSI Information Units. Transport of SCSI commands and data over the Fibre Channel protocol hierarchy is such a major part of the Fibre Channel usage that a particular acronym, "FCP," is used to denote the Fibre Channel Protocol for SCSI.

In this chapter, we will go into a bit more detail about how the FC-4 level is structured and what interfaces are provided before discussing the FC-4 implementations for the IP and SCSI interfaces.

FC-4 Control over Fibre Channel Operations

In implementing an FC-4 over Fibre Channel protocols, there is a requirement for the data and control Information Units to be passed "transparently" to the Fibre Channel levels, so that the exact contents of the ULP Payloads transmitted only affect specific Fibre Channel operations. There is also a requirement that the ULP be able to control the transport of the Information Units and to control the interactions between them, to form complete operations at the ULP level.

Fibre Channel provides several mechanisms for providing FC-4 control over data and control operations at the Fibre Channel level. First, the FC-4 can control which Information Category a particular Sequence or block of a Sequence is sent under. This allows signaling of the FC-4 level at the destination N_Port and possibly allows direction of the data to different buffer pools. Second, the F_CTL field options for initiating and terminating Sequences and Exchanges, transferring Sequence Initiative, and setting Exchange Error Policy can be used for implementing FC-4 and ULP-level functions. Third, Fibre Channel allows a Device Header to be included in Data Frames, for passing FC-4 level header information transparently through the Fibre Channel levels. Fourth, and most important, the Payload of the Sequences can contain FC-4-level control and data information.

Of these mechanisms, all but the Device Header optional Header mechanism are used extensively in implementing the SCSI and IP protocols over Fibre Channel. These two protocols provide examples of the usage of Fibre Channel constructs for implementing higher-level functions.

IP over Fibre Channel

IP operation over Fibre Channel is quite simple, since the only operation required is for the Fibre Channel levels to transport two types of Information

Units: IP and ARP datagrams. IP datagrams are used to communicate data between Nodes connected over networks implementing the Internet Protocol stack. ARP (Address Resolution Protocol) datagrams are used at network configuration to build a mapping between the 32-bit worldwide unique IP addresses and the addresses used for routing data on the network, which are termed MAC (Media Access Control) addresses.

The type of service requested for IP and ARP datagrams is connectionless datagram service, which means that each datagram is assumed to be routed through the network as a unit, without any guarantee of reliable delivery. Protocol operations above the IP layer do reassembly of IP datagrams and assurance of reliability and error recovery, if required.

IP and ARP datagrams are mapped as Sequences to be delivered to the proper destination over a Fibre Channel network. The connectionless datagram requirements of IP service can be met by using any of the Fibre Channel Classes of service. Logically, Class 3 service maps directly onto IP delivery requirements (non-guaranteed, without acknowledgment required), but due to the performance limitations of Class 3 implementation (described in the "Performance and Reliability with Class 3 Service" section, on page 190) and the low overhead of acknowledgments, Class 2 will generally be used. Class 1 service could be a reasonable option for large datagrams and Point-to-point topologies.

The IP protocol makes the assumption that data delivery is fully bidirectional, with datagrams being transmitted and received simultaneously. Since each Exchange is only allowed to have non-simultaneous bidirectional traffic (Sequences from A to B must be closed before Sequences from B to A can be opened), Exchanges are only used unidirectionally, with the Exchange Originator initiating all Sequences. Bidirectional IP communication between two N_Ports requires originating one Exchange in each direction. This enhances performance by allowing simultaneous bidirectional transmission. Transmission of IP datagrams to two different destination Ports requires establishment of two different Exchanges.

Transmission of IP and ARP Datagrams

When IP and ARP datagrams are sent over Fibre Channel networks, they are encapsulated into a Sequence of one or more Data Frames. IP and ARP are distinguished by setting the Routing Control field to "Unsolicited Device Data" (R_CTL[31-24] = b'0000 0100'), and the TYPE field is set to b'0000 0101' to indicate the 8802.2 LLC/SNAP Encapsulation for IP and ARP datagrams, without ordering. The Frames are allowed to contain an Association Header or an Expiration_Security Header but cannot contain a Network or Device Header. An Association Header is required for IP implementation on systems, such as mainframes, which require Initial Process Associators to associate the IP datagrams with a particular system image.

The Information Units sent over Fibre Channel consist of IP and ARP datagrams, preceded by the IEEE 802.2 LLC (Logical Link Control) and SNAP (SubNetwork Access Protocol) Headers. The full formats of the IP and ARP datagrams transmitted are shown in Figures 18.1 and 18.2. The base of the Relative Offset value in the Parameter field for each Frame is zero.

Figure 18.1
Format of an IP Information Unit.

Internet Protocol (IP) Information Unit

<table>
<tr><th>Bit / Word</th><th>33222222 10987654</th><th>22221111 32109876</th><th>11111100 54321098</th><th>00000000 76543210</th></tr>
<tr><td>0</td><td colspan="3">Logical Link Control (LLC)
DSAP = x'AA' SSAP = x'AA' Cntrl = x'03'</td><td>SNAP
0000 0000</td></tr>
<tr><td>1</td><td colspan="4">Sub-Network Access Protocol (SNAP) - cont'd.
0000 0000 0000 0000 ---Ethertype=x'0800'---</td></tr>
<tr><td>2</td><td>Vers/Hlen</td><td>Svc Type</td><td colspan="2">Total Length</td></tr>
<tr><td>3</td><td colspan="2">Identification</td><td colspan="2">Flags[3:0]|Fragment offset</td></tr>
<tr><td>4</td><td>Time to Live</td><td>Protocol</td><td colspan="2">Header Checksum</td></tr>
<tr><td>5</td><td colspan="4">Source IP Address</td></tr>
<tr><td>6</td><td colspan="4">Destination IP Address</td></tr>
<tr><td>7</td><td colspan="3">IP Options (if any)</td><td>Padding</td></tr>
<tr><td>8
:
:</td><td colspan="4">Data
:
:</td></tr>
</table>

The default Exchange Error Policy for IP and ARP is "Abort, discard a single Sequence." This ensures that the different datagrams are transmitted independently, so that an error in the delivery of one Sequence, or datagram, does not cause the destination to discard later Sequences in the Exchange unnecessarily. It is not prohibited to use either of the "Discard multiple Sequences" Exchange error policies, and on some systems the performance might be improved by exposing this capability of the Fibre Channel levels for preventing datagrams from being delivered out of order. On the other hand, with IP implementations, interoperability is generally more of a concern than performance. The "Stop Sequence" protocol is also allowed, although its use is limited, given the capabilities of the IP and higher levels.

IP Operations

For understanding the implementation of IP over Fibre Channel, it is helpful to understand the mapping of several other IP concepts over their Fibre Channel implementation.

Figure 18.2
Format of ARP request and ARP response datagrams or Frame payloads.

Address Resolution Protocol (ARP) Information Unit

Bit / Word	33222222 10987654	22221111 32109876	11111100 54321098	00000000 76543210
0	Logical Link Control (LLC) DSAP=x'AA' SSAP= x'AA' Cntrl=x'03'			SNAP 0000 0000
1	SubNetwork Access Protocol (SNAP) - cont'd. 0000 0000	0000 0000	---Ethertype=x'0800'---	
2	hardware:x'0012'		protocol: x'0800'	
3	hlen:5+ 3 × (# of Ports)	plen:x'04'	operation: rqst=x'0001,' response=x'0002'	
4	sha:validity	sender pa: Initial Process Associator[3-1]		
5	spa:IPA[0]	sender ha: N_Port Identifier(s) - size to 3 bytes × (number of Ports)		
:	spa: IP address of requestor (ARP request) or request target (ARP response)			
:	tha: validity	target pa: Initial Process Associator[3-1]		
:	tpa:IPA[0]	target ha: N_Port Identifier(s) - size to 3 bytes × (number of Ports)		
:	tpa: IP address of target (ARP request) or requestor (ARP response)			

Mapping between IP and Fibre Channel Addresses. At the IP level, there is a single layer of addressing, which is implemented using the IP address. This is a 32-bit field, which is effectively unique worldwide. Physically, there may be a division of the IP address space into different networks, with routers and so on, but the address space is flat, with one IP address mapping onto one network interface.

There are several parts of the Fibre Channel addressing which relate to the IP address. First, there is the conception that a single Node may have multiple Ports into the Fibre Channel network, each with its own Port identifier. Second, a system that uses multiple system images, such as a mainframe, or multiple separate processes uses the Initial Process Associator to map Sequences to those images. Each image on a host or Node may have a separate IP address. The Fibre Channel mapping under IP specifies that multiple N_Port Identifiers can map to the same IP address as long as they have the same Initial Process Associator. The Initial Process Associator can of course be absent on systems which don't require it. In further discussions here, references to N_Port Identifiers carry with them an implication of an Initial Process Associator on systems which require it.

Determining the mapping between IP addresses and N_Port Identifiers is carried out using the Address Resolution Procedures described in the "ARP Server and the Name Server Function" section, on page 322 below.

IP Subnetworks. In the IP addressing and routing algorithms there is the concept of an IP subnetwork. A subnetwork is a set of hosts which can directly communicate with each other. Communication between subnetworks is accomplished using a router (or gateway). A subnetwork mask can be used to route datagrams to a particular subnetwork, and then to route datagrams within the subnetwork, by separating the IP address bits.

The addressing and routing scheme implies that hosts within a subnetwork can directly communicate with each other and that hosts within different subnetworks may only be able to communicate by routing datagrams through a router. The router may do protocol conversion, speed conversion, or data formatting, etc., to route datagrams from one subnetwork onto another.

In a Fibre Channel network with multiple attached Nodes, even if all the Ports attached to the network are using Fibre Channel protocols, it is possible that they may not all be able to directly communicate. This is another consequence of the generality of the Fibre Channel protocol and the number of configuration options. If, for example, one group of Nodes only supports Class 1 communication, and another group of Nodes only supports Class 2 communications, they cannot directly communicate, even though they may be physically attached through a Fabric which supports both Classes.

In this case, each subset of N_Ports that can communicate is termed a "region," and all N_Ports within a region can communicate. In the above example, an N_Port that could support both Class 1 and Class 2 communications could be considered to be part of both regions.

There is a general correspondence between Fibre Channel regions and IP subnetworks. A Node connected by an N_Port which supports both Class 1 and Class 2 communications would be considered to be part of both Fibre Channel regions and could act as a router between the N_Ports acting as hosts on different IP subnetworks.

Maximum Transfer Unit (MTU). A significant factor affecting IP performance over different physical networks is the MTU, or maximum transfer unit. An MTU is the maximum size of a frame on the physical network. An IP datagram has a length field of 16 bits, so a datagram can be up to 65,535 bytes long. However, when the datagram is transported over a network, the largest datagram size that is allowed is limited by the MTU of the network. Each datagram is fragmented into datagrams of size determined by the MTU of the network, with an IP Header for each one. The fragmented datagrams are routed through the network independently and then reassembled, using a code built into the datagrams when they were fragmented, to rebuild the original IP datagram.

The fragmentation and particularly the reassembly are fairly slow operations, so it is generally a good policy to make the fragmented datagrams as large as possible, within the limits of the MTU of the network used.

On a token ring network, the MTU is roughly 4,000 to 4,500 bytes, while on an Ethernet network, the MTU is normally 1,500, 1,492, or 1,024 bytes. On these systems, a 65,535-byte IP datagram must be fragmented into a large number of fragmented datagrams for routing through the network and reassembly at the IP level.

Since the IP datagram is mapped as an Information Unit, or Sequence, at the Fibre Channel level, there is no MTU defined by hardware limitations. (A similar way to look at this is that the Fibre Channel has its own MTU of 2,112 bytes, but that fragmentation is hidden, so the MTU exposed to the IP level is unlimited.) The MTU is therefore limited by the requirements of fitting within the 16-bit IP length field, with space left over for intermediate headers. The choice for MTU over Fibre Channel therefore was 65,280 (x'FF00') bytes. This allows an IP datagram to fit in one 64 Kbyte buffer with up to 256 bytes of overhead. The only overhead currently defined is the 8-byte LLC/SNAP header, so there are 248 bytes left for expansion. This large MTU is a significant factor in allowing the speed-up of IP traffic over Fibre Channel networks, since fragmenting and reassembly require data copying and generation of new IP headers, which are relatively slow processes.

ARP Server and the Name Server Function. An ARP Server is an entity which provides a mapping between IP addresses and network addresses. This capability is somewhat different than implementation of the Address Resolution Protocol on networks which implement a broadcast function. On an Ethernet network, if a host wants to find the network address of a host with a particular IP address, it can do a broadcast of an ARP request, with the desired IP address in the contents. The IP layer on all connected hosts will read the ARP request, and the one host with a matching address will reply with an ARP Reply, containing its network address.

On a Fibre Channel network, there is limited support for a broadcast functionality. Broadcast Frames can only be sent in Class 3, which Fabrics and N_Ports are not required to support. To implement the mapping between IP addresses and Fibre Channel N_Port Identifiers, the concept of an ARP Server is defined. An ARP Server is expected to be a implemented as a Node, connected to the Fabric through a Port, which can respond to Frames directed to the x'FF FFFC' N_Port Identifier for the "Directory Server." A Node which can implement the ARP Server functionality can register with the network (Fabric, Point-to-point, or Arbitrated Loop) by performing a

Fabric Login with the requested N_Port Identifier of x'FF FFFC.' Following Fabric Login, N_Ports can register their IP Addresses by sending ARP Requests with their own IP addresses and N_Port Identifiers to the x'FF FFFC' Directory Server N_Port Identifier, which will be routed by the Fabric to the ARP Server N_Port and Node. The ARP Server Node can then reply to other ARP Requests, providing information for mapping IP addresses to N_Port Identifiers to other N_Ports and Nodes.

SCSI-3 over Fibre Channel: FCP Overview

The basic assumptions on physical topology for SCSI and Fibre Channel are different. Fibre Channel is logically a point-to-point serial data channel, with provisions for a Fabric topology and an Arbitrated Loop topology which allow physical connection among more than two N_Ports. SCSI differs from Fibre Channel in that it assumes a bus architecture.

Beyond the mechanics of bus arbitration vs. access to physical links, however, the architectures at a high level are fairly similar. They both involve an Initiator sending commands to target or Recipient Nodes, with multiple streams of data logically related together to perform communications operations. The Fibre Channel Fabric topology can be more efficient than the SCSI bus topology, since multiple operations can be simultaneously active, multiplexed over the network.

The general picture of FCP is of the Fibre Channel operating as a transport mechanism for transmitting SCSI-3 command, response, status, and data blocks. An N_Port which can implement the transmission and recognition of FCP Information Units and I/O operations is termed an "FCP_Port." FCP_Ports can be implemented on Point-to-point, Fabric, and Arbitrated Loop topologies, although by far the most prevalent usage is expected to be on the Arbitrated Loop topology.

A word of caution: The most current SCSI-3 standard documentation is roughly 3 in thick. There are options for implementation of SCSI-3 commands over a variety of parallel and serial interfaces, with great hope of conformity between the various interfaces. The discussion here does not attempt to give any more than the most general idea of how SCSI operations are implemented over Fibre Channel hardware.

Further reference documents for SCSI as well as for the FCP Fibre Channel protocol for SCSI are listed in Appendix A and include in particular the CAM, or *SCSI Common Access Method*, and SAM, or *SCSI-3 Architecture Model*. A number of SCSI-specific fields and command formats are described in these documents. They are not explicitly described here, since they are not directly related to SCSI implementation over Fibre Channel. Figure 18.3 shows the equivalence between SCSI and FCP terms.

Figure 18.3
Functional equivalence between SCSI-3 and FCP functions.

Functions		
SCSI Equivalent	**FCP Equivalent**	
I/O Operations	Exchanges	
Request/Response Primitives	Sequences	
Command service request	Unsolicited command IU	FCP_CMND
Data delivery request	Data description IU	FCP_XFER_RDY
Data delivery action	Solicited data IU	FCP_DATA
Command service response	Command status IU	FCP_RSP

Four types of functional management are defined for FCP:

Process Login/Logout Management: Defines initial communications which may or may not be required to set up FCP communications. This is a separate optional Login/Logout step beyond the standard Fibre Channel Fabric Login and Port Login which allows the definition of one or more virtual Initiators or Recipients supported over a single FCP_Port. This allows the management of multiple SCSI devices or multiple user-level processes by a single Fibre Channel Port. This is described in the "Overview of Process Login/Logout" section, on page 174.

Device Management: Defines communications for transferring data between Initiators and Recipients, using the FCP_CMND, FCP_XFER_RDY, FCP_DATA, and FCP_RSP IUs.

Task Management: Used when a task or group of tasks must be aborted or terminated. As shown in the "FCP_CMND Information Unit" section, on page 328 and Figure 18.5, various flags in the FCP_CNTL field of an FCP_CMND IU indicate which task management functions are to be performed.

Link Management: Defined under FCP to be the same as under standard Fibre Channel operation.

Following normal FC-4 practice, the interface between the SCSI ULP and the Fibre Channel levels is implemented through a set of Information Units, or IUs. These are the specific SCSI commands, data blocks, or status blocks transferred between initiators and targets, to complete I/O operations. Over Fibre Channel, they are each mapped as a separate Sequence. The set of IUs used by the FCP is shown in Figure 18.4. In this figure, the F/M/L column indicates whether the indicated IU is sent as the first, middle, or last Sequence of an Exchange, the SI column indicates whether the Sequence Initiative is held (H) or transferred (T) at the end of the Sequence, and the M/

O column indicates whether the indicated IU is mandatory (M) or optional (O). IUs T8-T11 and I6-I7 are examples of Sequences that are transmitted using two independent Information Categories, to optimize performance on systems which allow more than one Category per Sequence.

The Frames used for transferring FCP IUs basically follow the normal Fibre Channel rules, as described in Chapter 7. The SCSI Initiator is always the Exchange Originator and assigns the OX_ID values. Similarly, the SCSI Target is the Responder and can assign the RX_ID value to match the tag defined by SAM. The TYPE field has the value x'08,' as shown in Figure 7.3. Relative offset is not required, but it can significantly simplify reassembly of data transferred in FCP_DATA IUs.

FCP Information Unit Formats

This section describes formats of the Information Units transferred during FCP I/O operations.

The two types of Frames related to FCP are the Extended Link Service Command Frames related to Process Login/Logout, and the Information Unit Frames. The formats of all these Frames are shown in Figure 18.5, and are described below.

Process Login/Logout. The Process Login and Logout Sequences are Extended Link Service Commands, which can be used to build configuration information for multiple SCSI virtual initiators or virtual targets behind a single Fibre Channel Port. The processes map as images, as described in the "Overview of Process Login/Logout" section, on page 174.

In Process Login for FCP, word 4 of the payload contains 7 bits which specify capabilities and parameters specific to the operation of SCSI Targets and Initiators. These include bits specifying whether data overlay is allowed, whether the image can operate as a SCSI Initiator or Target, whether the CMND/DATA Sequences using two Information Categories per Sequence are usable, and whether the XFER_RDY Information Unit is used for reads or writes. The Process Login and Logout Information units also contain Originator and Responder Process Associators, which identify the image.

As stated earlier, it is not necessary to provide capability for transmission and reception of these Information Units if the same Login functions will always be carried out through implicit means.

Figure 18.4
Information Units (IUs) defined for FCP.

IU	SCSI SAM primitive	Info Cat.	Frame Content	F/M/L	SI	M/O
FCP Information Units sent to targets						
T1	Command Request	6	FCP_CMND	F	T	M
T2	Command Request	6	FCP_CMND	F	H	O
T3	Command Request (Linked)	6	FCP_CMND	M	T	O
T4	Command Request (Linked)	6	FCP_CMND	M	H	O
T5	Task Management	6	FCP_CMND	F/L	H	M
T6	Data Out action	1	FCP_DATA	M	T	M
T7	Data Out action	1	FCP_DATA	M	H	O
T8	Command request and Data Out action	6/1	FCP_CMND+FCP_DATA	F	T	O
T9	Command request and Data Out action	6/1	FCP_CMND+FCP_DATA	F	H	O
T10	Command request and Data Out action (linked)	6/1	FCP_CMND+FCP_DATA	M	T	O
T11	Command request and Data Out action (linked)	6/1	FCP_CMND+FCP_DATA	M	H	O
FCP Information Units Set to Initiators						
I1	Data delivery request	5	FCP_XFER_RDY (Write)	M	T	M
I2	Data delivery request	5	FCP_XFER_RDY (Read)	M	H	M
I3	Data In action	1	FCP_DATA	M	H	M
I4	Response	7	FCP_RSP	L	T	M
I5	Response (linked)	7	FCP_RSP	M	T	O
I6	Data In action and Response	1/7	RCP_DATA + FCP_RSP	L	T	O
I7	Data In action and Response (Linked)	1/7	RCP_DATA+FCP_RSP	M	T	O

Notes:

Information Category is indicated in R_CTL[27-24]. Values used in FCP are: x'1':Solicited Data, x'5':Data Descriptor, x'6':Unsolicited Command, and x'7':Command Status

F/M/L indicates whether the IU is the First, Middle, or Last Sequence of the Exchange.

H/T indicates whether the Sequence Initiative is Held or Transferred after the Sequence.

M/O indicates whether support for the IU is Mandatory or Optional.

T8, T9, T10, T11, I6, and I7 are only supported on N_Ports that support >1 Information Category per Sequence.

Figure 18.5 FCP Information Unit Formats.

PRLI Command

0	x'08'	x'10'	Payload length
1	x'08'	rsrvd	OR.. ....
2	Originator Process Associator		
3	.... .OIT CDRW		
:	More service pages may follow		

PRLO Command

0	x'08'	x'10'	Payload Length
1	x'08'	rsrvd	OR.. ....
2	Originator Process Associator		
3	reserved		
:	More service pages may follow		

PRLI ACC (Accept)

0	0x20	0x10	Payload length
1	0x08	rsrvd	OR.. ARc. ....
2	Originator Process Associator		
3	Responder Process Associator		
4	....IT CDRW		
:	More service pages may follow		

PRLO ACC (Accept)

0	0x20	0x10	Payload length
1	0x08	rsrvd	OR.. AR.. ....
2	Originator Process Associator		
3	Responder Process Associator		
4	reserved		
:	More service pages may follow		

FCP_CMND

0	FCP_LUN (Logical Unit #) high word
1	FCP_LUN (Logical Unit #) low word
2	FCP_CNTL (Control Field)
3	FCP_CDB (SCSI Cmd Dsc Blk) Word 0
4	FCP_CDB (SCSI Cmd Dsc Blk) Word 1
5	FCP_CDB (SCSI Cmd Dsc Blk) Word 2
6	FCP_CDB (SCSI Cmd Dsc Blk) Word 3
7	FCP_DL (Data Length)

FCP_XFER_RDY

0	DATA_RO (Rel. offset for FCP_DATA)
1	BURST_LEN (Length for FCP_DATA)
2	reserved

Relative offset and length refer to the FCP_DATA IU following the FCP_XFER_RDY

FCP_DATA

0	Data, Word 0
:	:
N	Data, Word N+1

Note:
The FCP_DATA Information Unit is a Sequence containing one or more Frames

FCP_RSP

0	reserved
1	reserved
2	FCP_STATUS (Field validity, SCSI stat)
3	FCP_RESID (Residual count)
4	FCP_SNS_LEN (Length of SNS_INFO)
5	FCP_RSP_LEN (Length of RSP_INFO)
6	FCP_RSP_INFO (FCP Response info)
:	:
:	FCP_RSP_INFO (FCP Response info)
:	FCP_SNS_INFO (SCSI Sense info)
:	:
:	FCP_SNS_INFO (SCSI Sense info)

FCP_CMND Information Unit. The FCP_CMND IU carries a SCSI command to be executed or a task management request to be performed. The payload contains the fields shown in Figure 18.5.

The FCP_LUN field contains the logical unit number, which is the address of the desired logical unit inside the SCSI target. A SCSI INQUIRY from LUN 0 can be used to determine the SCSI device type, manufacturer, and model of the logical unit. The FCP_CNTL field contains control flags which describe handling of tasks, and it contains bits to terminate tasks, reset targets, clear or abort task sets, and to indicate whether the operation is a read or write operation, among others.

The actual work of the FCP_CMND IU is specified in the FCP_CDB field, containing the SCSI Command Descriptor Block. The 16 bytes of the Command Descriptor Block field describe the SCSI command to be executed, as described in the SCSI documentation. Finally, the FCP_DL specifies the maximum number of data bytes expected to be transferred under the command, which can be used to prevent buffer overflow on receiving data.

FCP_XFER_RDY Information Unit. This IU is optionally implemented to indicate that the target is ready to perform all or part of the data transfer for a command. This IU can be used for handshaking to prevent buffer overflow when either the target or initiator may send data too early for the other Port to respond. For both a write operation and a read operation, the FCP_XFER_RDY IU is transferred by the target of the operation. The DATA_RO field duplicates the contents of the Relative Offset field in the Parameter field of the incoming operation. The BURST_LEN field is used on a data transfer from SCSI initiator to target to indicate the amount of buffer space reserved for the next FCP_DATA IU. For a data transfer from SCSI target to initiator, it indicates the exact length of the next FCP_DATA IU.

FCP_DATA Information Unit. The FCP_DATA IU contains the application data that is transferred for an FCP_CMND. More than one FCP_DATA IU may be used to transfer the data associated with a particular I/O command. The FCP_DATA IU should contain the same Relative Offset value as was in the FCP_XFER_RDY IU that preceded it, if XFER_RDY is implemented. The normal Fibre Channel mechanisms, depending on Class of service used and service parameters, in effect will apply for transfer of the IU.

FCP_RSP Information Unit. The FCP_RSP IU is used to return several types of status information relative to the CMND sent. The

FCP_STATUS field indicates the validity of the other fields, telling whether the FCP_RESID field contains a valid count indicating too many or too few bytes transferred and whether the FCP_SNS_LEN and FCP_RSP_LEN fields contain valid counts for the lengths of the FCP_RSP_INFO and FCP_SNS_INFO. These fields indicate protocol failure information and logical unit error information, respectively. For a successful operation as expected, the FCP_STATUS field and all other fields will be zero.

Sample I/O Operation under FCP

Following is a simple example of how the FCP Information Units described above can be used to execute an I/O operation. The Fibre Channel interface under SCSI is actually somewhat simpler than the native SCSI operation, since there is no concept of different bus phases, when different operations can occur. Most of the complexity of SCSI is retained in a Fibre Channel implementation.

An application begins an FCP I/O operation by requesting SCSI command service from the FCP system. This must contain all the information required, including data storage addresses, lengths, and data transfer characteristics. We'll consider two cases here, for a SCSI write and a SCSI read operation, as shown in Figure 18.6.

After receiving the request for a SCSI operation, the FCP_Port builds the necessary internal data structures describing the operation, then originates an Exchange for the Operation by building and initiating a Sequence containing the FCP_CMND Information Unit. This Sequence is transmitted over the attached link to the D_ID of the FCP_Port operating as SCSI target for the command. The FCP_CMND Sequence travels over the Fabric, Point-to-point link, or Arbitrated Loop topology to arrive at the FCP_Port of the Target Node. As indicated in Figure 18.4, the FCP_CMND uses the "Unsolicited command" Information Category.

On receiving the FCP_CMND Sequence, the target device determines what is required. In this case, the FCP_CMND requires a read or write data transfer. When it is ready to transfer the data, the target transmits a FCP_XFER_RDY IU, using the "Data Descriptor" Information Category. If the FCP_CMND operation indicates a read, it follows the FCP_XFER_RDY Sequence with an FCP_DATA Sequence. For a write, the Initiator will respond to the FCP_XFER_RDY Sequence by transmitting an FCP_DATA Sequence. As indicated in the figure, multiple FCP_XFER_RDY / FCP_DATA pairs might be required for the operation.

When the operation is completed, the SCSI target returns the completion status to the SCSI initiator in an FCP_RSP Sequence. For a successful operation, this Sequence will contain zeros, and the operation is over. The Sequence's last Frame will have the Last_Sequence and End_Sequence

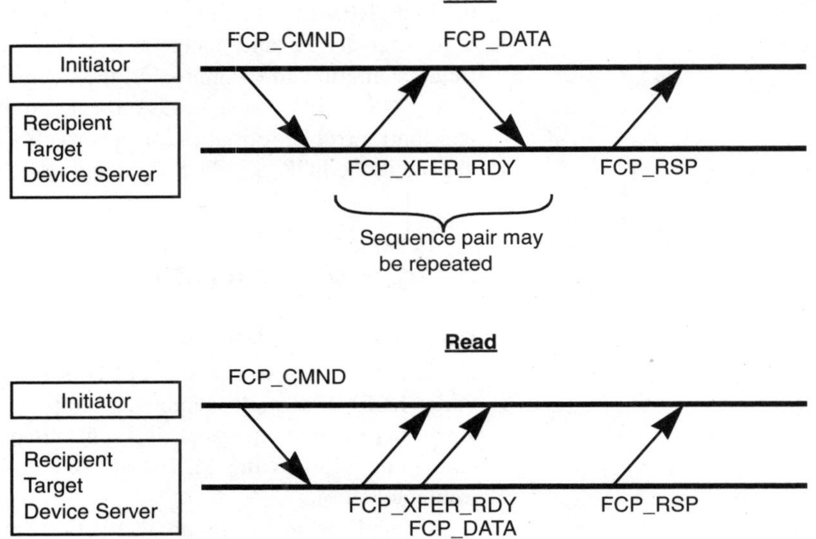

Figure 18.6 Information Unit transfers under FCP read and write operations.

F_CTL bits set, along with the other normal indications, and an **EOFt** delimiter (or possibly **EOFdt**, if Class 1 service was used for the Sequence or Sequences, and the target wants to remove the connection). The FCP_Port terminates the Exchange and indicates to the SCSI ULP that all Sequences terminated normally, and the SCSI level indicates to the process that made the SCSI I/O operation request that the operation completed normally and that it is ready for the next request.

Recovery from errors at the device level is carried out using the normal SCSI operations, with the SNS_INFO and RSP_INFO indicating any possible problems, and recovery occurring through transmission of other FCP Information Units. Recovery from any transmission errors at the Fibre Channel level uses the ABTS-LS procedure for aborting an entire Exchange, as described in the "Example 3: Aborting an Entire Exchange" section, on page 237. Once the Exchange has been unambiguously aborted, it can be retried, under control of the FCP ULP.

Note that there is no requirement for the operation to reserve all the transmission resources for the entire period of the operation. On typical disk drives, the time period between receipt of the FCP_CMND and transmission of the FCP_XFER_RDY on a SCSI read may be measured in milliseconds, for example. There is no architectural reason why the SCSI initiator may not be transmitting other FCP_CMND or FCP_DATA Sequences during this time. This multiplexing of Information Units allows a much higher utilization of the transmission and reception bandwidth resources than on a SCSI bus architecture, where the transfer of the SCSI command to the target generally reserves the bus, which must be idle until the target can fetch the data. Fabrics and Arbitrated Loops allow simultaneous operations between one or more FCP_Ports acting as SCSI Initiators and one or more FCP_Ports acting as SCSI targets.

Chapter 19

Future Work: FC-PH-?? and Beyond

Introduction

One of the truly huge risks in writing this kind of a book is in attempting to predict the future of the technology. I'm going to try, though, on the assumption that several years from now the predictions will either appear prescient, or at least very interesting, if perhaps not both.

There are several important trends driving the current development of Fibre Channel and similar networking technologies, including the following.

- The exponential rate of improvement in all technologies will continue. As applied to silicon chip area or processor performance, this is commonly referred to as Moore's Law, which predicts a performance doubling time of 18-24 months, but the same phenomenon can be observed in many other technologies as well, with comparable rates of improvement.
- There are different rates of exponential improvement for different technologies. Recently, disk drive density has been improving even faster than processor performance, and Gilder's law states that the doubling time for performance improvement in long distance networking technology is even faster, between 9 and 6 months (or even 3 months, in some areas such as Wavelength Division Multiplexing technology at some time). Sometimes these different rates can cause revolutionary changes in how different technologies are used.
- Further, there's sometimes even the observation of faster-than-exponential improvement, particularly in complex systems, as interrelated technological developments reinforce each other. An example of this is in Internet data caching technologies, where the usable bandwidth of WANs and servers can be tremendously increased, in effect, by caching frequently used information in the various intermediate routers in the internetwork.
- This exponential or faster-than-exponential improvement in technologies will be balanced by an equally fast increase in the rate of usage of those technologies, in most areas. That is, applications expand to fit the technologies available to them, getting progressively more complicated and distributed across more tightly coupled networks.

These trends all indicate that the need for widely distributed access across large-scale networks to vast amounts of data will increase extremely quickly for the foreseeable future (i.e, 3 to 5 years) — which is obviously a very positive set of trends for the future of Fibre Channel as a technology.

On the other hand:

- Network economy effects will accelerate. In some areas, this is phrased as Metcalfe's law, which states that the value of a network grows exponentially with the number of ports, while the cost per port stays constant or goes down. The most obvious consequence of this is a "winner-take-all" scenario, where the largest or most-accepted networks grow more quickly, where they compete with smaller or less-accepted networks. An important example here is IP, the Internet Protocol, which has accelerated in number of ports and wide acceptance over a number of other architectures which provided analogous capabilities. This effect will be particularly important as networks become more tightly integrated and compete more closely.

A few important practical observations are that:

- In the long term, many technologies are exhibiting a trend toward where the cost of hardware is a negligible fraction of the overall cost of a system. That is, administrators, software, and management, and particularly services, become increasingly important factors as time goes on. Phrased differently, this means that services and management will continue to get more difficult, and more important. In the Fibre Channel arena, this clearly includes network management, but more importantly includes data management, for consistency, location, status, and access rights. Many more resources will be shared, time-sliced, and programmable, with processors everywhere — so the whole concept of "data ownership" will become increasingly abstract, and file formats and file system structures will have to become increasingly interoperable and distributed.
- Linux will become more important, driven by these trends, less as an operating system technology than as a development methodology. That is, with the previous trends in mind, it will become increasingly straightforward for widely distributed people to pick up complex building blocks for little or no investment, incrementally improve them, and widely distribute the modified version for feedback and further fast improvement.

Finally:

- In the long run, a network technology succeeds less by the specific capabilities or performance it provides, than by the interoperability it offers, in the application area towards which it's targeted.

On balance, these are fairly conservative extrapolations of current trends. The remainder of this chapter describes several different possibilities for how these trends will affect the future of Fibre Channel technology over the next few years, in relation to similar technologies. Overall, these scenarios are extremely promising for Fibre Channel as a technology, especially over the next few years, but with some risks increasing in the longer term.

The Fibre Channel Future

In this possible future, Fibre Channel continues to be the de facto standard network architecture used for Storage Area Networks. There are some prerequisites for this to occur. Fibre Channel switch vendors must assure that switch interoperability issues are reasonably well resolved, so that customers have the option of selecting switches from multiple vendors. Also, a more clear definition of exactly what Fibre Channel options can reasonably be expected, and of how Fibre Channel networks interoperate on the Internet would be extremely useful. To meet data transmission requirements, in something less than 5 years, a 10 Gb/s version will have to be readily available. Finally, network management software tools will have to be developed that are as functional and simple to use as those for other networks, since the network management will be such a large part of the lifetime network cost.

This obviously the default case.

The "Ethernet Everywhere" Future

The Ethernet view of the future is driven by the observations that (a) Ethernet is the default network, for local area networks, with consequent large volumes, low hardware prices, and wide industry support, (b) performance concerns associated with Ethernet will be somewhat alleviated once Gigabit Ethernet and 10 Gigabit Ethernet become available, (c) the actual performance of the network in a storage area network environment is primarily driven by the NIC capabilities, and the host protocol stacks, which can be optimized over an Ethernet-based Fabric as well as over a Fibre Channel-based fabric.

The "Ethernet Everywhere" future is based on the idea that the network infrastructure (switches, links, NICs, and host protocol) would be based on IP protocols running over Ethernet hardware. Storage Area Network traffic, such as SCSI Command Data Blocks and responses, would simply be transported as a different kind of packet over the same network. This is reasonable because: (1) the links and switches for a SAN are not that different from a LAN, (2) the main hardware difference between the two are in management, usage, and HCAs or NICs, and (c) the main software difference between them is in what type of traffic they transport, and protocols they try to run — which can be modified.

The main area of controversy here is whether the TCP transport-level protocol, or a different protocol, would be used, across the IP network. TCP is a well-understood, very standard protocol, but its fundamental transport model (a uni-directional stream of bytes writable by one Host and readable by another) doesn't map particularly well to either packet-switched networks or to the SCSI command set, which is a request/response protocol using variable-sized command and data messages. This basic difference in fundamental transport models causes some pretty major inefficiencies in protocol implementation.

The model of transporting SCSI command blocks and responses over TCP sockets (termed "iSCSI") has been demonstrated, but it has not as of this writing demonstrated high efficiency, in terms of bandwidth transmitted per processor instruction. This model would be improved if more function were moved from the Host onto the NIC or HBA than is currently typical with Ethernet NICs. This TCP offload engine would ideally perform functions such as error checking, header parsing, and segmentation and reassembly, to alleviate the host processor from having to do these functions, and improve performance. Some efficiency issues can be resolved using TCP offload engines, but some of them are fundamental to the difference in transport models between TCP and the SCSI command set.

Alternatively, a new transport model could replace the TCP layer in the IP-based protocol stack, providing flow control and buffering on a per-message basis, with pre-allocated receive buffering, to better match the characteristics of SCSI traffic.

Another very interesting model is something of a "tunneling through IP" model. In this case, a small "storage gateway" device is used to translate between a common storage physical interface (SCSI or perhaps even IDE) and an Ethernet network link, such that an Ethernet could be used to link to widely separated storage interface buses. If two of these gateways are placed at each end of the host/device storage path, then the intervening path can be a normal Ethernet network, or, with routers, any other network that uses IP protocols. The advantage of this model is that the host and device interfaces are completely standard, highly optimized storage interfaces, and the gateway can be made as fast as any router. The primary issues here are (1) management of the gateway, since it has to act as a proxy for the host or storage device for the IP-based network, and (2) security, to assure that all hosts attached to the network access their devices coherently. These issues are analogous or identical to similar issues on any sophisticated LAN, so the gateway would be managed through its Ethernet port(s) similarly to a common Ethernet switch. Another common network administration issue which is particularly important here is network provisioning, to assure that each endnode sees adequate quality of service. For example, storage networking traffic could be restricted to a different LAN or VLAN than normal server/client traffic.

The advantages of any of these Ethernet-based strategies would be that the network links and switches would be high-volume units, which would presumably be less expensive than Fibre Channel components, and would be familiar to system administrators. Similarly, network management of such an Ethernet-based SAN would be the same as management of a LAN, and could use the same tools.

The "InfiniBand Grows Outward" Future

The InfiniBand^{TM} Architecture is a fundamentally disruptive technology in the whole area of server design, and connection of computers to their I/O, both networking and channel technologies. InfiniBand is a technology that brings networking technology, specifically IPv6 networking technology, directly into the heart of a server's I/O system.

The InfiniBand^{TM} Architecture addresses a problem which is present in both of the futures described above, which is that the network interface (the Fibre Channel Host Bus Adapter or Ethernet NIC) is attached onto an I/O bus, typically an I/O bus. Since I/O buses are not improving at the same rate as either processor/cache/memory subsystems, or as networks, the I/O bus, and PCI or PCI-X bus is becoming a bottleneck in server systems.

In a server system using an InfiniBand network I/O & IPC network, the point-to-point PCI-X bus would be replaced with a very high-efficiency network interface termed a "Host Channel Adapter," and a packet-switching network replaces the PCI bus as the I/O infrastructure for the server. This allows the server to be decoupled not only from the storage, but from the complete I/O infrastructure. Upgrading a processor in such a server would not require upgrading or even moving the I/O cards — the new server would just contain processor/cache subsystem, and memory, and would be attached to the I/O infrastructure through a link or a backpanel.

An InfiniBand network is termed a "System Area Network" (which unfortunately uses the same acronym as "Storage Area Network," although they are conceptually and practically quite different). The intention is for the network to be used to attach the system's I/O, and to attach to any other devices that the system needs to attach to. A sample diagram of an InfiniBand System Area Network is shown in Figure 19.1

Some of the innovative parts of the InfiniBand Architecture are that it combines IP-based network management with extremely efficient, low-latency switching, and very scalable bandwidth — each link comprises either 1, 4, or 12 lines transmitting 8b/10b coded data at 2.5 Gb/s per line to achieve 2, 8, or 24 Gb/s of user data on each link, with little or no change to the circuitry or packaging. Further the InfiniBand transport layer provides nearly equivalent functionality to the IP and TCP layers, but exploits assumptions on network performance and functionality within a well-controlled, localized network in order to highly optimize the overall communications protocol and move a large portion of it onto the HCA, offloading the host processor. Also, the InfiniBand^{TM} Architecture describes a complete set of packaging for I/O modules, based on horizontal plugging of EMI-shielded modules into the backplane of a rack-mounted system.

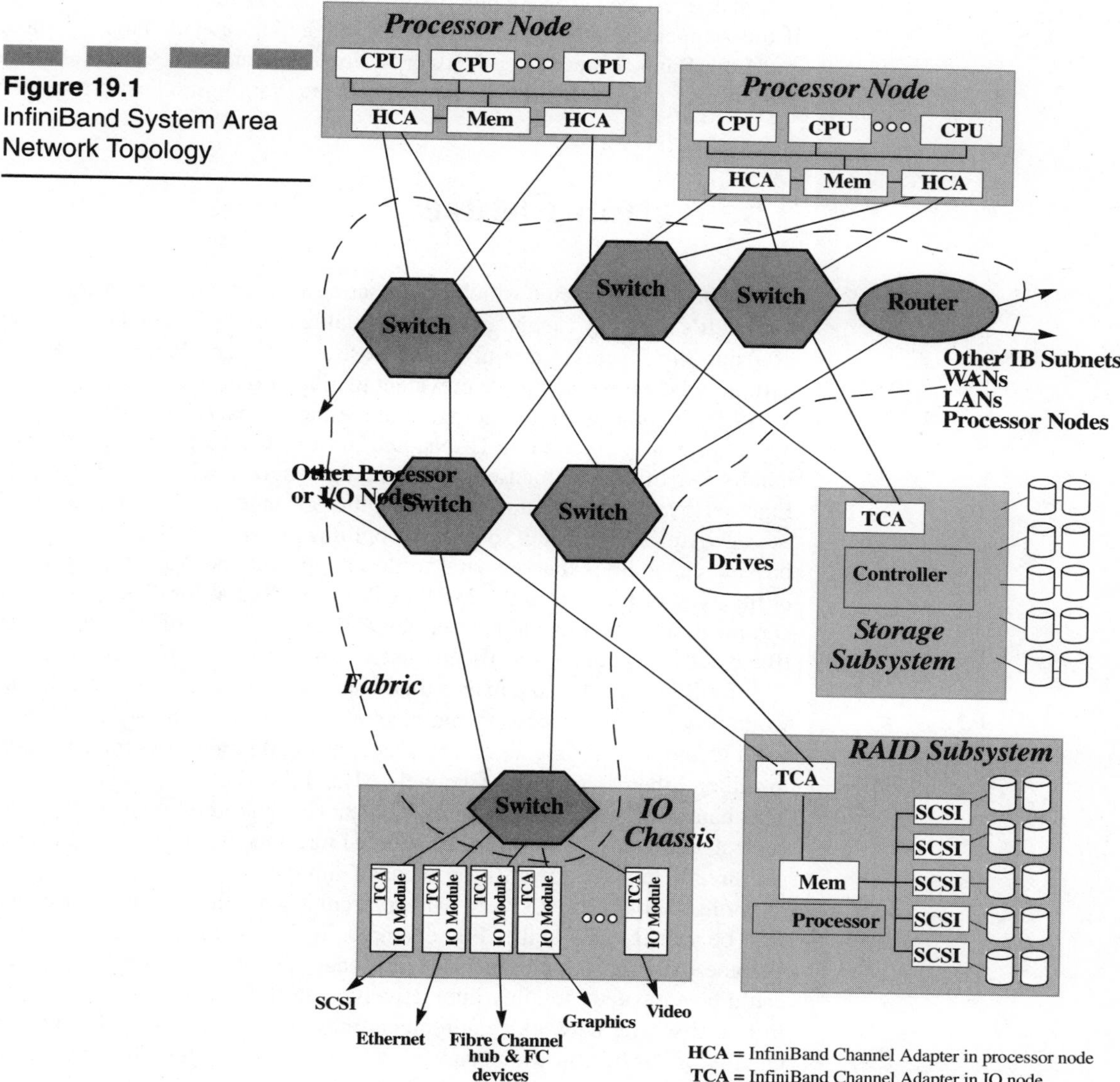

Figure 19.1
InfiniBand System Area Network Topology

The diagram shows two different mechanisms to attach storage to a server — either through a Target Channel Adapter bridging to disks or tape through SCSI or Fibre Channel links, or directly. The formats of the messages crossing the InfiniBand Fabric would naturally be different in the two cases.

The initial attachment of storage through InfiniBand will be through TCAs and SCSI or Fibre Channel links. There is, however, some possibility that the necessary software and device drivers will be written, and the necessary disk interface hardware built to allow disks to be directly attached to IB. If this happens, it would likely be a direct competitor with Fibre Channel, and InfiniBand System Area Networks would perform the set of functions that Storage Area Networks do, with tighter integration into host memory.

The Actual Future

Which of these three, or which combination of them, will form the actual future? It's probably fundamentally impossible to know, beyond a time scale of about 2 to 3 years. Certainly, over those 2 to 3 years, Fibre Channel will definitely become much more prevalent than it currently is.

Ethernet could work for storage, but the fact is that the Ethernet community is primarily focused on LANs, and most of the 10 Gb/s Ethernet community is particularly focused on extending Ethernet Wide-area LANs, i.e., Ethernet over distances up to 40 km. Fibre Channel is focused directly on Storage Area Networking, so it can be optimized to that application area, and provide capabilities that are vital to these applications. A possible example of this type of function is the MRK1x Primitive Signal for disk spindle synchronization, which would not normally be a part of either Ethernet or InfiniBand networks, and is specifically useful for a disk-interface network.

InfiniBand has the advantages that where it is installed, it will be the network closest to the processors and memory, so that attaching storage directly to IB would involve one less layer of adapters and fabric links than attaching to either Ethernet or Fibre Channel. Also, InfiniBand supports extremely high bandwidth over relatively short distances, and good quality of service control, which are both useful in a storage interface. However, InfiniBand is not directly targeted towards the needs of storage.

More importantly, it's very hard to predict what the traffic requirements will be (other than high). For example, real-time interactive video may become extremely important — every page, every application, every icon could be animated, possibly interactively, with full-motion video resolution. In this case, the network requirements may be biased much more towards Quality of Service, in which case either InfiniBand, with its fine granularity of flow control and sophisticated separation of traffic into Virtual Lanes and Service Levels, or Fibre Channel with its clear separation of Classes of Service, and video-oriented functionality in Class 1, Class 4, and Class 6, could be a better choice than Ethernet.

The most important point here is that the success of a particular technology depends somewhat on the quality of the actual technology, but it has a lot more to do with how much effort and investment are put into enabling the technology in various applications. The winning technology will be the one that the most people work the hardest to make into the best solution for the application.

In any case, we can again be sure that the future will see changes in the way computing will be done, and that Fibre Channel will assume a large and growing role in that future.

Appendix

References and Further Reading

This appendix contains references to documents available for further reference and detailed implementation.

Fibre Channel ANSI Standard Documents

Hardcopies of the standards documents may be obtained from Global Engineering Documents, An IHS Group Company, at http://global.ihs.com/. Also, electronic versions of most of the approved standards are also available from http://www.ansi.org, and at the ANSI electronic standards store at http://webstore.ansi.org.

Further information on ANSI standards and on both approved and draft international, regional and foreign standards (ISO, IEC, BSI, JIS, etc.) can be obtained from the ANSI Customer Service Department. References under development can be obtained from NCITS (National Committee for Information Technology Standards), at http://www.x3.org.

All documents referred to as drafts are working documents and are actively being modified. They are made available for review and comment only.

This is not a complete set, even at the time of writing, but this set of documents covers most of the essential parts of the technology.

- Physical and Signaling Interface
  - FC-PH: ANSI X3.230-1994, *Fibre Channel Physical and Signaling Interface (FC-PH)* — original Fibre Channel standard. Also amended by ANSI X3.230-1994/AM 1-1996, *Fibre Channel Physical and Signaling Interface (FC-PH) - Amendment 1.*
  - FC-PH-2: ANSI X3.297-1997, *Fibre Channel Physical and Signalling Interface-2 (FC-PH-2)* — Enhancements to FC-PH
  - FC-PH-3: ANSI X3.303-1998, *Fibre Channel Physical and Signalling Interface-3 (FC-PH-3)* — Enhancements to FC-PH and FC-PH-2. Also, see ANSI NCITS 321-1998, *Fibre Channel - Low-Cost 10-km Optical 1063-MBaud Interface.*
- Arbitrated Loop
  - FC-AL: ANSI X3.272-1996, *Fibre Channel Arbitrated Loop (FC-AL)* — Arbitrated Loop topology to support Fibre Channel
  - FC-AL-2: ANSI X3 NCITS 332-1999, *Fibre Channel Arbitrated Loop (FC-AL-2)* — Enhanced version of FC-AL.
- Switching and General Services
  - FC-FG: ANSI X3.289-1996, *Fibre Channel - Fabric Generic Requirements (FC-FG)* — topology-independent interconnecting Fabric to support the Fibre Channel standard
  - FC-SW: ANSI NCITS 321-1998, *Fibre Channel - Switch Fabric (FC-*

SW) — interoperable switch topology to support the Fibre Channel standard

- FC-GS-2: ANSI NCITS 288-1999, *Fibre Channel Generic Services (FC-GS-2)*

- FC-4 Mappings to Upper Level Protocols
  - FCP: ANSI X3.269-1996, *Fibre Channel Protocol for SCSI (FCP)* — Frame format and protocol definitions required to transfer command and data between a SCSI Initiator and Target using Fibre Channel
  - FC-LE: X3.287-1996, *Fibre Channel Link - Encapsulation (FC-LE)* — Encapsulation of IEEE STD 802.2 Logical Link Control Protocol Data Units over Fibre Channel.
  - FC-SB: ANSI X3.271-1996, *Fibre Channel Single-Byte Command Code Sets (SBCCS) Mapping Protocol (FC-SB).*
  - FC-SB-2: NCITS T11/Project 1357-D, *Fibre Channel Single-Byte Command Code Sets-2 Mapping Protocol (FC-SB-2).*
  - FC-FP: ANSI X3.254-1994, *Fibre Channel Mapping to HIPPI-FC (FC-FP).*
  - HIPPI-FC: ANSI C3.283-1996, *HIGH-PERFORMANCE PARALLEL INTERFACE — Encapsulation of Frames of the Fibre Channel Physical and Signaling Interface (FC-PH Encapsulation) (HIPPI-FC).*

Web Resources

The following web sites provide information on technology related to Fibre Channel, SANs and storage networking.

- http://www.fibrechannel.org
  Fibre Channel Industry Association
- http://www.snia.org
  Storage Networking Industry Association
- http://www.T11.org
  ANSI T11 Organization
- http://www.storageperformance.org
  Storage Performance Council
- http://www.iol.unh.edu
  University of New Hampshire InterOperability Laboratory

The following web site provides information on the objectives, history, and specification of the InfiniBand^{TM} fabric I/O technology.

- http://www.infinibandta.org
  InfiniBand^{TM} Trade Association

8B/10B Transmission Code

A. X. Widmer and P. A. Franaszek, "A DC-balanced, Partitioned-Block, 8B/10B Transmission Code," *IBM Journal of Research and Development*, 27, No. 5: 440–451 (September, 1983).

U. S. Patent 4,486,739. Peter A. Franaszek and Albert X. Widmer. Byte Oriented DC Balanced (0,4) 8B/10B Partitioned Block Transmission Code (December 4, 1984).

Storage Area Networks and Storage Networking

Clark, Tom, *Designing Storage Area Networks: A Practical Reference for Implementing Fibre Channel SANs*, Addison-Wesley Longman, Inc., Reading, Mass., 1999.

Farley, Marc, *Building Storage Networks*, The McGraw-Hill Companies, 2000.

Networking and Internetworking

Black, Uyless, *ATM — Foundation for Broadband Networks*, Prentice-Hall PTR, Englewood Cliffs, NJ, 1995.

Comer, Douglas E., *Internetworking with TCP/IP Vol. 1: Principles, Protocols, and Architecture*, Prentice-Hall, Englewood Cliffs, NJ, 1995.

Kumar, Balaji, *Broadband Communications: A Professional's Guide to ATM, Frame Relay, SMDS, SONET, and BISDN*, The McGraw-Hill Companies, New York, 1995.

McDysan, David E., and Spohn, Darren L., *ATM: Theory and Application*, The McGraw-Hill Companies, New York, 1995.

Miller, Mark A., P.E. *Internetworking: A Guide to Network Communications: LAN to LAN; LAN to WAN*, M&T Books, New York, 1991.

Partridge, Craig, *Gigabit Networking*, Addison-Wesley, Reading, MA, 1994.

Schwartz, M. *Computer-Communication Network Design and Analysis*, Prentice-Hall, Inc., Englewood Cliffs, NJ, 1977.

Tanenbaum, Andrew, *Computer Networks*, Prentice-Hall, Englewood Cliffs, NJ, 1989.

Washburn, K, and Evans, J. T., *TCP/IP — Running a Successful Network*, Addison–Wesley Publishing Company, Wokingham, England, 1993.

Index

F